W0112949

AI-Centric Smart City Ecosystems

Over the next few years, smart city technologies will be rolled out, and the IoT devices and AI-centric systems will provide even more far-reaching connectivity. This book presents various concepts in the design and development of a smart city and methodologies and solutions involved in designing contemporary infrastructure for building smart cities around the world. The book will focus mainly on six areas of smart city infrastructures: smart city entities, IoT-based solutions, AI-centric control systems, smart systems, cybersecurity mechanisms, data science, and cloud computing for the deployment of the smart ecosystem.

AI-Centric Smart City Ecosystem: Technologies, Design, and Implementation will discuss the role of AI-centric innovative systems and beyond intelligent solutions in the smart city framework. Readers will discover how to apply design principles and technologies for operating intelligent cities and develop an understanding of how to integrate AI-based control systems to make systems smarter. The book will present various concepts in the design and development of smart cities and methodologies and solutions involved in designing modern infrastructure. Also, readers can discover how to develop applications and connect the IoT devices for collecting and mining real-time data and uncover the challenges and techniques for improving the automatic operation in the smart city by using high-tech solutions.

This book is intended to serve the needs of the industry, engineers, professionals, researchers, and master's and doctoral students studying emerging technologies in smart city ecosystems.

AI-Centric Smart City Ecosystems

Technologies, Design, and Implementation

Edited by Alex Khang, Sita Rani,
and Arun Kumar Sivaraman

CRC Press

Taylor & Francis Group

Boca Raton London New York

CRC Press is an imprint of the
Taylor & Francis Group, an informa business

First edition published 2023
by CRC Press
6000 Broken Sound Parkway NW, Suite 300, Boca Raton, FL 33487–2742

and by CRC Press
4 Park Square, Milton Park, Abingdon, Oxon, OX14 4RN

CRC Press is an imprint of Taylor & Francis Group, LLC

ISBN: 978-1-032-17079-4 (hbk)
ISBN: 978-1-032-18028-1 (pbk)
ISBN: 978-1-003-25254-2 (ebk)

DOI: 10.1201/9781003252542

Typeset in Times New Roman
by Apex CoVantage, LLC

Contents

Contents

Preface

In the domain of city planning, smart cities made perceivable differences in easing city life, adding to its quality while keeping its people connected, engaged, and informed. This book aims to introduce the idea of smart cities comprehensively by covering the conceptual basis and the principles in practice systematically and sector-wise. The major domains focused on in the book are healthcare, education, transport, resource management, and supply management, among others.

The main aim of the book is to serve the needs of IT engineers, industrial engineers, professionals, researchers, and master's and doctoral students studying emerging technologies in the smart city ecosystem by focusing on six areas of infrastructure: IoT-based solutions, AI-centric control systems, smart systems, cybersecurity mechanisms, data science, and cloud computing for the deployment of the smart ecosystem.

Written lucidly, covering both possible new attempts and retrofitting options in turning smart, the book is a handy all-in-one volume for beginners, from city enthusiasts to advanced learners. Readers who have a basic idea of computer science, data science, and AI skills can benefit completely from this book.

The authors are grateful to CRC Press for giving the opportunity to present this book to the readers to cater to their different types of requirements. Last but not least, the authors express their heartfelt gratitude to their families and friends for their continuous support in completing this project.

Happy reading!

The Editors,
Alex Khang
Sita Rani
Arun Kumar Sivaraman

Acknowledgments

Nowadays, the concept of smart cities is rapidly developing in urban areas, which significantly enhances the living conditions of humans. But unlike several other engineering disciplines, the smart city ecosystem lacks well-defined design principles and research strategies. So this edited book will present various concepts in the design and development of smart cities and the methodologies and solutions involved in designing contemporary infrastructures for building smart cities around the world.

This book is to serve the needs of the IT engineers, industrial engineers, professionals, researchers, and master's and doctoral students studying emerging technologies in the smart city ecosystem by focusing mainly on six areas of the infrastructure of smart city entities: IoT-based solutions, smart systems, AI-centric control systems, data-centric systems, cybersecurity mechanism, data science, and cloud computing for the development of the smart city ecosystem.

Having an idea of a smart system, turning it into a chapter, and then sharing it with us is as hard as it sounds, but your effort and experience are both internally challenging and rewarding for the academic world. Without the support and contribution from friends and colleagues in the world, this book would not exist. So I especially want to say complete thanks to the individuals that contributed make this book successful.

To all the reviewers we have had the opportunity to collaborate with and watch their hard work from afar, we acknowledge their tremendous support and valuable comments. We would like to say a big thank you for being the inspiration and foundation for the project's success.

We thankfully acknowledge all the advice, support, motivation, sharing, and inspiration we received from our faculty and academic colleagues.

We express our grateful thanks to our publisher CRC Press (Taylor & Francis Group) and the entire editorial team who spent their wonderful support for making sure the timely processing of the manuscript and bringing out of the book to readers.

Finally, this is all about trust, honor, and respect for the contribution of all those who submitted the book chapters. We will always welcome the chance to represent you.

Thank you for all.

Editorial Team

Editor Biographies

Alex Khang, Professor in Information Technology, Doctor of Computer Science, Universities of Science and Technology in Vietnam and United States, Software Industry Expert, AI Specialist and Data Scientist, Workforce Development Solutions Consultant, and Chief of Technology (AI and Data Science Research Center) at the Global Research Institute of Technology and Engineering, NC, USA, ORCID: 0000-0001-8379-4659. He has 28+ years of teaching and research experiences in Information Technology (Software Development, Database Technology, AI Engineering, Data Engineering, Data Science, Data Analytics, IoT-based Technologies, and Cloud Computing) at the Universities of Technology and Science in Vietnam, EU, India, and United States. He is the chair session for 20+ international conferences; international keynote speaker for 25+ international conclaves; Expert Tech-talks for 100+ seminars and webinars; international technical board member for 10+ Intl. organizations; international editorial board member for 5+ ISSNs; reviewer and evaluator for 100+ the journal papers; international examiner and evaluator for 15+ the Ph.D. thesis in Computer Science field. He is the receiver of the Best Professor of the Year 2021, Researcher of the Year 2021, the Global Teacher Award 2021 (AKS), the Life Time Achievement Award 2021, the Leadership Award 2022 (Educacio World), and many other reputed awards. He has been contributing to the various research activities in the fields of AI and Data Science, while publishing many International articles in the renowned journals and conference proceedings. He has published 52 authored books (in Computer science year 2000–2010 in Vietnam), two authored books (Software development), four edited books, ten book chapters, and two edited books (calling for book chapters) in the fields of AI, Data Science, Big Data, IoT, Smart City Ecosystem, Healthcare Ecosystem, Fintech technology, and Blockchain technology (since year 2020). He has over 28 years of working in the field of software productions, data engineering, and specialized in database technology for foreign corporations from Germany, Sweden, the United States, Singapore, and multinationals (former CEO, former CTO, former Engineering Director, and Senior Software Production Consultant).

Sita Rani Dr. Sita Rani is presently working as Assistant Professor in the Department of Computer Science & Engineering, Guru Nanak Dev Engineering College, Ludhiana, India. Earlier, she has served as a Professor and Deputy Dean (Research) at Gulzar Group of Institutions, Khanna (Punjab), India. ORCID: 0000-0003-2778-0214. She has completed her B. Tech and M. Tech degrees in the Faculty of Computer Science and Engineering from Guru Nanak Dev Engineering College, Ludhiana. She obtained her PhD in computer science and engineering from IK Gujral Punjab Technical University, Kapurthala, Punjab, in the year 2018. She is an active member of IEEE and IAEngg. She is the recipient of ISTE Section Best Teacher Award and International Young Scientist Award. She has contributed to various research activities while publishing articles in renowned journals and conference proceedings. She has published three international patents also. She has delivered many expert talks in AICTE-sponsored faculty development programs and organized many international conferences during the 18 years of her teaching experience. She is a member of editorial board of four reputable international journals.

Arun Kumar Sivaraman Arun Kumar Sivaraman is currently working as Assistant Professor (Sr. Grade) in VIT University, Chennai Campus, India, ORCID: 0000-0003-0514-484X. He obtained his bachelor's degree in computer science and engineering from Anna University, Chennai, and master's in computer science and engineering in 2010 from the College of Engineering Guindy (CEG), Chennai Campus. He was awarded a PhD in computer science and engineering in 2017 from Manonmaniam Sundaranar University (Govt.), Tirunelveli, India. He received his master's in business administration (MBA) in 2020 and master's in education management from Alagappa University, Karaikudi, India. He has more than a decade of professional experience in the industrial, R&D, and academic sectors. He has worked as a lead data engineer for top MNCs, like Cognizant, Standard Chartered, and Gilead Life Sciences. He worked as a project consultant for a healthcare research (R&D) project, the Research Council, Sultanate of Oman, funded by the Omani Ministry of Health. He published a book on machine learning titled *Image Processing for Machine Learning* (ISBN: 978-93-5445-509-4). He published many research papers in a reputable Scopus-indexed journal, and he also has two Indian patents and got one grant in international patent. For his merit, he got an offer as the lead data engineer in Tata Consultancy Services (TCS) in 2011, an Employment Pass Eligibility Certificate (EPEC) from the government of Singapore in 2012, and a Young Scientist Award from the government of Oman in 2018. He is an active co-editor of a couple of special issues in Tech Science Press (Computer, Materials, and Continua—IF 3.77). His academic and research expertise covers a wide range of subject areas, including data engineering, data analytics, data science, and machine learning.

Contributor Biographies

Abhishek Vishnoi, Assistant Professor, Department of Electrical & Electronics Engineering Kanpur Institute of Technology, Kanpur, Uttar Pradesh, India. ORCID: 0000-0002-1052-4890.

Abuzarova Vusala Alyar, Assistant Professor Abdullayev Vugar, Azerbaijan State Oil and Industry University, Baku, Azerbaijan.

Ahmed Muayad Younus, Senior Lecturer, Faculty of Management (information technology), Limkokwing University, Selangor, Malaysia. ORCID: 0000-0002-5686-3505.

Alex Khang, Professor in Information Technology, Doctor of Computer Science, Universities of Science and Technology in Vietnam and United States, Software Industry Expert, AI Specialist and Data Scientist, Workforce Development Solutions Consultant, and Chief of Technology (AI and Data Science Research Center) at the Global Research Institute of Technology and Engineering, NC, USA. ORCID: 0000-0001-8379-4659.

Aman Kataria, Head Tracking, Artificial Intelligence and Neural Networks, Thapar Institute of Engineering and Technology, Patiala. Project Associate, CSIR-Central Scientific Instruments Organization, CSIR-CSIO, Chandigarh, India.

Ananthi Sadhasivam, Assistant Professor in Computer Science, PPG College of Arts and Science—Coimbatore, Tamil Nadu, India. ORCID: 0000-0003-2070-8236.

Ankit Jain, Assistant Professor, Department of Electronics and Communication Engineering, Indore Institute of Science and Technology (IIST), Indore (Madhya Pradesh), India.

B. Prabu, Assistant Professor, Department of Civil Engineering, Sona College of Technology, Salem, Tamil Nadu, India. ORCID: 0000-0003-2030-7862.

Eugenia Litvinova, Doctor of Science, Professor of Computer Engineering Faculty, Design Automation Department, Kharkiv National University of Radio Electronics, Kharkiv Oblast, Ukraine. ORCİD: 0000-0002-9797-5271.

Gardashova Latafat Abbas, Doctor of Sceinces on Engineering, Vice-Rector for Science and Technology (for Scientific Affairs) of Azerbaijan State Oil and Industry University, Baku, Azerbaijan. ORCID: 0000-0003-3227-2521.

Humberto Martínez-Camacho, Universidad Panamericana, Biblioteca, Zapopan, Jalisco, México. ORCID: 0000-0001-5430-0648.

Kashish Wilson, Assistant Professor at MM College of Pharmacy, Maharishi Markandeshwar (Deemed to be University) Mullana, Ambala, Haryana, India.

Kumar Guarve, Principal at Guru Gobind Singh College of Pharmacy a NBA Accredited College, Haryana, India.

Leelavathy Vellingiri, Assistant Professor in Computer Science, PPG College of Arts and Science, Coimbatore, Tamil Nadu, India. ORCID: 0000-0002-9845-2288.

M.P. Karthikeyan, Assistant Professor, School of CS and IT, Jain (Deemed-to-be University), Bengaluru, Karnataka, India. ORCID: 0000-0002-2346-0283.

Manish Kumar Mukhija, Ph.D. in Computer Science & Engineering, Sangam University, Bhilwara, Rajasthan, Jaipur, India. ORCID: 0000-0003-1816-6511.

Meetali Chauhan, Assistant Professor, Computer Science & Engineering at Guru Nanak Dev Engineering College, Ludhiana, Punjab, India.

M.N.A. Gulshan Taj, Associate Professor, Department of civil Engineering, Sona College of Technology, Salem, Tamil Nadu, India. ORCID: 0000-0002-6581-297X.

Mohanad S.S. Abumandil, Senior Lecturer, Faculty of Hospitality, Tourism and Wellness, Universiti Malaysia Kelantan, Kelantan, Malaysia. ORCID: 0000-0003-0121-306X.

Mukesh Patidar, Assistant Professor, Department of Electronics and Communication Engineering, Indore Institute of Science and Technology (IIST), Indore (Madhya Pradesh), India. ORCID: 0000-0002-4401-8777.

N. Madhan, Intern, Cognizant Technology Solutions, Chennai, India.

N. Muthumani, Principal, PPG College of Arts and Science, Coimbatore, Tamilnadu, India.

N. Vijayaraghavan, Assistant Professor of Mathematics in KCG College of Technology, Chennai, India. ORCID: 0000-0002-4424-5048.

Namit Gupta, Director of Shri Vaishanv Institute of Technology and Science (SVITS), Indore (M.P.), Shri Vaishnav Vidyapeeth Vishwavidyalaya (SVVV), Indore (Madhya Pradesh), India.

Nazila Ali Ragimova, Head of "Computer Engineering" Department at the Azerbaijan State Oil and Industry University, Baku, Azerbaijan.

Nilesh Patidar, Assistant Professor, Electrical and Electronics Engineering, SVITS, Shri Vaishnav Vidyapeeth Vishwavidyalaya, Indore (Madhya Pradesh), India. ORCID: 0000-0001-8712-5938.

Pankaj Bhambri, Assistant Registrar (Academics), Assistant Professor (Senior Scale), I.K.G. Punjab Technical University, Jalandhar at GNDEC, Ludhiana, Assistant Registrar (Academics), Assistant Professor (Senior Scale), Department of Information Technology, Ludhiana, Punjab, India.

Pooja Singh, M. Tech* in Computer Engineering, Arya Institute of Engineering & Technology, Jaipur, Rajasthan, India. ORCID: 0000-0002-8285-7408.

Prerna Sharma, Assistant Professor in Guru Gobind Singh College of Pharmacy, Haryana, India.

R. Dhanalakshmi, Professor and Head of the Department of Computer Science Engineering in KCG College of Technology, Chennai, India. ORCID: 0000-0003-0233-5202.

R. Malathy, Professor and Head, Department of Civil Engineering, Sona College of Technology, Salem, Tamil Nadu, India. ORCID: 0000-0002-7604-6289.

R.K. Tailor, Associate Professor, Department of Business Administration, Manipal University, Jaipur, Rajasthan, India.

S.K. Chaya Devi, Assistant Professor in the Information Technology Department at Vasavi College of Engineering, Hyderabad, India.

Satish Kumar Alaria, M. Tech. in Computer Engineering, SKIT, Jaipur (RTU, Kota), India. ORCID: 0000-0001-8298-1364.

Shashi Kant Gupta, Researcher, Integral University, Lucknow, Uttar Pradesh, India. ORCID: 0000-0001-6587-5607.

Sita Rani, Assistant Professor, Department of Computer Science and Engineering, Guru Nanak Dev Engineering College, Ludhiana, Punjab, India.

Sumeet Gupta, Dean and Principal in M. M. College of Pharmacy constituted college of (M. M. Deemed to be University) (NAAC "A ++ "), Mullana, Ambala, Haryana accredited by Scientific and Industrial Research Organization under Department of Science and Technology, Ministry of Science and Technology, Government of India, New Delhi, India.

Veer P. Gangwar, Professor, Mittal School of Business, Lovely Professional University Phagwara, Punjab, India. ORCID: 0000-0001-8617-5511.

Vikas Verma, Assistant Professor, Department of Electrical & Electronics Engineering Kanpur Institute of Technology, Kanpur, Uttar Pradesh, India. ORCID: 0000-0003-2910-5255.

Vladimir Hahanov, Doctor of Science, Professor of Computer Engineering Faculty, Design Automation Department, Kharkiv National University of Radio Electronics, Kharkiv Oblast, Ukraine. ORCİD: 0000-0001-5312-5841.

Vugar Abdullayev Hajimahmud, Doctor of Technical Sciences, Assistant Professor, Azerbaijan State Oil and Industry University, Baku, Azerbaijan. ORCID: 0000-0002-3348-2267.

1 Smart City Ecosystem
Concept, Sustainability, Design Principles, and Technologies

Sita Rani, Pankaj Bhambri, Aman Kataria, Alex Khang

CONTENTS

1.1 INTRODUCTION

The smart city is a complex ecosystem of processes, people, policies, technology, and other enablers that work collaboratively to accomplish a set of goals (Rani & Kumar, 2022). The smart city isn't really "owned" by the city alone, but other value makers are also engaged, sometimes in partnership with one another and sometimes on their own. In a smart city, numerous electrical systems, sensors, and voice activation methods are used to collect data (Silva, Khan, & Han, 2018).

The management of the assets, resources, and services may be made more efficient using the data obtained from them, and this, in turn, improves city operations. This

DOI: 10.1201/9781003252542-1

contains data collected from devices, citizens, buildings, and assets that is processed and analyzed for the purpose of monitoring and managing power plants, traffic and transportation systems, utilities, waste, crime detection, water supply networks, and information systems, as well as libraries, schools, hospitals, and other community services. Smart cities are those in which the government uses technology to monitor, analyze, plan, and govern the city, among other things.

1.1.1 CONCEPT

A smart city refers to a city's utilization of intelligent technologies and data to address its environmental challenges. The use of information and analytics to enhance transportation, health, energy use, and air quality, as well as drive economic growth, is becoming increasingly common in many cities.

Others are intelligently constructed from the ground up (Law & Lynch, 2019). There are various organizations working to investigate different communication and networking technologies that facilitate the functioning of smart cities. Various communication technologies are also compared and contrasted in terms of their most important similarities and differences. Additionally, taxonomy is developed by categorizing the literature according to future and developing technologies, contemporary communication technologies, objectives, network classes, and modes of operation.

Additionally, several case studies from various cities are provided, including Barcelona, Stratford, Singapore, and Porto. Finally, several research issues, including interference management, customizable wireless solutions, support for interoperability among diverse wireless networks, vehicular networks, and high energy consumption, are debated as future research for enabling unimpeded connectivity in smart cities.

In the near future, 5G (fifth generation) internet will revolutionize how we interact with daily devices and network infrastructure. In order to connect a larger number of people and provide better and quicker communications, wireless technologies, such as 5G, are widely employed to transport data between devices. Boosting the network's capacity, reliability, and responsiveness are the primary goals here, as are reducing network latency and increasing global network coverage.

The primary goal of 5G and wireless communications is to reduce load and stabilize technological solutions so that they can be used via cutting-edge business applications and apps for improved mobile communication. For many of today's unified communication devices, the existing long-term evolutionary advanced (4G) network, which requires low latency QoS, extremely high data speeds, increased bandwidth, and interference immunity, is simply not adequate (LTE-A).

Mobile apps and networks like smart cities, health apps, beacons, smart watches, and smart cards will continue to benefit from the Internet of Things (IoT) and 5G (Zhao, Askari, & Chen, 2021).

1.1.2 SMART CITY FRAMEWORK

The smart city environment is a straightforward decision-making platform that helps both the public and commercial sectors to more efficiently formulate and execute smart city initiatives. The majority of cities actually go through this process intuitively rather than in an organized fashion. A systematic approach will not only permit more efficiency in city infrastructures and also greater transparency into how cities operate.

A four-layer smart city framework is shown in Figure 1.1 (Falconer & Mitchell, 2012). It creates a logical flow that enables stakeholders to "push" ideas forward and test them. For instance, suppose a city leader is committed to improving sustainability, which later becomes a Layer 1 high-level objective.

Additionally, let us assume that the city has determined that its bus system's journey times are not among the best in the world through international transportation indices (Layer 2). After collecting this data, stakeholders can consider a city proposal for a "connected bus fleet" (Layer 3) and the needs for building and executing the system.

From there, municipal leaders can look for best practices in similar efforts around the world, including how a system got financed and administered, as well as the appropriate regulatory and policy frameworks (Layer 4).

The cyclical flow of information inside the smart city framework creates a feedback loop that enables stakeholders to learn about other smart city initiatives' best practices (Baig, Himarish, Pranaya, & Ahmed, 2018).

Not only does a smart city architecture provide a complete insight into how cities operate, but it also permits three significant outcomes: taxonomy/typology allows cities to benchmark the pertinent content in relation to the hierarchy underlying physical city components; stakeholder roles that determine who does what.

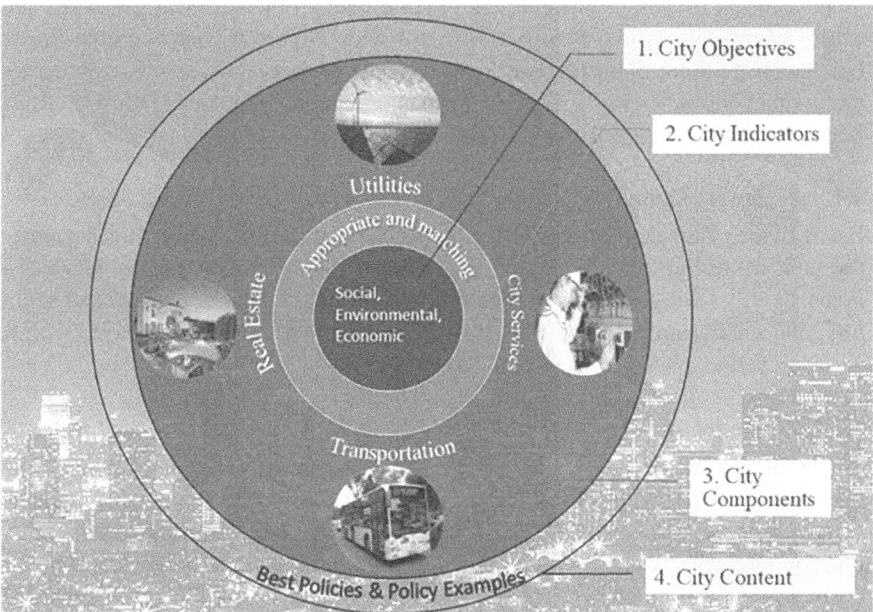

FIGURE 1.1   Smart city: a four-layer framework.

Unfortunately, this section is frequently omitted from municipal talks; its exclusion leads to a lack of understanding regarding how to adopt smart public solutions, such as a catalog system of conveniently available city content.

1.1.3 SMART CITY APPLICATION DOMAINS

1.1.3.1 SMART ENERGY

Smart energy is used in sensors for electric vehicles, smart grids, transmission, and distribution of energy. The advantage of using smart energy is to reduce energy consumption. In addition, it prevents power grid failure in terms of energy.

1.1.3.2 Smart Environment

An environment that is supportive of a comfortable and sustainable climatic condition is termed a smart environment. It depends upon various factors, such as air, forest conditions, water pollution, waste of gas, greenhouse gas, noise in the city, and climatic conditions. If all the conditions are satisfied, then the city can approach sustainable development.

1.1.3.3 Smart Industry

Industrial production with surplus products and resources is one of the key factors in developing a city into a smart city. A smart industry has brought a revolution in the field of the industrial sector. It has rolled out surplus production of resources with minimal cost and labor.

In addition to this, it prevents the environment from industrial gas waste and excess heat transmission. Precise operations with a span of control are an important factor in a smart industry. For precise operations, actuators, servers, robots, and motors play key roles in performing various operations in a smart industry (Rana, Khang, Sharma, Goel, & Dubey, 2021).

1.1.3.4 Smart Living

Smart living offers comfortable homes to residents of the smart city. This has raised the standard of living of the people with services such as remote control services at home, energy saving, education, climatic adjustments, automatic sensors and actuators, and entertainment. Smart living also includes various other factors, such as waste recycling, parking, and social networking.

1.1.3.5 Smart Services

Smart services facilitate the citizens in various aspects. For example, considering the case of traffic patterns on roads, traffic analysis can be easily done with the help of sensors, cameras, GPS, and smartphones. Efficient management of transportation can be done to avoid road accidents, help travelers to avoid traffic congestion, and help travelers to opt for a suitable route by enabling route navigation. In addition to this, traffic lights and public transportation can be used to reduce traffic.

Another example is smart healthcare; e-medical checkup facilities, e-medical prescriptions, and e-healthcare monitoring facilities have been provided to patients. This has brought a revolution in the development of the smart healthcare sector. All the basic services in various sectors are getting advanced and smarter to provide a comfort zone to the citizens (Zhang et al., 2017; Banerjee et al., 2022).

1.1.4 A MULTI-SCALAR APPROACH TO "SMART CITY" PLANNING

Cities, without question, are complicated systems. The "smart" challenge presents a fresh perspective on the city system, implying the ability to take meaningful actions in response to specific challenges. Indeed, rethinking city planning through the lens of "smart rules" necessitates novel planning practices.

The key to "smart city" planning is to incorporate objectives and collaboration among the critical parts that contrast the challenges.

The authors' concept of smart planning a district implies that smart planning entails optimal planning, which takes into account factors such as resource efficiency, technology use, and sustainability objectives while also respecting human capital and investing in the life quality of its users. The study takes a multi-scalar approach to the notion of smart planning, from the regional (district) to a global (city) (shown in Figure 1.2).

1.2 SUSTAINABLE DEVELOPMENT: KEY OBJECTIVE

Everyday activities, such as transportation, governance, agriculture, maintenance, logistics, education, and healthcare, may now be remotely handled and controlled via smart devices, thanks to the widespread adoption of technology in these areas.

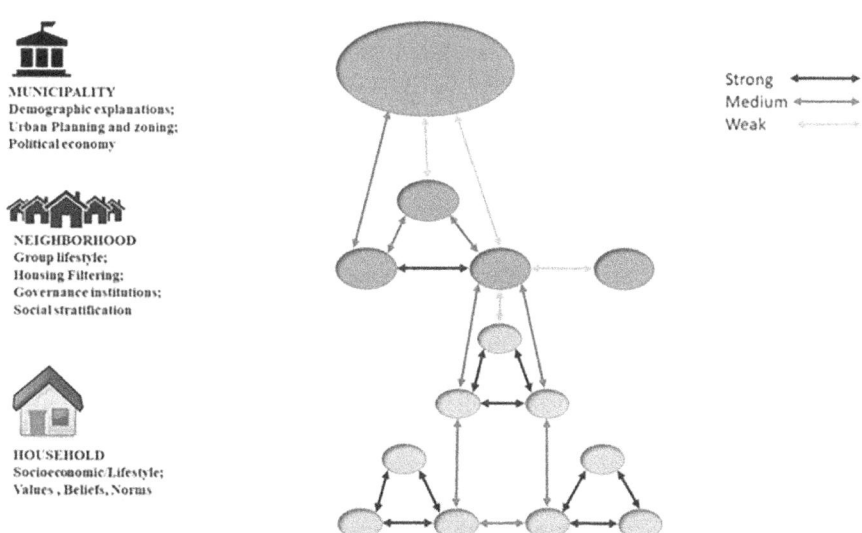

MUNICIPALITY
Demographic explanations;
Urban Planning and zoning;
Political economy

NEIGHBORHOOD
Group lifestyle;
Housing Filtering;
Governance institutions;
Social stratification

HOUSEHOLD
Socioeconomic Lifestyle;
Values, Beliefs, Norms

Strong
Medium
Weak

FIGURE 1.2 Smart city planning: a multi-scalar approach.

As a result, the idea of smart cities was born, in which existing traditional infrastructure is combined with modern information and communication technology, and the system is then digitally coordinated and regulated as a whole. A growing number of nations around the globe are embracing this idea and developing their own form of a smart city (Khang & Khanh, 2021).

The sensors and actuators built into smart gadgets are at the core of the smart city, allowing people to make better decisions about their environment. Using sensor data, the microcontrollers in these gadgets are programmed to make autonomous judgments. These technologies include artificial intelligence (AI), protocol-based communication and data exchange, the IoT, wireless sensor networks, and more (Zhang, Pee, Pan, & Cui, 2022).

Many researchers investigate the potential of smart city concepts and mechanisms to promote ecologically sustainable urban development. Recent studies have emphasized the importance of conducting a more systematic examination of the relationship between smart and sustainable cities, concentrating on real-world applications that might facilitate a stronger insight into the included domains, typologies, and design concepts, and this article aims to fill that research gap. Simultaneously, they examined whether these apps are capable of contributing to the "zero vision" strategy, a very ambitious goal in the context of smart cities (Silva, Khan, & Han, 2018).

In recent years, the idea of smart and sustainable cities has attracted a lot of attention. And it is fast gathering pace and global attention as a potentially transformative approach to the urban sustainability dilemma. This is especially true for environmentally and advanced technological nations.

Smart *and* sustainable cities are examined, covering their underlying assumptions and foundations, present state-of-the-art research and development, research vistas and opportunities, rising scientific and technical trends, and future planning practices.

Many authors have demonstrated that how smart and sustainable cities were discursively interpreted and literally created through socially defined understandings, socially anchored, and institutionalized ICT practices associated with the new generation of urban sustainability computing (Silva, Khan, & Han, 2018). Thus, such cities are medicated and located among cultures that are ecologically and technologically advanced. Additionally, as urban representations of scientific innovations and professional innovation, they are formed by sociocultural and politico-institutional institutions, as well as shaping them.

Additionally, the research confirms that the success and development of smart, sustainable cities are due to the transformative power, knowledge/power relationship, productive and constitutive force, and legitimization capacity inherent in ICT of the latest influx of processing for urban sustainability, as a result of its association with scientific discourse and its societal implications.

City governments throughout the world have recently taken a conscious step toward making their cities smarter and more sustainable through the implementation of big data technologies and their applications in many urban domains in order to achieve and improve inhabitants' living standards.

In response to the need for sustainability and urbanization issues, smart and smarter cities are progressively adopting sophisticated forms of ICT in order to increase their productivity in accordance with sustainability goals and urban expansion needs. It is one of these kinds that has great potential to boost urban activities, operations, designs, services, strategies, and policies in this regard. This is because big data computing enables the kind of informed decision-making and better insights made possible by applied intelligence.

Today's study of smart cities' big data techniques and approaches is heavily weighted toward economic growth and life quality improvements in terms of service efficiency but ignores and hardly explores the enormous potential for such applications to advance sustainable development. Indeed, the creation and implementation of smart and smarter cities in such a sustainable manner raises a variety of concerns and presents significant challenges.

1.3 COMPUTATIONAL TECHNOLOGIES

The "smart city" concept represents an integrated environment that is difficult to describe with a single definition (Rani, Mishra, et al., 2021). But many authors gave different definitions of the most developing concept from time to time from different aspects.

In 2008, Hollands described smart city architecture by considering data transfer techniques, political changes, and social development (Hollands, 2008). In the year 2009, Caragliu explained this concept using the parameters of optimal resource usage, ICT infrastructure, economic development, and quality of life (Caragliu, Del Bo, & Nijkamp, 2013). Fundamentally, the concept of a smart city is realized by integrating a variety of technologies to provide better and sustainable solutions and better health and economic development for the residents.

The major focus in developing smart cities is to develop and modernize urban infrastructure to aid social development. A variety of technologies are amalgamated across different functional areas of a smart city to meet the fundamental objectives which are discussed in the following sections.

1.3.1 INTERNET OF THINGS (IoT)

Smart/IoT-equipped devices have become a very important part of our day-to-day life. IoT devices have captured our routine activities and facilitates our communication with the surrounding environment. IoT-supported systems consist of data collection, storage, processing, and analysis, along with users and other technologies (Rani, Khang, Chauhan, & Kataria, IoT & Healthcare, 2021).

The main idea behind the enormous development of the IoT framework is to support anytime-anywhere connectivity (Rani, Khang, Chauhan, & Kataria, 2021). Increasing usage of IoT devices is helping to provide solutions for many complex problems in an urban environment.

The fast growth of IoT devices is causing the evolution of many advanced concepts, among which smart cities are the prominent ones. In smart cities, IoT devices cater to a number of services to enhance the efficiency of the processes and provide quality of life to urban residents (Rani & Kumar, 2022; Rani et al., 2022).

A number of smart city application domains, like smart homes and buildings, smart healthcare, smart energy management, transportation management, and so on

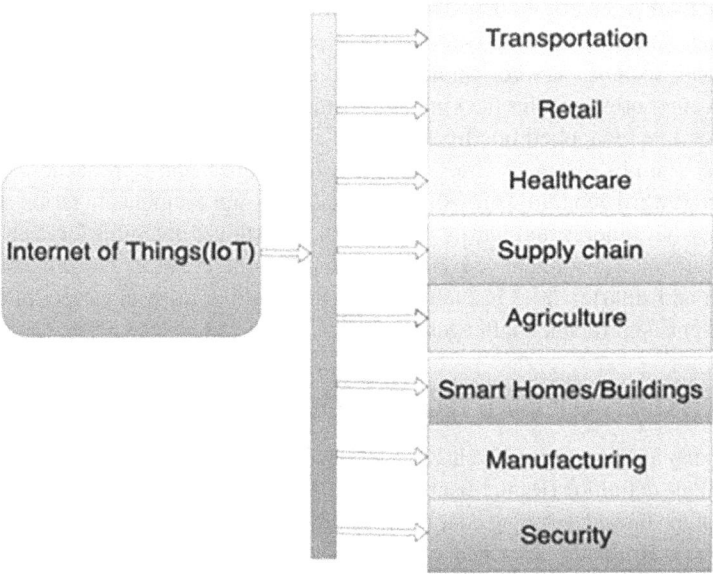

FIGURE 1.3 IoT applications in smart cities.

(shown in Figure 1.3), are adding comfort to the life of the citizens (Arya, Rani, & Choudhary, 2022; Rani, Mishra, et al., 2021). Along with providing a better quality of life, smart devices also provide sustainable solutions for different application areas.

It has been analyzed that to fulfill various requirements, the volume of smart devices is increasing abruptly in the urban environment, both in number and variety.

Consequently, managing this huge network of IoT devices is one of the biggest challenges. There are many other challenges experienced in different IoT-supported smart city services:

- Data is generated from different sources, in different formats, making them difficult to join and analyze.
- Data is gathered using different types of sensors with varying ranges, speeds of transfer, surface areas, and so on. Data gathered from a versatile environment becomes difficult to process and manage.

1.3.2 ARTIFICIAL INTELLIGENCE (AI)

Artificial intelligence (AI) plays a very important role in various smart city domains (Voda & Radu, 2018). It is contributing enormously to the rapid development of the smart city ecosystem.

The term "artificial intelligence" was introduced by John McCarthy in 1979 (Srivastava, Bisht, & Narayan, 2017). Initially, it was introduced with a vision to portray human behavior using neuroscience, mathematics, and psychology.

But over the years, it has captured many domains of real life. When it comes to smart cities, it is facilitating various domains like education, safety, healthcare, agriculture, industry, transport, urban planning, tourism, waste management, and many

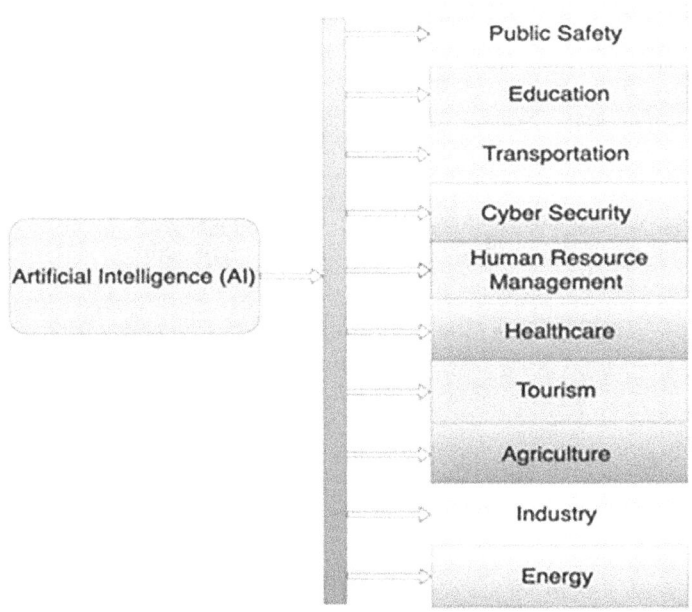

FIGURE 1.4 AI applications in smart cities.

more to cater to convenient and sustainable solutions (shown in Figure 1.4) (Rani, Bhambri, & Chauhan, 2021).

Along with benefits, the usage of AI in smart city applications faces the following issues (Inclezan & Pradanos, 2017):

- Its usage becomes a challenge in the domains where data is not readily available.
- Heterogeneity of data is another major challenge.
- AI applications are controlled by social, ethical, and legal policies.

1.3.3 BLOCKCHAIN

Blockchain is a distributed ledger technology that can be practiced with any class of transaction (Berdik, Otoum, Schmidt, Porter, & Jararweh, 2021). It is a public database to execute secure and reliable transactions (Namasudra, Deka, Johri, Hosseinpour, & Gandomi, 2021).

It plays a very important role in banking, finance, cryptocurrency, organization of medical data, supply chain, and so on. A voluminous amount of data is generated in a smart city ecosystem, which requires reliable, fast, and secure data transfer framework (Deepa et al., 2022). The Validity of the sender is also important in such kind of environment.

As sensitive data is generated from many applications, so privacy is also a major concern. Old methods are facing a few issues like dependency on data sources, lack of privacy, and centralized data storage (Hasan, Sivakumar, & Khang, 2022), to validate data integrity:

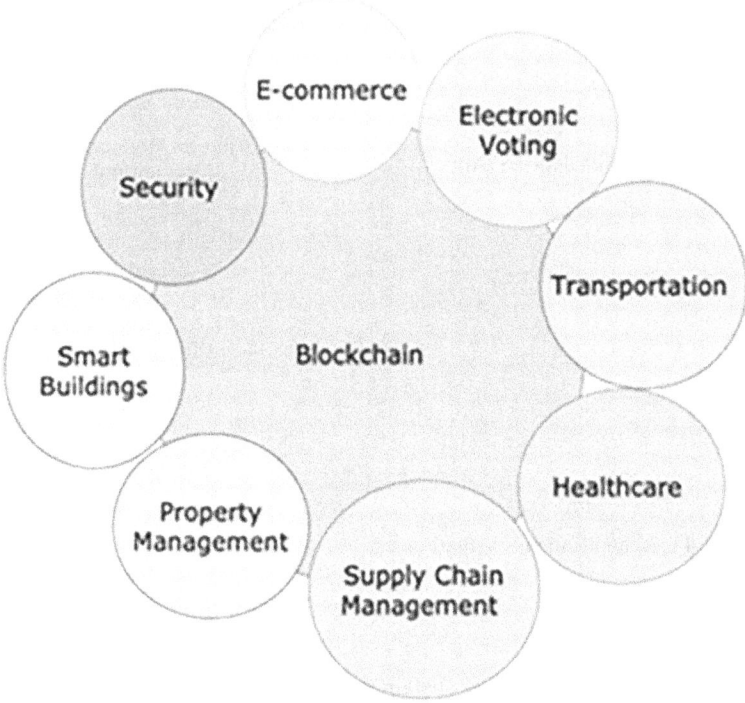

FIGURE 1.5   Applications of blockchain technology in smart cities.

Blockchain has evolved as e very fruitful technology to manage various data security issues in a smart city ecosystem (Bhardwaj, Negi, Nagrath, & Mittal, 2022). Although blockchain caters to the security requirements of a wide variety of infrastructure used in various smart city applications (Biswas & Muthukkumarasamy, 2016) (shown in Figure 1.5), the deployment of these solutions faces the following issues:

- Techniques used to generate and verify the keys are threat prone.
- Communication lag is one of the major issues faced in the implementation of distributed blockchain applications.
- The role of artificial intelligence in blockchain applications must be clearly identified (Khang & Khanh, AI & Blockchain, 2021).
- Difference in the computational competence of various IoT devices is also a major challenge in deploying blockchain solutions for smart infrastructure.
- Reliability of the different nodes is also a major point of concern for distributed blockchain applications.

1.3.4 BIG DATA

A huge volume of data, in the size of petabytes and exabytes, is generated regularly in various smart city subsystems. This voluminous amount of data is known as big data.

The key characteristics of big data are volume, velocity, and variety (Rodríguez-Mazahua et al., 2016). This huge volume of data is administered using a variety of data analytics and management tools, which aid many smart city processes (shown in Figure 1.6) (Alshawish, Alfagih, & Musbah, 2016).

- Data are gathered using different types of sensors deployed in various applications, so they are available in various formats and, consequently, difficult to manage.
- Data is highly prone to cyberattacks as communicated in cyberspace, which initiates the requirement of standard security protocols.
- Most of the data gathered through various applications is unstructured (Khang, Chowdhury, & Sharma, Data-Driven Blockchain Ecosystem, 2022).

But big data analytics also face a number of issues (Bhadani & Jothimani, 2016).

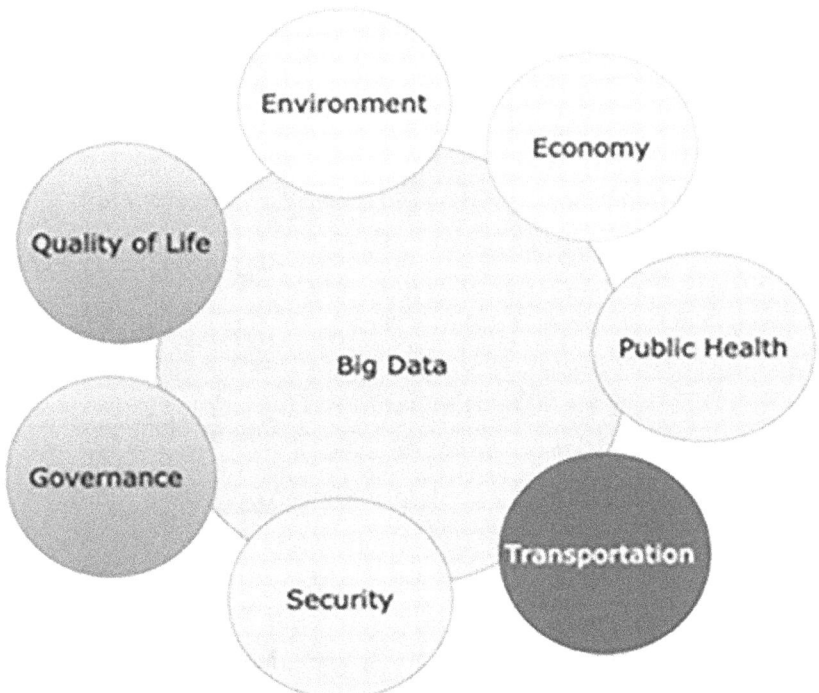

FIGURE 1.6   Big data applications in a smart city environment.

1.3.5 CLOUD COMPUTING

One of the most powerful technologies to store and manage a huge volume of data gathered through various smart city applications is cloud computing, which has become very popular in usage in the last few years (Kaur, Kaur, & Rani, 2015). It causes optimal resource utilization by putting them in cyberspace, which is available on the need of the application (Bhambri, Rani, Gupta, & Khang, 2022).

In a smart city ecosystem, cloud frameworks offload the burden of storage and processing from diverse local infrastructure and enhance the quality of the services. Various smart city services are characterized using the features size of data, storage requirement, computational needs, communication delays, and throughput (Marinescu, 2022). Cloud computing contributes to a number of smart city services, as shown in Figure 1.7.

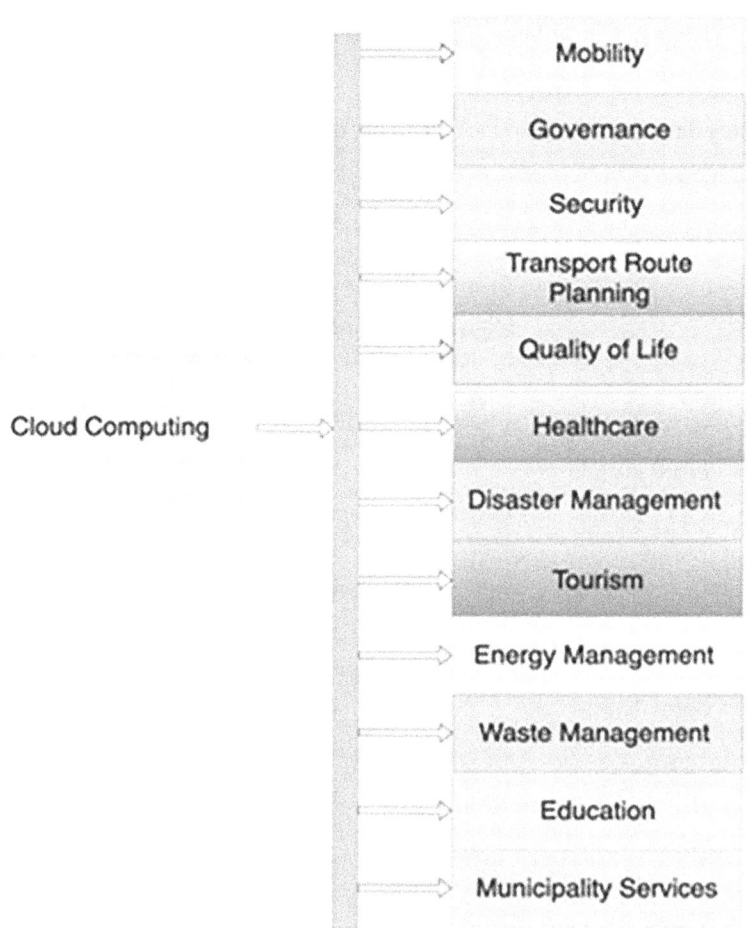

FIGURE 1.7   Cloud-computing applications.

Cloud deployment for smart city applications faces a few challenges:

- Security of residents' data gathered through various applications may be compromised.
- Clod platforms provided by different companies are not scalable (Khang, Kataria, Bhambri, & Rani, Big Data, Cloud Computing & IoT, 2022).
- Data are gathered through various smart devices and equipment, so they are diverse in nature.

1.4 COMMUNICATION TECHNOLOGIES

Along with computational technologies, communication technologies are also playing a very critical role in the rapid development of sustainable smart cities. These technologies support all-time, everywhere connectivity among various entities in a smart city environment. These technologies also provide a platform to take maximum benefit from other technologies (Rani et al., 2021).

Different types of technologies are deployed in various applications as per the requirement of communication speed, coverage, and infrastructure (Yaqoob et al., 2017).

Some of the advanced communication and network technologies employed in a smart city environment are as follows:

- IPv6 low-power wireless personal area network (LowPAN)
- Visible light communication (VLC)
- Green communication (GC)
- Neul
- NFC (near field communication)
- Cognitive radio networks (CRN)
- Software-defined wireless networks (SDWN)
- Thread (IP-based IPv6 networking protocol)
- Sigfox
- Network function virtualization (NFV)

1.5 APPLICATIONS

Mariem and Mahmoud (2015) provided a method for designing a network based on the most recent WiMax standard, 802.16m, and for dimensioning and planning such a network in the Sidi-Bouzid region in order to reduce interferences, noise, and cut-out.

Many network topologies, including pure wireless, wireless-optical/optical-wireless, and pure optical, were simulated. The results confirm a decrease in bandwidth blockage ratio across all networks studied, as well as optimal network resource usage (Naik & De, 2018).

Zemrane, Abbou, Baddi, and Hasbi (2018) chose metropolitan area networks, such as the WiMax-Mobil protocol, to connect sensors with a 50-kilometer range. The simulation is demonstrated using the OPNET simulator, which simulates protocol performance at the data link level when used with a voice application and when used with an HTTP application.

Lee and Kim (2021) examined the difficulties and issues (such as wireless coverage, mobility, bandwidth, and signal strength) that 4G wireless networks confront when delivering quality service.

A framework enabling future 4G wireless connections based on existing wireless connectivity in smart cities was also provided in the study. In the Sultanate of Oman, advanced services—including smart parking systems, smart transportation systems, and pervasive computing, which will improve people's social lives, the sultanate's economy, and the overall well-being with better quality of life—will be made available as a result of this proposed solution.

Dong and Wang (2016) advocated that the logical nodes of a Long-Term Evolution (LTE) system be allocated the responsibility of providing immediate proximity solutions to user equipment (UE). It demonstrates the recommended solutions through the use of a restaurant recommendation system that is based on the availability of adjacent parking.

The proposed methods facilitate collaboration and interaction between numerous public services in smart cities, as well as a large reduction in network traffic generated by an evolved packet core (EPC) and the Internet.

Polese, Centenaro, Zanella, and Zorzi (2016) offered a patch to ns-3, one of the most widely used open-source network simulators, in order to improve the accuracy of the procedure that replicates an LTE random-access channel (RACH). The updated version of the multiple-access technique is contrasted to the default one, and using a simulation campaign, the challenges associated with enormous levels of connectivity through wireless connections in LTE are explored.

Raza (2016) conducts a quick examination of public safety service use cases and requirements, addresses radio and core technology advancements, and outlines the dangers associated with partnership-based methods.

Ali, Hossain, and Kim (2017) explained the current state-of-the-art approaches for limiting MTC devices' massive random access in LTE/LTE-A networks. Five important measures are used to compare the proposals: access success rate, access delay, QoS guarantee, power efficiency, and influence on HTC.

Additionally, a unique collision resolution, multiple access framework for massive MTC over LTE/LTE-A, was proposed. This model resolves preamble collisions rather than avoiding them and is geared toward managing massive and busty access attempts.

The proposed model's simulations demonstrate significant improvements in the random-access success rate when compared to standard slotted Aloha-based models. Additionally, the new model can coexist with current LTE/LTE-A MAC protocols, ensuring high network reliability and efficiency.

Samoilenko, Accurso, and Malandra (2020) presented SimuLTE software used to study how IoT traffic and users could cohabit in a smart-city LTE infrastructure. We used publicly available actual geographic data on the positions of LTE base stations and IoT devices.

Key network metrics, including user throughput and cell utilization, were used to evaluate both network and user performance. When IoT traffic is introduced into the network, major performance deterioration is observed, according to the simulation results.

Cherkaoui, Keskes, Rivano, and Stanica (2016) made significant contributions to the modeling and computation of the LTE-A Random Access Channel's (RACH) capacity in terms of concurrent successful access. It tested the hypothesis of piggybacking machine-type communications payloads from M2M devices within the RACH and demonstrated that M2M densities perceived as realistic for smart city applications are tough to maintain with the current LTE-A architecture.

Along with significantly lowering the power consumption of UEs running video-streaming services and diverse data applications (DDA), the suggested HPC-DRX/TRTS also maintains the video playback quality at UEs (Li & Chen, 2018).

Kumar, Bhagyalakshmi, Lavanya, and Gowranga (2016) proposed a system to reduce Bluetooth's power consumption via Bluetooth Low Energy (BLE) technology while also preventing fraud and favors in an attendance system through the use of a Bluetooth smart beacon.

To facilitate real-time communication with smart city items, personal area networks (PANs) have been developed to include BLE as the underlying technology. A smart gate prototype was built on the college campus to demonstrate the viability of the suggested approach (Bhattacharjee et al., 2017).

Hasan and Hasan (2020) created FinderX, a Bluetooth beacon-based technology for locating nearby trash bins and bathrooms. The system delivers real-time location information for adjacent amenities depending on the position of the associated beacons.

The system is self-contained and does not require Internet or communication facilities to operate. It also operates indoors, where a GPS signal is not available. To illustrate FinderX's practicality, a testbed has been established, and the system is extensively assessed in an urban context. It was determined that FinderX displayed appropriate usability and feasibility and a reduction in the average time required to locate a garbage bin of 18.98%.

Hasan, Hasan, and Islam (2021) created Insight, a Bluetooth beacon-based system that employs a smartphone application to detect signals from beacons indicating danger zones. Because the system is not dependent on the Internet or communication infrastructure, it is resistant to communication disruptions during disasters. To validate Insight's feasibility, a study was conducted on an urban university campus. The system was found to be sufficiently usable and feasible.

Ali, Chew, Khan, and Weller (2017) proposed a cloud-based automatic meter reading (AMR) WSN system that enables utilities to improve demand side management and control of electricity demand while also enabling customers to monitor and optimize their electricity usage and contribute to peak load reduction. The experimental results obtained with the proposed AMR testbed illustrate the system's significant potential for use in smart cities.

Naik, Das, and Bindiya (2018) discussed the development of a unit that detects the presence of hazardous gases, uploads the data to a website, and also sends alerts to affected individuals. The vehicle monitoring unit, which is installed in a vehicle, tracks the vehicle's location, detects vehicle accidents, and monitors the engine's temperature and the presence of toxic gases in the exhaust.

If the vehicle is stolen, it also includes a feature that locates the vehicle and prevents it from moving until the owner sends a message. Vehicles that perform signal jumping are detected and penalized during implementation. This feature is critical in nations such as India, where traffic restrictions are often ignored (Tailor, Ranu Pareek, & Khang, 2022). The density of traffic at traffic intersections is measured, and the webpage is updated with the results. An Android application is designed to ensure that all necessary information is conveniently accessible.

Chowdary, Abd El Ghany, and Hofmann (2020) presented an IoT-based smart metering solution for smart cities. Data from smart meters is transmitted to the energy/utility provider via a network.

As a result, a smart meter network is constructed using star and mesh topologies. App, website, and database for electricity supplier company are all part of this proposed system. There is a 35% savings over the worldwide rival with the system architecture that is being supplied. A 25% reduction in power usage was achieved by following the recommended design.

Yu, Zhu, and Fan (2017) created a LoRaWAN simulator in MATLAB to provide a software-based method for performance evaluation. The implementation of a practical LoRaWAN network encompassing the Greater London area. Its performance is assessed using two representative city monitoring apps.

Economic analysis and business models for networks were developed to serve as a guide for commercial network operators, city planners, and IoT vendors investigating future LoRaWAN deployments for smart city applications.

1.6   WORLD'S TOP SMART CITY MODELS

Any smart city is an amalgamation of a wide variety of concepts, entities, subsystems, and infrastructures. Two fundamental objectives behind the development of any smart city are sustainability of the services and quality of life (QoL).

Another major motivation is the optimal use of resources. With the same objective, 181 cities in the world are listed as the smartest cities. Some of the famous smart cities are presented in Table 1.1.

TABLE 1.1
Top Smart Cities in the World

Smart City	Name of the Country
London	UK
San Francisco	USA
Barcelona	Spain
Stander City	Spain
Nice	France
Padova	Italy
Singapore	Singapore
Stratford	Canada
Jakarta	Indonesia
Moscow	Russia

1.7 CONCLUSION

It has been observed that the need for sustainable solutions is the key requirement in various smart city application domains. Both computational and communication technologies play a very vital role in achieving this objective. But there is a requirement for more standardized protocols to administer the data gathered through various applications using various technologies. A standardized protocol will contribute toward the sustainability of the solutions and more optimal usage of the existing infrastructure.

REFERENCES

Ali, H., Chew, W., Khan, F., & Weller, S. R. (2017). *Design and implementation of an IoT assisted real-time ZigBee mesh WSN based AMR system for deployment in smart cities.* Paper presented at the 2017 IEEE International Conference on Smart Energy Grid Engineering (SEGE). doi: 10.1109/SEGE.2017.8052810

Ali, M. S., Hossain, E., & Kim, D. I. (2017). LTE/LTE-A random access for massive machine-type communications in smart cities. *IEEE Communications Magazine, 55*(1), 76–83. doi: 10.1109/MCOM.2017.1600215CM

Alshawish, R. A., Alfagih, S. A., & Musbah, M. S. (2016). *Big data applications in smart cities.* Paper presented at the 2016 International Conference on Engineering & MIS (ICEMIS). doi: 10.1109/ICEMIS.2016.7745338

Arya, V., Rani, S., & Choudhary, N. (2022). *Enhanced bio-inspired trust and reputation model for wireless sensor networks.* Paper presented at the Proceedings of Second Doctoral Symposium on Computational Intelligence. doi: 10.1007/978-981-16-3346-1_46

Baig, M. N., Himarish, M. N., Pranaya, Y., & Ahmed, M. R. (2018). *Cognitive architecture based smart homes for smart cities.* Paper presented at the 2018 2nd International Conference on Trends in Electronics and Informatics (ICOEI). doi: 10.1109/PGSRET.2017.8251799

Banerjee, K., Bali, V., Nawaz, N., Bali, S., Mathur, S., Mishra, R. K., & Rani, S. (2022). A machine-learning approach for prediction of water contamination using latitude, longitude, and elevation. *Water, 14*(5), 728. doi: 10.3390/w14050728

Berdik, D., Otoum, S., Schmidt, N., Porter, D., & Jararweh, Y. (2021). A survey on blockchain for information systems management and security. *Information Processing & Management, 58*(1), 102397. doi: 10.1016/j.ipm.2020.102397

Bhadani, A. K., & Jothimani, D. (2016). Big data: challenges, opportunities, and realities. In *Effective big data management and opportunities for implementation* (pp. 1–24): IGI Global. doi: 10.4018/978-1-5225-0182-4.CH001

Bhambri, P., Rani, S., Gupta, G., & Khang, A. (2022). *Cloud and fog computing platforms for internet of things*: CRC Press. ISBN: 978-1-032-101507. doi:10.1201/9781032101507

Bhardwaj, J., Negi, R., Nagrath, P., & Mittal, M. (2022). Applications of blockchain in various domains. In *Smart IoT for research and industry* (pp. 1–30): Springer. doi: 10.1007/978-3-030-71485-7_1

Bhattacharjee, A. K., Bruneo, D., Distefano, S., Longo, F., Merlino, G., & Puliafito, A. (2017). *Extending bluetooth low energy pans to smart city scenarios.* Paper presented at the 2017 IEEE International Conference on Smart Computing (SMARTCOMP). doi: 10.1109/SMARTCOMP.2017.7947007

Biswas, K., & Muthukkumarasamy, V. (2016). *Securing smart cities using blockchain technology.* Paper presented at the 2016 IEEE 18th International Conference on High Performance Computing and Communications; IEEE 14th International Conference on Smart City; IEEE 2nd International Conference on Data Science and Systems (HPCC/SmartCity/DSS). doi: 10.1109/HPCC-SmartCity-DSS.2016.0198

Caragliu, A., Del Bo, C., & Nijkamp, P. (2013). Smart cities in Europe. In *Smart cities* (pp. 185–207): Routledge. doi: 10.1080/10630732.2011.601117

Cherkaoui, S., Keskes, I., Rivano, H., & Stanica, R. (2016). *LTE-A random access channel capacity evaluation for M2M communications*. Paper presented at the 2016 Wireless Days (WD). doi: 10.1109/WD.2016.7461480

Chowdary, S. S., Abd El Ghany, M. A., & Hofmann, K. (2020). *Iot based wireless energy efficient smart metering system using zigbee in smart cities*. Paper presented at the 2020 7th International Conference on Internet of Things: Systems, Management and Security (IOTSMS). doi: 10.1109/IOTSMS52051.2020.9340230

Deepa, N., Pham, Q.-V., Nguyen, D. C., Bhattacharya, S., Prabadevi, B., Gadekallu, T. R., . . . Pathirana, P. N. (2022). *A survey on blockchain for big data: approaches, opportunities, and future directions*. Future Generation Computer Systems. https://www.science-direct.com/science/article/abs/pii/S0167739X22000243

Dong, L., & Wang, G. (2016). *Enable close proximity services for smart cities with information centric LTE system*. Paper presented at the 2016 IEEE 83rd Vehicular Technology Conference (VTC Spring). doi: 10.1109/VTCSpring.2016.7504212

Falconer, G., & Mitchell, S. (2012). Smart city framework. *Cisco Internet Business Solutions Group (IBSG), 12*(9), 2–10. http://www.cisco.com/web/about/ac79/docs/ps/motm/Smart-City-Framework.pdf

Hasan, R., & Hasan, R. (2020). *Towards designing a sustainable green smart city using Bluetooth beacons*. Paper presented at the 2020 IEEE 6th World Forum on Internet of Things (WF-IoT). doi: 10.1109/WF-IoT48130.2020.9221118

Hasan, R., Hasan, R., & Islam, T. (2021). *InSight: A bluetooth beacon-based ad-hoc emergency alert system for smart cities*. Paper presented at the 2021 IEEE 18th Annual Consumer Communications & Networking Conference (CCNC). doi: 10.1109/CCNC49032.2021.9369621

Hasan, S., Sivakumar, T. B., & Khang, A. (2022). Cryptocurrency methodologies and techniques. In *The data-driven blockchain ecosystem: fundamentals, applications, and emerging technologies* (1st ed.) (pp. 27–37): CRC Press. https://doi.org/10.1201/9781003269281

Hollands, R. G. (2008). Will the real smart city please stand up? Intelligent, progressive or entrepreneurial? *City, 12*(3), 303–320. doi: 10.1080/13604810802479126

Inclezan, D., & Pradanos, L. I. (2017). A critical view on smart cities and AI. *Journal of Artificial Intelligence Research, 60*, 681–686. doi: 10.5555/3207692.3207707

Kaur, G., Kaur, R., & Rani, S. (2015). Cloud computing-a new trend in it era. *International Journal of Science, Technology, and Management*, 1–6. https://www.researchgate.net/publication/326439556_Cloud_Computing-A_new_Trend_in_IT_Era

Khang, A., Chowdhury, S., & Sharma, S. (2022). *The data-driven blockchain ecosystem: fundamentals, applications and emerging technologies*: CRC Press. ISBN: 978-1-032-21624. doi:10.1201/9781003269281

Khang, A., Kataria, A., Bhambri, P., & Rani, S. (2022). *Big data, cloud computing and internet of things*: CRC Press. ISBN: 978-1-032-284200. doi:10.1201/9781032284200

Khang, A., & Khanh, H. H. (2021). The role of artificial intelligence in blockchain applications. In *Reinventing manufacturing and business processes through artificial intelligence* (pp. 19–38). https://doi.org/10.1201/9781003145011

Kumar, B. A., Bhagyalakshmi, K., Lavanya, K., & Gowranga, K. (2016). *A bluetooth low energy based beacon system for smart short range surveillance*. Paper presented at the 2016 IEEE International Conference on Recent Trends in Electronics, Information & Communication Technology (RTEICT). doi: 10.1109/RTEICT.2016.7808018

Law, K. H., & Lynch, J. P. (2019). Smart city: technologies and challenges. *IT Professional, 21*(6), 46–51. doi: 10.1109/MITP.2019.2935405

Lee, H., & Kim, J. (2021). *Trends in blockchain and federated learning for data sharing in distributed platforms.* Paper presented at the 12th International Conference on Ubiquitous and Future Networks, ICUFN 2021. doi: 10.1109/ICUFN49451.2021.9528593

Li, M., & Chen, H.-L. (2018). Energy-efficient traffic regulation and scheduling for video streaming services over LTE-A networks. *IEEE Transactions on Mobile Computing, 18*(2), 334–347. doi: 10.1109/TMC.2018.2836421

Mariem, F., & Mahmoud, P. A. (2015). *4G WiMax network for smart-Sidi Bouzid area communication.* Paper presented at the 2015 SAI Intelligent Systems Conference (IntelliSys). doi: 10.1109/IntelliSys.2015.7361257

Marinescu, D. C. (2022). *Cloud computing: theory and practice*: Morgan Kaufmann. doi: 10.1016/C2012-0-02212-0

Naik, D. L., & Kiran, R. (2018). Naïve Bayes classifier, multivariate linear regression and experimental testing for classification and characterization of wheat straw based on mechanical properties. *Industrial Crops and Products, 112*, 434-448. doi: 10.1016/j.indcrop.2017.12.034

Naik, D. R., Das, L. B., & Bindiya, T. (2018). *Wireless sensor networks with Zigbee and WiFi for environment monitoring, traffic management and vehicle monitoring in smart cities.* Paper presented at the 2018 IEEE 3rd International Conference on Computing, Communication and Security (ICCCS). doi: 10.1109/CCCS.2018.8586819

Naik, D. R., & De, T. (2018). *Congestion aware traffic grooming in elastic optical and WiMAX network.* Paper presented at the 2018 Technologies for Smart-City Energy Security and Power (ICSESP). doi: 10.1109/ICSESP.2018.8376693

Namasudra, S., Deka, G. C., Johri, P., Hosseinpour, M., & Gandomi, A. H. (2021). The revolution of blockchain: state-of-the-art and research challenges. *Archives of Computational Methods in Engineering, 28*(3), 1497–1515. doi: 10.1007/s11831-020-09426-0

Polese, M., Centenaro, M., Zanella, A., & Zorzi, M. (2016). *M2M massive access in LTE: RACH performance evaluation in a smart city scenario.* Paper presented at the 2016 IEEE International Conference on Communications (ICC). doi: 10.1109/ICC.2016.7511430

Rana, G., Khang, A., Sharma, R., Goel, A. K., & Dubey, A. K. (2021). *The role of artificial intelligence in blockchain applications.* Reinventing Manufacturing and Business Processes Through Artificial Intelligence. doi:10.1201/9781003145011

Rani, S., Bhambri, P., & Chauhan, M. (2021). *A machine learning model for kids' behavior analysis from facial emotions using principal component analysis.* Paper presented at the 2021 5th Asian Conference on Artificial Intelligence Technology (ACAIT). doi: 10.1109/ACAIT53529.2021.9731203

Rani, S., Kataria, A., Chauhan, M., Rattan, P., Kumar, R., & Sivaraman, A. K. (2022). Security and privacy challenges in the deployment of cyber-physical systems in smart city applications: state-of-art work. *Materials Today: Proceeding.* doi: 10.1016/j.matpr.2022.03.123

Rani, S., Khang, A., Chauhan, M., & Kataria, A. (2021). *IoT equipped intelligent distributed framework for smart healthcare systems.* Networking and Internet Architecture. https://arxiv.org/abs/2110.04997v2, doi:10.48550/arXiv.2110.04997

Rani, S., & Kumar, R. (2022). Bibliometric review of actuators: Key automation technology in a smart city framework. *Materials Today: Proceedings.* doi: 10.1016/j.matpr.2021.12.469

Rani, S., Mishra, R. K., Usman, M., Kataria, A., Kumar, P., Bhambri, P., & Mishra, A. K. (2021). Amalgamation of advanced technologies for sustainable development of smart city environment: a review. *IEEE Access, 9*, 150060–150087. doi: 10.1109/ACCESS.2021.3125527

Raza, A. (2016, August). LTE network strategy for smart city public safety. In *2016 IEEE international conference on emerging technologies and innovative business practices for the transformation of societies (EmergiTech)* (pp. 34–37): IEEE. doi: 10.1109/EmergiTech.2016.7737306

Rodríguez-Mazahua, L., Rodríguez-Enríquez, C. A., Sánchez-Cervantes, J. L., Cervantes, J., García-Alcaraz, J. L., & Alor-Hernández, G. (2016). A general perspective of Big Data: applications, tools, challenges and trends. *The Journal of Supercomputing, 72*(8), 3073–3113. doi: 10.1007/s11227-015-1501-1

Samoilenko, R., Accurso, N., & Malandra, F. (2020). A simulation study on the impact of IoT traffic in a smart-city LTE *network*. In *2020 IEEE 31st annual international symposium on personal, indoor and mobile radio communications* (pp. 1–6): IEEE. doi: 10.1109/PIMRC48278.2020.9217179

Silva, B. N., Khan, M., & Han, K. (2018). Towards sustainable smart cities: a review of trends, architectures, components, and open challenges in smart cities. *Sustainable Cities and Society, 38*, 697–713. doi: 10.1016/j.scs.2018.01.053

Srivastava, S., Bisht, A., & Narayan, N. (2017, January). Safety and security in smart cities using artificial intelligence—a review. In *2017 7th international conference on cloud computing, data science & engineering-confluence* (pp. 130–133): IEEE. doi: 10.1109/CONFLUENCE.2017.7943136

Tailor, R., Ranu Pareek, K., & Khang, A. (Eds.). (2022). Robot process automation in blockchain. In *The data-driven blockchain ecosystem: fundamentals, applications, and emerging technologies* (1st ed.) (pp. 149–164): CRC Press. https://doi.org/10.1201/9781003269281

Voda, A. I., & Radu, L. D. (2018). Artificial intelligence and the future of smart cities. *BRAIN. Broad Research in Artificial Intelligence and Neuroscience, 9*(2), 110–127. https://lumenpublishing.com/journals/index.php/brain/issue/view/154

Yaqoob, I., Ahmed, E., Hashem, I. A. T., Ahmed, A. I. A., Gani, A., Imran, M., & Guizani, M. (2017). Internet of things architecture: recent advances, taxonomy, requirements, and open challenges. *IEEE Wireless Communications, 24*(3), 10–16. doi: 10.1109/MWC.2017.1600421

Yu, F., Zhu, Z., & Fan, Z. (2017, October). Study on the feasibility of LoRaWAN for smart city applications. In *2017 IEEE 13th international conference on wireless and mobile computing, networking and communications (WiMob)* (pp. 334–340): IEEE. doi: 10.1109/WiMOB.2017.8115748

Zhang, D., Pee, L. G., Pan, S. L., & Cui, L. (2022). Big data analytics, resource orchestration, and digital sustainability: A case study of smart city development. *Government Information Quarterly, 39*(1), 101626. doi: 10.1016/j.giq.2021.101626

Zhang, K., Ni, J., Yang, K., Liang, X., Ren, J., & Shen, X. S. (2017, January). Security and privacy in smart city applications: challenges and solutions. *IEEE*, 122–129. doi: 10.1109/MCOM.2017.1600267CM

Zhao, X., Askari, H., & Chen, J. (2021). Nanogenerators for smart cities in the era of 5G and Internet of Things. *Joule, 5*(6), 1391–1431. doi: 10.1016/j.joule.2021.03.013

2 Cyber-Physical-Social System and İncident Management

Alex Khang, Vladimir Hahanov, Gardashova
Latafat Abbas, Vugar Abdullayev Hajimahmud

CONTENTS

2.1 INTRODUCTION

The creation of robots (artificial intelligence [AI]) that could think like humans and judge and decide in real and uncertain conditions was the first major step in breaking down the wall between humans and robots (Khang et al., 2021). Although this is event happened almost half a century ago, it can be considered a great event to connect the virtual world with the physical world. It is known that in many parts of the world today, robots are trusted as much as humans and are given "human tasks," sometimes even more.

DOI: 10.1201/9781003252542-2

One of the main problems today is that two different races (human and robot) live in the same world in "peace"—in other words, making robots our most loyal helpers (Arunachalam et al., 2021). It is known that, with the exception of many economically developed countries, the integration of robots into human life in many countries is still not up-to-date. However, experts are constantly working to change this in the future to create conditions for robots and humans to "live" together (Bhambri et al., 2022). In Japan, for example, research is underway to make autonomous robots closer to humans and access them to our homes.

However, it was the concept of the cyber-physical system (CPS) that "built" the main bridge between the virtual world and the physical world (Gupta and Rani, 2013). Despite being a relatively new concept, CPS has, to some extent, completed its development. Work is still underway to develop it.

2.2   CYBER-PHYSICAL SYSTEM REVIEW

2.1.1   CYBER-PHYSICAL SYSTEM DEFINITION

CPS are systems that connect the physical world with the virtual processing world. CPS has different ingredients. These components include modern preventive information technologies.

In other words, CPS collects information in the real (physical) world, analyzes this information using digital technology in the virtual (cyber) world, facilitates the use of information and knowledge, and returns that information to the physical side to create added value. In CPS-Toshiba, the main component of CPSs (as well as, in a sense, the Internet of Things [IoT]) are sensors (Kaur et al., 2015). We will talk about this in detail in the "CPSS Security" section.

In a CPS, various information in the real world (physical space) is collected by IoT devices and processed and analyzed in the virtual world (cyberspace) (Bhambri et al., 2022).

CPS can be encountered in many areas to some extent. Mainly, on the eve of the fourth industrial revolution, CPS began to be applied in production and other fields. For example, the leading areas where CPS is applied are aviation, automotive industry, chemical processes, civil infrastructure, energy, health, manufacturing, transport, entertainment, and consumer goods.

2.1.2   CPS AND OTHER TECHNOLOGIES

CPS and IoT are three concepts that are closely intertwined in a new generation of cooperative solutions in which people, autonomous devices, and the environment interact to achieve specific goals (Ochoa et al., 2017).

The closest concept to the concept of CPS is IoT. The IoT ecosystem can be seen as a subsystem of the CPS. The basis of the CPS is the IoT, or more precisely, its virtual part is this technology. Suppose we need to create a bridge between the virtual world and the physical world (Rani, Bhambri et al., 2021).

Of course, for this to happen, the two technologies need to have similar aspects in order for them to interact with each other. While the virtual part is made up of interacting smart devices that create IoT technology, the physical part is made up of interacting (smart) people (Rani and Gupta, 2017). The common denominator is that both sides are "smart." Let us take a closer look at the concept of IoT technology, which is the basis of CPS.

2.1.2.1 Internet of Things and CPS

"It would be even more useful if one smart device was connected to another smart device." This idea is a transition to the concept of the IoT in a broad sense. That is, the IoT is an ecosystem created by the interaction of smart devices. In other words, the IoT is not necessary for the existence of smart devices, but smart devices are necessary for the existence of the IoT (Rani et al., 2022). Rather, smart devices with Internet access and even AI are key factors in the availability of this technology.

The IoT is an environment where objects, each with a unique identifier, can send information over a network. There is no need for a human-to-human or human-computer relationship to exchange this information (Rani, Khang et al., 2021). The IoT is created by combining wireless technologies, micro-electronic systems, and the Internet. Therefore, this concept can be called the Internet of Everything (Singh et al., 2017).

2.1.2.2 The Difference between the Internet of Things and the Cyber-Physical System

IoT guarantees that any smart device with Internet access is part of the IoT ecosystem. In other words, any smart device with Internet access can create an IoT ecosystem. This is the basis of the IoT technology.

One of the goals of CPS, unlike the İoT, is to make not only devices with Internet access but also devices without the Internet interact with each other.

It is this factor that makes the fundamental difference between CPS and İoT. There are many areas where CPS and IoT are used together. The basics are as follows:

- Smart cyber-physical factories
- Smart cities
- Smart cyber universities

2.1.2.3 Smart Cyber-Physical Factories

A smart factory is an IT-based factory. A smart factory collects information, such as factory equipment and work content, making full use of IoT. By transmitting the analyzed data to in-house and out-of-factory systems and to the head office, it is possible to optimize production efficiency and provide high-quality products.

Implementing smart cyber-physical factories is a rather difficult process. Because these systems are expensive. Nevertheless, many economically developed countries are investing heavily in this area. There are several reasons for this:

- Smart factories are expected to solve the shortage of workers by replacing manual work with machines such as robots.

- It is possible to reduce costs by reducing energy consumption and increasing work efficiency by optimizing work.

However, it should be noted that smart factories may not show results immediately after presentation (NETA122, 2018). Of course, this factor applies to every new project used.

2.1.2.4 Smart Cities

There are many different systems within smart cities—smart homes, smart transportation, smart healthcare, smart management, smart store chains, and more (Rani, Khang et al., IoT & Healthcare, 2021). The interconnectedness of smart devices within the smart city, the availability of not only machine-to-machine but also human-to-machine connections is a particular impetus for the development of İoT ecosystems and CPSs.

In terms of the function and structure of the ICT system that builds a smart city, it can be said that it is the CPS itself. CPS is a system that optimizes activities in real space by collecting and analyzing information in virtual space that reflects the state and movement of real space, such as residential areas and streets.

In addition to various data terminals, IoT devices must be used to collect data. Big data analysis using cloud or AI is required for analysis and optimization in virtual space (Khang and Khanh, AI & Blockchain, 2021).

Moreover, a data terminal is required to receive the analysis results and provide some information to the residents. In the future, control data will be transferred to infrastructure equipment, self-propelled (driverless) cars, robots, and so on will be returned and will become an autonomous system that operates optimally according to the conditions and movements of the city.

2.1.2.5 Smart Cyber University

A smart cyber-physical system (SCPS) is a combination of virtual and real components related to communication in the digital metric space with adequate real-time physical monitoring and optimal cloud management functions to achieve the set goals.

The essence of cyber-physical systems is important in the management of a smart university.

A smart cyber university is a unit to conduct relevant scientific research through adequate monitoring and cloud management (Khang et al., Big Data, Cloud Computing & IoT, 2022) of digital science and education processes and events in order to attract investment and achieve high-quality living conditions for employees and to train specialists with academic and scientific degrees required by the market is a metric culture that connects personnel and smart infrastructure to the network as Figure 2.1.

These examples are just three of the many areas of joint use of CPS and İoT, such as digital twin technologies and autonomous robots. Areas such as İoT and CPS are also areas where they are applied jointly.

FIGURE 2.1 Smart cyber university structure.

2.3 SECURITY: A GLOBAL PROBLEM

Security is a global problem. And it is not only available on Internet networks. Security as a global problem can be characterized by such types as the following:

- Global terrorism
- Government instability
- External conflict
- Kidnapping and usurpation
- Serious and organized crime
- General and low-level crime
- Spying and cyberattacks
- Civil unrest and disorder

One of today's biggest security challenges can be seen as "data loss as a result of a cyberattack."

Although the increase in information has solved many problems: finding a lot of information on any topic has also created many problems. Compared to previous years, it is much more difficult to distinguish between valuable and invalid information. Anyone has the right to submit an article in any form of "authorship."

The biggest problem is data protection. While it is now possible to obtain any information at any time, very important personal information can be lost in the blink of an eye. This information can be stolen by a group of people—cybercriminals. The fact that large companies lose the information of their customers is a big problem. Although efforts are being made to prevent this, it is impossible to completely prevent data theft.

Information security is a situation where legitimate users can use information securely at any time. While this is legally true, there are certain exceptions where the situation is violated. In particular, there are four exceptions:

1. Information theft
2. Falsification of information
3. Personalization as a result of unauthorized use of identification information
4. Destroying/breaking the network (cyberattack)

3.3.1 INFORMATION THEFT

From the moment of birth, a person has a unique database in the state database. For example, time of birth, weight and height over time, vaccinations, place of study, place of residence, family, relatives, and everything else related to a person are stored in this database. Then, the most dangerous, which can be considered, is his card number, account, and more personal information, which are ideal information for cybercriminals.

The first step in a cyberattack is to steal information. Data theft can be classified as "unauthorized access to information." A third party accesses the user's information without their permission. The most stolen pieces of information are the following:

* Personal information, such as name and address
* Personal information of customers and general users trusted by companies
* Important information within the company, such as projects and products

Cybercriminals can use this information for various purposes. Certain regional and global crimes can be processed with your personal information without your knowledge—for example, using your name to deceive other users to obtain their information. These are, in a sense, the most harmless crimes. In addition, advanced crimes are possible.

2.3.2 FALSIFICATION OF INFORMATION

Falsification changes part of the research process, often allowing the results to appear more sensational and relevant than reality. In addition to fabrication and plagiarism, fraud is a serious investigative violation. In other words, the data is manipulated with the intention of creating a false impression. This includes image distortion, removal of extraneous indicators or "unfavorable" results, changing or adding data points, and so on.

For example, criminals can make changes to the code written on a site by accessing it and going to its "background." Also, they may add or delete any information that does not belong there.

2.3.3 PERSONALIZATION

Personalization is a result of unauthorized use of identification information—in other words, confronting a criminal who imitates you. The simplest case of this may be the theft of your social network profile. Many people who use social networks may encounter this. Celebrities can especially have more problems with this. To be sure that you are not facing this danger, you need to take certain steps.

You can check your email addresses, inboxes, and spam or trash folders, and see whom you last spoke to. You can also look at your favorite posts on other social networks, such as your Instagram profile, whom you talked to in your DM box, whether someone sent a message on your behalf, and so on. In the simplest case, you will be notified when another person logs into your account (Khang et al., Data-Driven Blockchain Ecosystem, 2022).

One of the most advanced levels of this threat is the threat of a cyberattack by a cybercriminal who can imitate you, use your card accounts, or make certain decisions on your behalf.

2.3.4 DESTROYING/BREAKING THE NETWORK (CYBERATTACK)

Destroying or disrupting a network is a threat commonly referred to as a cyberattack.

Unlike the previous three threats, this is a deliberate attack on a computer or network over the Internet. This is an attempt to use a compromised computer system to shut down computers, steal data, or launch additional attacks. Cybercriminals use a variety of methods to carry out cyberattacks, including malware, phishing, ransomware, DDoS, or other methods.

In the simplest case, you can take the following steps to protect yourself from cyberattacks:

1. Check if you are already involved in data corruption. As mentioned earlier, you can take a look at the trash or spam folder in your email account.
2. Check the strength of your passwords. Passwords should be long and consist of different letter-number-character combinations.
3. Avoid passwords like this in particular. The following passwords are the most common and the easiest to crack:
   123456 (or any chronological sequence numbers)
   987654321
   123123
   QWERTY
   111111
   Password
4. Don't reply to every message your emails receive or open them at that moment. Messages can infect your computer. We need to be very careful in this regard.
   • Do not open emails from unknown email addresses.
   • Throw attachments in the trash in unexpected emails.
   • Avoid risky clicks—enter the address in your browser instead.
5. You can protect your devices by doing the following:
   • Install antivirus software.
   • Set up a password, face scanning, or fingerprint reading that must be used to unlock your device.
   • Set the device to request a password before installing applications.
   • Hide your Bluetooth when not in use and turn off the automatic connection to networks.
   • Activate remote locking and/or deletion functions if your device supports them.

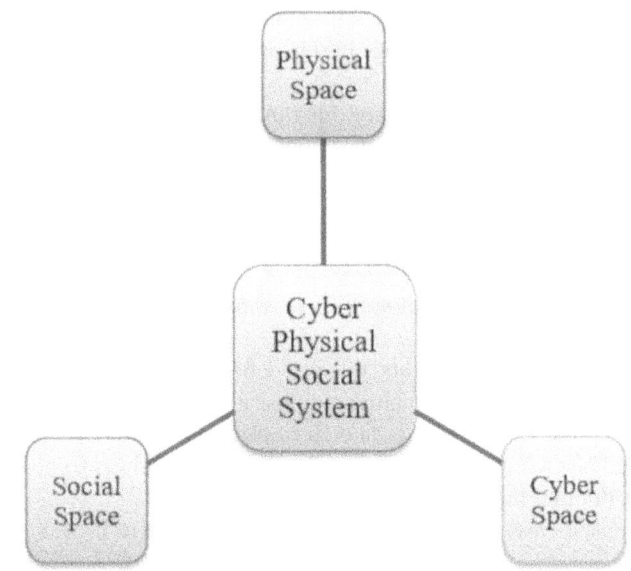

FIGURE 2.2 Cyber-physical-social system components.

2.4 CYBER-PHYSICAL SOCIAL SYSTEM

Cyber-physical-social systems (CPSS) is a continuation of cyber-physical systems (CPS) that perfectly combines cyberspace, physical space, and social space. CPSS promotes information resources from one place to three places to cause a revolution in data science (Wang et al., 2019).

The CPSS system perceives human factors as part of the system instead of placing them outside the system. Such systems consist not only of cyberspace and physical space but also of human knowledge, mental capabilities, and socio-cultural elements.

Information from cyberspace interacts with physical and mental spaces in the real world, as well as with artificial space that maps various aspects of the real world. CPSS affects parallel execution, self-synchronization, and the physical, informational, cognitive, and social spheres. CPSS provides synchronization with the real world, as well as self-synchronization of the human and physical system (Liu et al., 2020)

Although the concept of CPSS has been around for almost ten years, research on it has received more attention in recent years. The main reasons for this are the widespread use of smart devices, other research results obtained in the IoT ecosystem, and the fact that smart devices that communicate with each other are now more in contact with people.

CPSS is a key step toward the concept of a smart society. This concept, introduced in Japan, is presented as Society 5.0.

The concept of human-in-the-loop (HITL) is also included in CPSS. HITL is defined as a model that requires human interaction. This model can also be defined for CPSSs. The reason is that instead of keeping a person out of the system, CPSSs make it an integral part of themselves.

Although the concept of CPSS is, in a sense, a future-oriented concept, there are certain problems associated with it. The main issue between man and the cyber world is still security issues. The extent to which the relationship between two different races (human and machine) can be safe is better understood after a certain level of application of CPSS. However, in theory, it is possible to say some things about it outside of practice.

2.5 CPSS SECURITY

At the heart of CPSS, sensors always play the role of hardware support that connects the data of the physical world to man. However, both security guarantees and power consumption limit the development of sensors in CPSS, whereas privacy information can cover—for example, camera and e-health (Rani, Khang et al., IoT & Healthcare, 2021).

2.5.1 SENSORS AS A KEY COMPONENT OF CPS

The reason that sensors are a key component of CPSs is that they are the main ancillary technologies to achieve the main objectives of CPSs. Thus, sensors collect information from the physical world and transmit it to the virtual world.

CPS collects information from the real (physical) world, analyzes this information using digital technology from the virtual (cyber) world, facilitates the use of information and knowledge, and returns that information to the physical side to create added value.

Sensors are devices that measure an event or change in the environment and convert it into an electronic signal that can be read and calculated. Sensors can detect any aspect of the physical environment and turn it into useful information. Thus, sensors convert stimuli, such as heat, light, sound, and movement into electrical signals. These signals pass through an interface that converts them into binary code and transmits them to a computer for processing (Tailor et al., 2022).

A sensor network is a group of sensors where each sensor monitors information in a different location and sends that information to a central location for storage, viewing, and analysis. There are two types of sensor networks, wired and wireless. Sensor network components include sensor nodes, sensors, gateways, and control nodes. The four topologies of sensor networks are point-to-point, star, tree, and mesh.

Wireless sensors are used in CPSs. The architecture of the wireless sensors is as follows:

2.5.2 WIRELESS SENSOR NETWORK SECURITY

Inside the wireless sensors, there are original sensory nodes (nodes) distributed in space, which can work in different conditions: noise, temperature or pressure. As can be seen from the previous image, the sensor nodes must be directly connected to the base (ie, the destination) in order to transmit information. Among them, the router acts as a router interface. In general, such networks—which are router nodes—are known as multi-hop networks (Alotaibi et al., 2019).

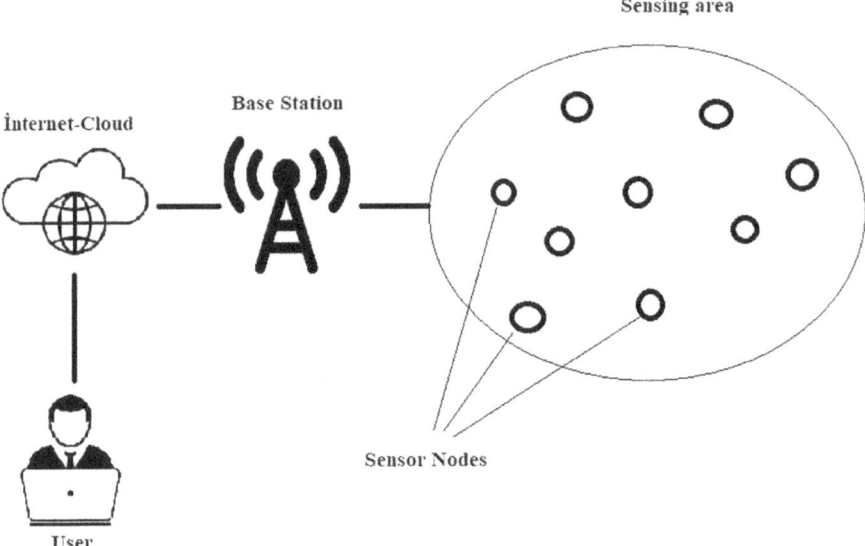

FIGURE 2.3 Wireless sensor architecture.

WSN provides a gateway that acts as an interface between the end user to process the data transmitted by the sensor nodes. These types of networks place certain limitations. As nodes become more widespread, Wireless Sensor Networks may be subject to certain cyberattacks.

For example, a malicious node can easily access the network, and the rival node is disguised as one of the network nodes, misdirecting other nodes and network density that exist on the network. Due to the large number of nodes in the WSN, security must be provided at different levels, which is also complicated.

2.5.3 Security Solutions: Cryptography

The ideal way to prevent attacks on wireless sensor networks is with cryptography. Cryptography can be presented as the science of encryption in a narrow sense. The best protection for systems is to encrypt all transmissions, making it difficult for the intruder to understand the nature of the information being exchanged.

In general, some intensive measures must be taken to ensure the safety of any sensor devices. And strong encryption methods should be used. For example, ECDH-256 and AES-128 encryption can be used between the sensors, end users, and gateways.

Many researchers find it undesirable to use public key algorithm methods such as the Diffie-Hellman key agreement protocol or RSA signatures in WSNs for code size, data size, processing time, and power consumption. Public key algorithms, such as RSA, typically take up to ten seconds or minutes to perform encryption and decryption on wireless devices with limited resources, which makes them vulnerable to DoS attacks. In contrast, symmetric key cryptographic algorithms and hash functions consume less computing power than open key algorithms (Hasan et al., 2022).

As shown in the example, it is more appropriate to use symmetric key cryptographic mechanisms such as AES in wireless sensor networks.

Basically, many experts prefer to study and apply cryptography, which is a security solution in CPSSs.

2.6 INCIDENT MANAGEMENT (SYSTEMS)

In the narrow sense, incident management is a sequence of processes carried out to detect, analyze, and resolve critical events.

2.6.1 INCIDENT MANAGEMENT DEFINITION

Incident management describes the steps the system takes to analyze, identify, and correct problems while taking measures to prevent future incidents. If critical events are not identified in time, it will be inevitable that more dangerous situations will arise later.

Critical incidents can disrupt operations, cause temporary outages, and lead to data and productivity losses. In this regard, the incident management process is very important. This can be done in different areas, such as business, work, and IT.

2.6.1.1 IT Incident Management

An IT incident is generally considered to be a breach of any of the IT services affecting the user or organization.

Incident management is the process of managing an IT service breach under the agreed service level agreements (SLAs) framework and restoring services.

This process begins when the user (or organization) reports the error and ends with the team member who solves the problem.

2.6.1.2 The Life Cycle of an IT Incident Management

The life cycle of an IT incident management can be presented as follows:

1. Incident logging: This is the process of recording incidents (phone calls, emails, or even events automatically generated by the system) reported to the service desk with a date and time stamp.
2. Incident categorization: Classification is the process of classifying incidents and problems into classes or categories. The main goal is to understand what kind of incident happened.
3. Incident prioritization: It is a process of clarifying how quickly an incident can be resolved. Four types of priority are possible: critical, high, medium, and low. While critical-priority incidents are very important, low-priority events are problems that do not stop the user or organization from working and are easily solved.
4. Incident assignment: Incident routing deals with the correct grouping of incidents. Thus, once an incident has been identified, it is the process of routing it to that group after determining which group is responsible for the problem.
5. Task creation and management: Depending on the complexity of the incident, it can be divided into sub-activities or tasks. Tasks are usually created

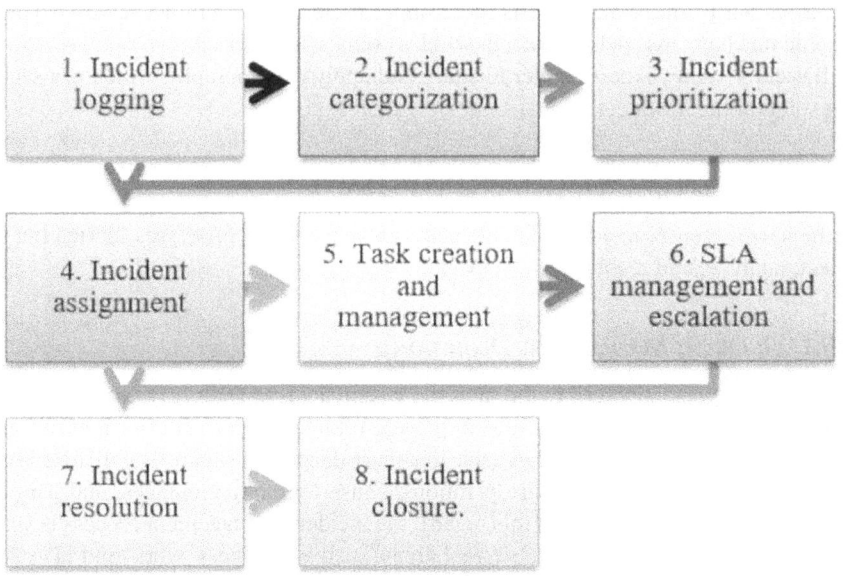

FIGURE 2.4 IT incident management life cycle.

when the resolution of an incident requires the assistance of a large number
of technicians from different departments.

6. SLA management and escalation: Escalations monitor processes to ensure
   that service providers meet their obligations under service-level contracts.
   When you create an escalation for a service-level contract, the escalation is
   applied to the business object (such as a service request or incident) used in
   the service-level contract.

7. Incident resolution: It is the process of resolving the incident by the technical staff.

8. Incident closure: Closing an incident is the eighth step in the incident management process. This is the process of resolving and restoring the incident.

2.6.2 INCIDENT RESPONSE PLAN

The concept of an incident response plan also has a special place in incident management to ensure security.

An incident response plan is a set of instructions for IT staff to detect, respond to,
and then recover from network security incidents. These types of plans address issues
such as cybercrime, data loss, and service outages that threaten day-to-day operations.

Everyone should have some knowledge about this because anyone can be a victim
of cybercrime. The incident response plan consists of six phases, outlined in Figure 2.5.

1. Preparation
2. Identification

FIGURE 2.5   Six phases of the incident response plan.

3. Containment
4. Eradication
5. Recovery
6. Lessons learned

In summary, all these phases can be characterized as follows:

- Preparation: It is the process of planning in advance how security incidents will be managed and prevented.
- Identification: It covers all phases, from monitoring potential attack vectors to searching for and prioritizing incident signs.
- Containment, eradication, and recovery: Developing a security strategy is the process of identifying and mitigating attacks on hosts and systems and having a recovery plan.
- Lessons learned: It is the process of reviewing the lessons learned and having a plan for storing evidence.

2.7   CONCLUSION

One of the most pressing requirements of the time is to be able to create the right interaction between humans and robots. Research to achieve this by creating humanoid robots has led to other technological innovations (Rani and Gupta, CLUS_GPU-BLASTP, 2017).

Today, various studies are being conducted to create a "bridge" between two different worlds (physical and virtual). Although robots are created by humans, their world is completely different from the human world. People should try to integrate their machines into the real world and learn to live with them.

One project that aims to do this is the cyber-physical system and its expanded application, the cyber-physical-social system.

These systems include many other projects to help humans and machines coexist—smart cities, smart universities, smart hospitals, and so on. It is possible to continue this list to some extent.

The following results can be noted:

1. Cyber-physical-social systems are an ideal project designed to create a smart future.
2. The main function of these systems is not only to establish a connection between two different spaces but also to properly manage and improve these connections.
3. Although cyber-physical-social systems is a creatively designed "perfect" project, it faces some challenges. The main problem is security.
4. Security problems can be classified as unauthorized access to the virtual environment from the physical environment (cybercriminals). This is especially true of sensor security, a key component of CPSS/CPS.
5. Experts recommend cryptography as the key to security solutions.
6. Any threats can manifest themselves in the work of an organization or an ordinary user. In this case, certain steps must be taken to eliminate them. Here we are talking about incident management.
7. Everyone should have information on the incident response plan, which is a set of instructions to protect their devices.

REFERENCES

Alotaibi, Majid, "Security to wireless sensor networks against malicious attacks using Hamming residue method," *EURASIP Journal on Wireless Communications and Networking*, 2019, 8 January 2019. doi:10.1186/s13638-018-1337-5.

Arunachalam, P., N. Janakiraman, A. K. Sivaraman, A. Balasundaram, R. Vincent, S. Rani, B. Dey, A. Muralidhar, and M. Rajesh, "Synovial sarcoma classification technique using support vector machine and structure features," *Intelligent Automation & Soft Computing*, 32(2), 2021, 1241–1259, doi:10.32604/iasc.2022.022573.

Bhambri, Pankaj, Sita Rani, Gaurav Gupta, and Khang, A., *Cloud and fog computing platforms for internet of things*, CRC Press, 2022. ISBN: 978-1-032-101507, doi:10.1201/9781032101507.

Селянская, Г. Н. *SMART*-университет—ответ на вызовы новой промышленной революции [Электронный ресурс], 5 May 2016, https://bgscience.ru/lib/9080/.

Gupta, O. P., and Sita Rani, "Accelerating molecular sequence analysis using distributed computing environment," *International Journal of Scientific & Engineering Research—IJSER*, 10, 2013. https://www.ijser.org/paper/Accelerating-Molecular-Sequence-Analysis-using-Distributed-Computing-Environment.html.

Hasan, S., T. B. Sivakumar, Khang, A., "Cryptocurrency methodologies and techniques," in *The data-driven blockchain ecosystem: Fundamentals, applications, and emerging technologies* (1st ed., pp. 27–37), CRC Press, 2022. https://doi.org/10.1201/9781003269281.

"Japan's Smart Cities Solving Global Issues Such as the SDGs, etc. *Through Japan's Society 5.0*" | PDF. https://www.kantei.go.jp/jp/singi/keikyou/pdf/Japan's_Smart_Cities-1(Main_Report).pdf.

Kaur, Gurleen, Ranjit Kaur, and Sita Rani, "*Cloud computing-a new trend in IT era*," *International Journal of Science, Technology and Management*, 2015, 1–6. https://www.research-gate.net/publication/326356549_Cloud_Computing-_A_new_Trend_in_IT_Era.

Khang, A., Pankaj Bhambri, Sita Rani, and Aman Kataria, *Big data, cloud computing and internet of things*, CRC Press, 2022. ISBN: 978-1-032-284200, doi:10.1201/9781032284200.

Khang, A., Subrata Chowdhury, and Seema Sharma, *The data-driven blockchain ecosystem: fundamentals, applications and emerging technologies*, CRC Press, 2022. ISBN: 978-1-032-21624, doi:10.1201/9781003269281.

Khang, A., and Haru Hong Khanh, "The role of artificial intelligence in blockchain applications," *Reinventing Manufacturing and Business Processes Through Artificial Intelligence*, 2021, doi:10.1201/9781003145011.

Liu, Zhong, Dong-sheng Yang, Ding Wen, Wei-ming Zhang, and Wenji Mao, "Cyber-physical-social systems for command and control," *IEEE Intelligent Systems,* 26(4), 2020, doi:10.1109/MIS.2011.69.

Ochoa, Sergio F., Giancarlo Fortino, and Giuseppe Di Fatta, "Cyber-physical systems, internet of things and big data," *Future Generation Computer Systems*, 75, October 2017, 82–84. doi:10.1016/j.future.2017.05.040.

Rani, Sita, P. Bhambri, and M. Chauhan, "A machine learning model for kids' behavior analysis from facial emotions using principal component analysis," in 2021 5th Asian Conference on Artificial Intelligence Technology (ACAIT) (pp. 522–525). IEEE, October 2021. doi:10.1109/ACAIT53529.2021.9731203.

Rani, Sita, and O. P. Gupta, "CLUS_GPU-BLASTP: accelerated protein sequence alignment using GPU-enabled cluster," *The Journal of Supercomputing*, 73(10), 2017, 4580–4595. doi:10.1007/s11227-017-2036-4.

Rani, Sita, A. Kataria, and M. Chauhan, "Fog computing in industry 4.0: Applications and challenges—a research roadmap," in *Energy conservation solutions for fog-edge computing paradigms* (pp. 173–190), Springer, Singapore, 2022. doi:10.1007/978-981-16-3448-2_9.

Rani, Sita, Khang, A., Meetali Chauhan, and Aman Kataria, "IoT equipped intelligent distributed framework for smart healthcare systems," *Networking and Internet Architecture*, 2021, https://arxiv.org/abs/2110.04997v2, doi:10.48550/arXiv.2110.04997.

Singh, P., O. P. Gupta, and S. Saini, "A brief research study of wireless sensor network," *Advances in Computational Sciences and Technology*, 10(5), 2017, 733–739.

Tailor, R. K, Ranu Pareek, and Khang, A., (Eds.), "Robot process automation in blockchain," in *The data-driven blockchain ecosystem: Fundamentals, applications, and emerging technologies* (1st ed., pp. 149–164), CRC Press, 2022. https://doi.org/10.1201/9781003269281.

Terreno, Saratu, Abiola Akanmu, Chimay J. Anumba, and Johnson Olayiwola, "Cyber-physical social systems for facility management," *Cyber-Physical Systems in the Built Environment*, 28 May 2020, 297–308.

Wang, Puming, Laurence T. Yang, Jintao Li, Jinjun Chen, and Shangqing Hu, "Data fusion in cyber-physical-social systems: state-of-the-art and perspectives," *ELSEVIER*, 51, November 2019, 42–57. doi:10.1016/j.inffus.2018.11.002.

3 Smart City Models Using Applications of the Internet of Things

M.P. Karthikeyan, N. Muthumani, Ananthi Sadhasivam, Leelavathy Vellingiri

CONTENTS

DOI: 10.1201/9781003252542-3

3.1   INTRODUCTION

The Internet of Things (IoT) is gaining traction in a variety of fields, including the transformation of beautiful urban groupings, the organization of key resources and frameworks, adaptability, transportation, partnerships, and so on. The growth in the applicability and importance of this concept results in an increasing volume of structured data being researched, safeguarded, and communicated in various settings (Agarwal et al., 2019; Bibri, 2018).

In the IoT, devices gather and share data directly with one another using web-based means, and the cloud understands how to assemble and examine data squares (Rani, Bhambri et al., 2021). In the 21st century, we need to be related to anything, at any time and in any place, which is currently occurring in many locations throughout the world (Khang et al., Big Data, Cloud Computing & IoT, 2022).

The city could be thought of as a service organization that provides services to its citizens (Bhambri et al., 2022). There is a desire for more brilliant, viable, productive, and practical urban communities, which can improve the capacity to conjecture and oversee urban streams and coordinate the measurements of a provincial agglomeration's physical, advanced, and institutional spaces (Padhi and Charrua-Santos, 2021).

The focus of urban development and change has shifted to innovation. Smart cities make use of a variety of data and communication technologies (Rani, Chauhan et al., 2021; Rani et al., 2022). An arrangement naturally integrates various components of a city's biological system, such as a strong foundation, astute administration and industry, astute instruction frameworks, and astute security frameworks (Dhanalakshmi et al., 2022).

The concept of a smart city unifies the measurements of an agglomeration's physical, institutional, and digital spaces.

The method presents viewpoints such as connectivity, input, self-association, and adjustment with the primary purpose of providing insight into the relatively natural creation, operation, and advancement of cities (Evers, 2008).

Cities are transitioning from computerized to smart urban communities and sophisticated or smart urban places that are more innovation-oriented counterparts of brilliant city ideas.

When a city is instrumented, interconnected, adaptable, self-ruling, learning, self-repairing, and powerful, it develops toward becoming smart (Rani, Kataria et al., 2021; Rani and Kumar, 2022). Parts of its foundation and offices are meticulously connected and renovated using information and communications technology (ICT) to provide services to its residents and other partners (Rani, Mishra et al., 2021).

Within the more comprehensive digital, advanced, sharp, smart urban areas writings, the idea of smart urban communities understood from the perception of innovations and segments has some good characteristics (Pawar et al., 2021).

3.1.1   OVERVIEW OF SMART CITIES

Smart cities, or clever spaces as they are more commonly known, refer to a wide range of electronic and computerized applications associated with advanced spaces of groups and urban communities, such as smart building, smart home, smart light system, smart meters, and other power, water, and waste management frameworks (Kataria et al., 2022).

Advanced urban regions, derived from computerized depictions of urban areas, refer to a computerized illustration of urban areas, while insightful urban areas, derived from new urban community information, refer to aggregated and appropriated understanding (Sterling and Kittross, 2001).

So despite the fact that a Telegram TV screen would have initially physically charged a receiving area, not only are current level screen sheets more traditionalist, but the development is pervasive to the point where a high-assurance screen fit for demonstrating TV substance can be incorporated into a door frame or a kitchen unit, and clearly, impressively tinier screens can be incorporated into music players and mobile phones.

Likewise, with PCs, it has ended up being so decrepit to make a comprehensively valuable microchip in contraptions that you're garments washer may contain a PC running Linux, the cash enrolled at the general store may continue running on Windows, and your video player may run an interpretation of Apple's OS 3.

Nevertheless, as we've quite recently demonstrated, basic figuring power isn't a sufficient precondition for the IoT. Or on the other hand, perhaps we are looking at force associated from one point of view to electronic sensors and actuators that coordinate with this present reality and on the other to the Internet.

Surprisingly, swift exchange and preparation of information with businesses or other customers is a huge differentiation. A radical transformation of the current Internet into a network of interconnected items that not only gathers data from the earth (detecting) and collaborates with the physical world (incitation/order/control) but also uses existing Internet standards to provide data exchange, investigation, applications, and correspondence administrations (Caputo et al., 2016).

As specified before, brilliant urban areas end up noticeably smarter due to the advanced idea of computerized innovation, wherein the smart city is prepared with various electronic gears used for different applications, such as road cameras for the reconnaissance framework, sensors for the transportation framework, and so on, even though there are similar activities that utilize items to give distinctive value-added administrations such as Google Street View, Global Positioning System (GPS), and so on.

Moreover, the improving nature likewise develops toward the use of individual cell phones, contributing to the said situation. Having said that, in this heterogeneous condition in term of articles, highlights, donors, inspirations, security rules, and so forth, unique inquiries emerge from a city domain that need answers (Edensor and Jayne, 2012):

- How do we handle vulnerability prompted because of the ongoing and disconnected elements and guarantee the nature of data?
- How do we make existing items more astute?
- How do we outline new questions more intelligently in light of the client's decision?
- How do we empower items to respond as needs be as for setting?

- How to limit the cost of information gathering that is being created by a few gadgets?
- How to get an understanding into the information if information is gathered and going to preparing stage in a continuous, In view of the inquiries specified over, the keen city idea uses ICT in a way that could help the subjects in an extremely day life inside constrained assets.

In addition, different association plan to build up a framework that utilization propelled innovation by giving the productive administrations to their residents.

3.1.2 TECHNOLOGIES FOR SMART CITIES

The larger part of these current innovations comprises of cutting-edge detecting abilities, stockpiling capacity for the remarkable volume of information, lastly, to get an understanding into the voluminous information (Mallo et al., 2020). As a result, a framework is necessary that can absorb the majority of current advancements in the field of ICT, resulting in a spectacular development in the not-too-distant future.

Every one of the capacities for detecting the earth and breaking down the detecting data are required for the design of this framework (Ramana et al., 2018). As a result of these mechanical assets, several ongoing activities may be invited.

Furthermore, it can be observed that combining a large amount of data to carry out a successful investigation is already being done. In any event, when working on a large scale, it's inescapable that a large chunk of data becomes disjointed. As a result, such knowledge cannot provide us with a better understanding of the situation so that we can prepare for the future.

Consequently, urban arranging and creating give a better approach to the field of the IoT, in which gadgets are coordinated by methods for their geographic area, and they are broke down by methods for a recently planned framework for different administrations in a city (Mohammed et al., 2020).

Obviously, there are a slew of different benefit space applications that make use of the smart city IoT framework to plan operations in the areas of air, commotion, pollution, vehicle portability, and reconnaissance in urban areas. There aren't many research findings in the topic of smart cities and urban regions in the current research (Khang et al., IoT & Healthcare, 2021).

Similarly, a smaller, more versatile, and effective structure has yet to be built (Sharma, 2019). The following are some of the explanations for the IoT's significant changes: Everything information—enables information on what clients truly want to know. The IoT's vibrancy takes shape as the focus shifts from the thing to the system. The unique association of items in a condition that is always associated (Vermesan and Friess, 2014).

Convergence is the process of arranging people, things, places, and information. Corporate at the next level—The IoT is transforming existing business capabilities by providing a mechanism to manage interfaces, measure, work, and analyses data. The IoT is the culmination of several dreams, including a Things- organized vision, an Internet-organized vision, and a Semantic-organized vision. The IoT supports various related physical contraptions over the Internet as an overall correspondence organize.

A cloud-driven vision for compositional thinking of an unavoidable recognizing condition is given by the common outline of the IoT consolidates other than various passing on devices a cloud-based server plan, which is required to relate and perform remote data organization and tallies.

Sensors, actuators, gadgets, and humans and programming administrators connect and grant data in order to carry out specified tasks or modify a company's or organization's structures.

IoT maps and organizes real-world objects in the virtual world, extending the relationship with adaptive systems, as well as structures and organizations for massive data and cloud environments (Mathur and Arora, 2020).

This research consists of different article analyses. In these articles, various methods for IoT application for various areas of applications are proposed, such as home automation, which is dependent on smart appliances (kitchens, refrigerators, laundry, etc.), smart buildings, smart healthcare, smart traffic monitoring, smart toll collection, smart highway light monitoring system, smart irrigation system, smart fire detection, and smart weather-monitoring system.

IoT applications will bring quick and bother-free administration. The administration relies on quick web association with the Internet service provider (Kalimuthu et al., 2021).

IoT-based technologies, as solutions, look to a great degree empowering in light of headway of 100 Smart urban groups wander in India (Kalimuthu et al., 2021). If such propel transforms into a reality, the gauges of IoT use in India may work out everything considered.

In this work, we propose how a smart city could be different in the future. Each chapter has its own importance. In the introduction, brief ideas of applications are listed. In the literature review, different views are analyzed for different applications.

In the model of a smart city, the detailed view of a smart city and its infrastructure are discussed. In the architecture and framework, explicit functions of different layers of architecture are discussed, and the technologies used for the framework are also discussed.

3.2 LITERATURE REVIEW

Innovation can simply provide us with new opportunities. It is up to us to make use of these in all of our development strategies.

Traditional business structures must be reconsidered. Other research fields must incorporate the IoT into their routine considerations.

The topic of urban manageability has gone to the front of research in the previous decade. It is coming about because of the extent that the urban population consistently increases; per the estimations, 70% of the population will be urban dwellers by 2050. On that time, these urban areas will be assumed a huge part in manageability, atmosphere insurance, and the lessening of unsafe discharges.

A few creative activities were performed in the region of the foundation of carbon-neutral urban areas of zero vitality utilization and maintainable districts (Angel et al., 2011).

The motivation behind smart city ideas is to boost urban life quality, to expand vitality effectiveness, to diminish unsafe outflows, and to improve the nature of urban administrations. To solve these expectations, the smart city framework must be included using the potential outcomes given by the accessible ICT gadgets (Info Communication Technology).

These days, the state of the urban condition and urban issues can always be estimated and observed by new methods and emerging instruments and, with the assistance of investigation and processing figures, we can present up-to-date answers for the smart gadgets have gone to the concentration while examining city associations (Sta, 2017).

Nowadays, it can be seen that a few urban areas as of now apply savvy arrangements and these have thoughts regarding the future advancement headings, yet a large portion of the urban communities have the term just the plan on paper.

In the following years, noteworthy sources from the European Union will be accessible for the advancement of urban administrations and the upgrade of vitality productivity. It is importance that the administration of urban communities advises itself and opens toward shrewd application as could be expected under the circumstances (Bradford, 2005) for developing the smart city. Thus, the significance of examinations inquiring about issue discernment does not just intend to demonstrate the urban legislative issues and how the neighborhood populace thinks about some urban inquiries related to the implementation of new smart infrastructure. However, the serves of public service providers can be additionally advanced, and general society engagement can be likewise developed identified with an issue separately.

In my examination, I have concentrated on the point of recognition as a matter of first importance, having attention the finish aims to put the weight and judgment of natural and get-togethers among urban issues.

Also, in light of the fact that in the European Union and in the enormous urban areas of the world as per the soul the city administration applying brilliant data correspondence innovation-based arrangements joining supportability and life quality purposes come increasingly to the front, I endeavored to overview the supposition of the populace about this as well (Rath and Pattanayak, 2019).

The challenges are: the foundation of ICT conditions is genuinely costly, the arrival time frame is by and large long, and so littler urban areas cannot be really incorporated into the line of huge urban areas applying genuine savvy arrangements (Checchi et al., 2002).

Furthermore, cost and maintenance address the level of awareness and receptiveness of urban nationals are additionally predominant factors in the usage of smart arrangements, on the grounds that more urban dwellers use the Internet today or use diverse cell phone applications. The greater space in the city can be widely used in e-administration, Internet-based correspondence channels, or even GIS-based cell phone estimations.

So it can be expressed that smart urban citizenship is a basic factor in such advancements. It likewise implies that the regions should make noteworthy strides to urge the populace to routinely utilize open administrations and e-correspondence channels (Davies et al., 2008; Cottrill et al., 2013).

As they have stated, there are numerous challenges in IoT security that demand solutions, such as RFID label security, remote security, organized transmission security, security insurance, and data preparation security.

The ebb and flow of system security innovation are examined in this research. It also provides a new method to work with experts in specific IoT applications and frameworks by examining and improving IoT security in a variety of ways (Singhet al., 2016).

Through universal question-enabled systems, the IoT provides clients with unique ways to communicate with the Web. Distributed computing allows a system to access a shared pool of configurable registering assets at a cost-effective, on-demand, and adjustable rate. The focus of this chapter is on a common way to combine IoT and cloud computing under the umbrella of Cloud Things design. We polled the best of the best when it comes to coordinating cloud computing with the IoT (Zhou et al., 2013).

To break down the IoT application requirements, we look at an IoT-enabled smart house scenario. We also recommend the Cloud Things design, a cloud-based IoT platform that supports Cloud Things IaaS, PaaS, and SaaS for faster IoT application development and management (George and Nazeh, 2019).

The architecture suggested in this research is a mobile reaction for assessing the surrounding conditions in a specific area and making the records accessible globally.

The development at the back of the smart city framework is net of factors (IoT), which is an advanced and successful response for associating matters to the Internet and interface the whole IoT in a framework. The important matters might be a few elements like virtual devices, sensors, and car digital equipment (Tripathi et al., 2022).

The device oversees searching and controlling the commonplace conditions like temperature, relative moistness, slight strength, and CO degree with sensors and sends the information to the Internet page and, a short time later, plot the sensor records as graphical bits of expertise. The information revived from the realized gadget can be opened within the net from anywhere and anytime (Tayyaba et al., 2020).

The planning of the infrastructure and transportation system of a nation plays extremely essential or fundamental parts being developed and development in different field like financial, social, social and so on. So, for better and fast development for any nation we should concentrate on street condition, plan of street, offices for client and terrible street organize overview in light of the fact that there are numerous issues faces by buildup individuals in their everyday routine life.

Some advanced systems and development strategies of the modern city are already acquainted with make or develop street more astute and resolve all issue likes automobile overloads, mischances, wounds and deferral because of terrible street condition and so on.

In this chapter, we are dealing with those strategies and systems that are already present but do not get such a large amount of support, as we acknowledged. So, there is a need to act automatically in strategies and techniques and use some more methods to solve as many issues as possible (Ruiz and Guevara, 2020; Shaikh and Chitre, 2017).

An IoT-based sensor-based computerizing plan technique for a smart physical well-being framework creates a recuperation approach and allows you to modify pharmaceutical assets according to the patient's specific requirements rapidly and repeatedly (Hosseinian et al., 2019).

The IoT is a popular articulation in the field of information technology. What's next is the IoT, which will transform current reality objects into smart virtual ones.

The IoT aims to bring everything in our lives under one roof, allowing us not only to manage the things around us but also keep us informed about their status (Ross and Gross, 2017).

The primary goal of this chapter is to provide an overview of the IoT, its plans, and its basic advancements, as well as their applications in our daily lives. This organization will provide exceptional guidance to new examiners who must do research on the subject of IoT and competently support data collection.

They attempt to coordinate the organizations provided by IoT in this article, with the ultimate goal of empowering application architects to construct a foundation organization (Botta et al., 2016).

In the next part, smart city infrastructure leveraging the IoT is addressed.

3.3 SMART CITY INFRASTRUCTURE USING IOT

Difficulties in the planning, development, and operation of urban settlements are enabling innovative thinking in a variety of fields. Experts from a variety of fields, including design, urban planning, construction, development, data innovation, frameworks, natural science, property development, and municipal government, gain a better understanding of partners and learn how to engage with them.

For intelligent city developments, framework models capable of seeing deeply into how urban areas function, how people use the city, how they feel about it, where the city faces challenges, and what types of progress can be connected could be used. Different partners—residents, workers, understudies, analyzers, financial specialists, or business people—use a similar structure 24 hours a day, seven days a week (Tailor et al., 2022).

A city can be thought of as a collection of stakeholder slider groups and group associations, business groups, and business people, nearby individuals and occupants, neighborhood specialists and neighborhood administrations, metro associations, scholastic groups, and instructive foundations, all of which compete for resources.

In addition, with neighborhood on-screen individuals, smart city layouts must fulfill community requirements within a close situation (Gittell and Weiss, 2004). Civil initiatives, IT and broadcast communications organizations, utilities, district-specialized administrations, and network foundation specialist organizations are all notable stakeholders in the smart city.

Group interests can be seen at a variety of institutional levels, including family units, neighborhoods, cities, corporations, and administrations, both public and private. These dispositions will be used to indicate the level of interest and influence of partners in relation to the planned venture, as well as to integrate diverse partners into the smart city planning method and adjust interests.

To solve above limitations, stakeholders need to share research and development resources, such as emerging ICT apparatuses, techniques, and know-how, test innovation stages, and client groups for experimentation on e-benefit applications and future web advances, associations and solid joint effort methodologies and strategies among key partners are required (Lewis, 2002).

As data frameworks have become more prevalent in urban settings, they have framed opportunities to catch data that was never previously available. An overview of the city's frameworks and key viewpoints.

This diagram is necessary for understanding the present relationships. Huge amounts of data on what happens in the city are available, and they can be used to develop and update smart arrangements in linked areas of e-administrations applications.

There is a variety of information and data obtained from the city administration and many partners, such as open data on transportation, energy use, water conditions, information entered at the source by shippers, and a variety of other data.

Understanding the city frameworks would be aided by knowing what information and data frameworks the city possesses (Goodwin, 1999).

A smart city is defined as an urban city that monitors and organizes the state of the majority of its important establishments, such as lanes, ranges, tunnels, rail/cable cars, aircraft terminals, seaports, correspondences, water, control, and even significant structures, in order to better overhaul its advantages, outline its defensive upkeep actions, and screen safety points while boosting organizations to its residents.

Alternative reply organization to both trademark and also manufactured challenges to the system can be locked in and fast. With cutting-edge monitoring systems and sensors, data can be accumulated and evaluated logically, updating the city's essential organizational authority (Pawar et al., 2021).

This city's smart city infrastructure should include the following. In terms of services, security, and safety, this infrastructure will make life easier. As a result, both the service provider and the city's citizens will gain more benefits.

The suggested infrastructure for the smart city model is as follows. IoT applications serve as the foundation for this strategy.

1. Smart apartments
2. Smart weather-monitoring system
3. Smart environment and pollution control
4. Smart power grids
5. Smart education system
6. Smart platform ticketing system
7. Smart healthcare
8. Smart transportations
9. Smart garbage disposal system
10. Smart logistic system
11. Smart home
12. Smart railway ticketing system (SRTS)
13. Smart industries and manufacturing (SM&I)
14. Smart highway systems

15. Smart water distributions and purification (SWDP)
16. Smart traffic control
17. Smart banking
18. Smart retail system
19. Smart library system
20. Smart entertainment system

Here are some explanations of the IoT applications used in a smart city.

3.3.1 SMART APARTMENT

In the current scenario in apartments, IoT devices measure the level of comfort in real estate sites. The IoT sensor can record audio and video while also detecting temperature, humidity, light, and acceleration. Despite the fact that the two apartments are in the same building, an upper-floor apartment receives two more hours of sunlight than a lower-floor apartment.

Furthermore, it was determined that apartment complexes near a train or motorway produce very little noise when facing away from the source of the noise. We would be able to avoid biased apartment evaluations and establish more objective assessments using this knowledge.

3.3.2 SMART WEATHER-MONITORING SYSTEM

The proposed system is a cutting-edge weather-monitoring system that makes real-time data publicly accessible across a wide range of devices using the IoT.

The system monitors weather and environmental changes using a variety of sensors, including temperature, humidity, wind speed, wetness, light intensity, UV radiation, and even carbon monoxide levels in the air. The data collected by these sensors is uploaded to a web page and plotted as graphical statistics.

The information on the web page is accessible from any location on the planet. This smart city application can help meteorological offices, weather stations, the aviation and maritime industries, and even the agricultural industry.

3.3.3 SMART ENVIRONMENT AND POLLUTION CONTROL

The IoT has a huge part to play in future smart cities. The IoT can be utilized as a part of essentially all situations for open administrations by governments. IoT-sensor-empowered gadgets can help screen the ecological effect of urban communities and gather insights about sewers, air quality, and rubbish. Such gadgets can likewise help screen woods, streams, lakes, and seas.

Numerous ecological patterns are complex to the point that they are hard to conceptualize. The primary goal of the IoT air and sound monitoring system is to address the fact that air and sound pollution is a growing concern in today's world.

It is critical to keep an eye on air quality and monitor it for a healthy future and present for everyone. We propose an air feature and sound contamination monitoring approach that allows us to monitor and present real-time air quality as well as strong tainting in a district using the IoT.

The structure employs air sensors to detect the proximity of toxic gases/blends that are perceptible all around it and to continuously relay this information. Similarly, the structure continues to estimate and report sound levels. The sensors communicate with the Raspberry Pi, which processes the data and sends it to the app.

Experts would be able to screen for air pollution in various districts and take action as a result of this license. Experts can also keep an eye on the sound pollution near schools, patching workplaces, and no-sounding spaces, and if the system detects air quality or commotion issues, it alerts professionals so they can take steps to address the problem.

3.3.4 SMART POWER GRID

The Internet of energy is a novel concept that has evolved as a result of the rapid growth of ICT and energy systems (IoE). The IoT is a subset of this notion, which refers to the use of powerful digital controllers, sensors, actuators, and meters that can send data over IT infrastructures. In this context, the term "Internet" encompasses any server-based or peer-to-peer network, not just the World Wide Web.

Data for local analysis and control, as well as analyzed commands, raw data, and instructions, can all be received by these communicational instruments. Although incorporating the IoT into power systems has a bright future in terms of overcoming power system operation difficulties and environmental challenges, it does require a significant initial capital investment portfolio.

However, it is expected that advances in present technologies, as well as the introduction of new technologies at lower prices, will assist in expanding the reach of these cutting-edge technologies.

3.3.5 SMART EDUCATION SYSTEM

In the current COVID-19 situation, education in the entire world is online-based. Smart city infrastructure is much needed in this education system. IoT is needed by the educator to understand and fulfill the students' requirements according to their needs.

In the future also, IoT-based solutions, built into online classes and examinations, are a focus of the smart education system. If we launch this approach in the smart city, it should be a less time-consumption process to the educators and teachers, also citizen can easily participate the interested classes.

It is demonstrated that the principles of the quality management system, whose essence consists of achieving, maintaining, and continuously improving the level of knowledge and general and professional learning culture and of continuously improving educational process quality by all its members, should be harmoniously integrated into smart education.

3.3.6 SMART PLATFORM FOR A BUS TICKETING SYSTEM

Using an efficient new algorithm and the estimated waiting time, the user can check platform ticket availability, book tickets, and receive a seat automatically. If there are no seats available on the platform, our system will allocate the seat that becomes available in the least amount of time.

The user will only be able to purchase the ticket if they connect to the device installed at the stop and pay digitally through our portal, resulting in a very comfortable and smart platform ticket service.

Users who do not have access to a smartphone will be able to do all of the aforementioned actions using the platform's gadget. The ticket booking confirmation will function as an e-ticket, which will be confirmed by the bus conductor.

3.3.7 Smart Healthcare System

The market for prosperity checking devices is shown in software-based healthcare plans, which are usually non-interoperable and are included in different models. While solitary things are expected to cost centers around, the whole deal target of achieving lower development costs transversely finished present and future divisions will unavoidably be outstandingly trying unless a more solid approach is used in the national medical system.

The IoT can be used as a part of medical care service in which hospitalized patients whose illnesses require close monitoring can be continually checked by IoT-guided noninvasive monitoring. This expects sensors to assemble thorough physiological information and usages entries and the cloud to investigate and save the information and after that ship the separated statistics remotely to doctor's figures for energize exam and evaluation (Bhambri et al., Cloud & IoT, 2022).

By using advances in medical technology, hospital will offer the care fee cheaper than in-person care by wiping out the necessity for discerning the care types to viably take part in strategies of the socialization of national medical examination and treatment.

3.3.8 Smart Transporation

The first smart application of transportation in smart city framework is how to manage car parkings in the urban. Tenants are encouraged to recognize and reserve nearby available parking by constantly analyzing the availability of parking places in the city.

The second smart application is how to monitor moving vehicles in the urban. Reduced peak hour jam blocks and increased compensation from dynamic pricing could be some of the benefits, as well as a less demanding task for action administrators who monitor secure utilization.

3.3.9 Smart Garbage Disposal System

The increase in population has resulted in a considerable degradation in the waste management system's hygiene. The pollution of the environment in the surrounding areas is caused by the spillover of garbage in public spaces. It has the potential to aggravate a number of serious illnesses in the area. As a result, the damaged area's reputation will be tarnished.

To eliminate or mitigate garbage and maintain cleanliness, a smart waste management system is required. Sensor systems are used to assess the waste level in the bins in a smart waste management system based on IoT. The system was adjusted to allow concern via GSM/GPRS as soon as it was discovered.

In this system, a microcontroller was used as an interface between the sensor system and the GSM/GPRS system. An Android application is being developed to track and combine the necessary data about the varying levels of rubbish in various places.

3.3.10 Smart Logistics

Today, logistics systems are critical for many sectors and business solutions to solve transportation issues and reduce costs. Many applications for logistics systems integrate many systems, but they cost a lot of money and aren't always used for what they're supposed to be used for. As a result, this study proposes a new way to solve some difficulties by utilizing the notion of IoT technology.

A model is created for users to access online interactions, and then the model is evaluated based on the satisfaction created by users. Users will be able to easily trace their materials and secure their products using this method. Both the user and the administrator have secure conversations.

3.3.11 Smart Home

Homes in the 21st century will become increasingly automated and robotized, particularly in private homes, due to the convenience it provides, and security has now become a major concern in both public and private organizations, with various security frameworks proposed and created for some significant procedures.

Data, property, and preventing burglary or wrongdoing in the home all require security frameworks. A smart home is a network of sensors and controllers that work together to provide a client with IoT-based remote control of various items in their home.

The sensors sense different changes, screen them, store the information, and show them all together for investigation and control. This encourages us to tweak our home to fit each family's lifestyle. This is a savvy framework produced using locally accessible segments like PIC controllers, light sensors, and PIR sensors, which permit us to control the machines of our homes.

The IoT is associating ordinary entertainment keenly and intensel to the Web to empower correspondence among the individuals and between things themselves. The current smart home system also has various inbuilt entertainment systems, so that home users are kept happy and entertained anytime.

3.3.12 Smart Railway Ticketing System

In the current train ticket providence scenario, biometric-based ticketing is used, although it is not restricted to that. The proposed IoT-based system evaluates each person's fingerprint straight away from registration and the purchase of tickets and authenticates the fingerprint on the day of the journey in order for him or her to go on a certain day and on the desired train to his or her preferred destination.

This technology is both secure and time-consuming, so it is required for smart city design.

3.3.13 Smart Highway System

The next generation of smart highways has become an irreversible global trend in transportation efficiency and safety. The study of smart highway characteristics and frameworks can aid in the modernization of existing transportation infrastructure.

In keeping with the popularity of automated and connected vehicles, the key technical components of the proposed are comprehensive elements sensing, cyber-physical systems, cooperative vehicle-infrastructure applications, and 5G mobile communication technology.

This smart highway system offers new perspectives on how to reform highways in a sustainable and repeatable manner, with implications for smart city design.

3.3.14 Smart Manufacturing and Industries

IoT technologies promises to have a deep impact on any industry and manufacturing businesses, allowing investors or managers to plan, control, integrate, analyze, and optimize processes in a better manner by creating a IoT-powered networks of connected machines, systems, devices, and humans.

Nowadays, the area of the IoT is completion up clearer in allowing access to gadgets in life what's more, technologies, which in assembling outlines, were covered up in all-around made-up manufacturing and warehouses. This development will empower IT to advance to change-making structures for every factory and industry.

IoT-based technologies, such as Industrial IoT, Factory IoT, can help monitoring the machines, workers, assemblies, and work spaces in real-time and generates accurate reports for better decisions. This helps business stakeholders in making business strategies better and quality-centric products. Thus, IoT always lays an impact on increasing the quantity and revenue of your manufacturing business. In earlier sense, the creation structure could be seen as one of the various IoT, where another natural framework for all the further clever and additional powerful conception could be described in industry revolution 4.0.

3.3.15 Smart Water Purification and Distribution

Because water is one of the most basic requirements for human living, some systems to monitor water quality must be included. Contaminated water is responsible for around 40% of all deaths worldwide. As a result, it is critical to guarantee that people in both cities and villages have access to clean drinking water.

Water quality monitoring is a cost-effective and efficient solution that uses IoT technology to monitor drinking water quality. Several sensors are used to monitor various characteristics, such as pH, turbidity in the water, tank level, temperature, and humidity in the surrounding environment.

Furthermore, these sensors are connected to the microcontroller unit, with further processing taking place on the personal computer. The collected data is uploaded to the cloud via an IoT-based application to monitor the quality of the water delivered to the people.

3.3.16 SMART TRAFFIC CONTROL

IoT technology is a development that utilizes the web to control physical things. Utilizing IoT-based technologies, we can get results that are more exact, faster, and correct. In IoT ecosystem, all databases will be fed in a PC and Cloud storage; this stockpiling is done through the web. Then, this database is utilized in like manner to their analytics and visualization and applications.

3.3.17 SMART BANKING

Smart cities can provide a variety of smart applications, such as intelligent transportation, smart industry, and smart banking, to improve people's quality of life. One of the most significant difficulties facing a smart city is security.

Smart cities can benefit from blockchain technology by keeping transactions in a secure, transparent, decentralized, and unchangeable ledger. These digital banking systems provide for one-on-one connection with customers and their needs, allowing for data gathering, processing, analytics, and decision-making based on that data (Khang et al., Blockchain Ecosystem, 2022).

In smart cities, an IoT and blockchain strategy will lead to safe banking applications (Rana et al., AI & Blockchain, 2021).

3.3.18 SMART RETAIL SYSTEM

Recent IoT retail systems and databases keep the user purchase details and user details. Smart shopping carts that communicate directly with customers through real-time shopping feedback are included in this category.

The ability to track the total amount of a shopping cart during the transaction is a key feature of such shopping carts, and it provides considerable benefits to both customers and retailers.

RFID sensors allow objects to be detected as soon as they are added to or withdrawn from a shopping cart. The cart total is then shown on an LCD screen affixed to the cart. If a consumer then proceeds to the checkout process, the billing information has already been sent to the checkout and allocated to the appropriate shopping cart ID.

As a result, customers save time and effort while paying by not having to unload and reload their whole shopping basket at the checkout for scanning. Additionally, because the checkout procedure requires less work, the retailer's staffing requirements are lowered.

Furthermore, technology can sometimes provide customers with additional information about products or promotions, make individualized product recommendations or use voice control to aid blind individuals when shopping.

3.3.19 SMART LIBRARY SYSTEM

Children's reading habits are dwindling in this digital era. Parents must take their children to the children's section in order to encourage them to read. All library information has recently been placed in a database, which protects the information from intruders.

The collecting and administration of information from one or more sources, as well as the presentation of that information to one or more audiences, is what information management is all about. Those with a stake in or right to that information are sometimes involved. The organization of and control over the structure, processing, and distribution of information is referred to as management.

In a nutshell, information management is the process of organizing, retrieving, obtaining, and managing data. It is linked to and overlaps with the practice of data management.

3.3.20 SMART ENTERTAINMENT SYSTEM

Because preventive security policies are unlikely to be effective enough to block evasive cyberattacks, cyber-resilience for home IoT networks is a critical concern. The music entertainment system is controlled through a user-friendly smartphone application.

In this aspect, the envisioned approach's situational awareness could be critical in limiting damage by giving data about an IoT network's security state and prompting or advising corrective measures to a better-informed end user.

3.4 CONCLUSION

This investigation begins by setting the stage for rapid urbanization, which continuously places pressure on urban areas to deliver great solutions in order to satisfy manageable development needs.

The examination gives a blueprint of smart physical establishments, accompanied by relevant investigations related to (1) smart water administration, (2) mobility and transport, (3) smart grid, (4) waste administration, (5) smart buildings, (6) smart healthcare, (7) smart shopping, (8) smart banking service, as well as the specific layers of smart systems required to maintain a well-working city. It focuses on the requirement for a consolidated approach to handle a magnificent establishment diagram.

The last part examines five key challenges of smart cities: (1) alteration of smart city perceptions by resident conditions, (2) public services, (3) abilities gap, (4) economic restrictions applying appropriate governance models, and (5) building inclusive smart city programs.

Individuals centered and inclusive infrastructure for modern smart city: It should be perceived that while innovation as brilliant city framework is a necessary piece of a shrewd city, it should just go about as an empowering agent to address the issues of the general population of the city.

Keen framework advancement ought to in this manner take after a "people-driven" approach which reacts to the manageable improvement needs of individuals and ought to stay away from an "innovation driven" approach for building the smart city model.

REFERENCES

Agarwal, P., Singh, V., Saini, G. L., & Panwar, D. (2019). Sustainable smart-farming framework: smart farming. In *Smart farming technologies for sustainable agricultural development* (pp. 147–173). IGI Global. doi: 10.4018/978-1-5225-5909-2.ch007.

Angel, S., Parent, J., Civco, D. L., Blei, A., & Potere, D. (2011). The dimensions of global urban expansion: estimates and projections for all countries, 2000–2050. *Progress in Planning*, 75(2), 53–107. doi: 10.1016/j.progress.2011.04.001

Bhambri, P., Rani, S., Gupta, G., & Khang, A. (2022). *Cloud and fog computing platforms for internet of things*. CRC Press. ISBN: 978-1-032-101507, doi: 10.1201/9781032101507.

Bibri, S. E. (2018). The IoT for smart sustainable cities of the future: an analytical framework for sensor-based big data applications for environmental sustainability. *Sustainable Cities and Society*, 38, 230–253. doi: 10.1016/j.scs.2017.12.034.

Botta, A., De Donato, W., Persico, V., & Pescapé, A. (2016). Integration of cloud computing and internet of things: a survey. *Future Generation Computer Systems*, 56, 684–700. doi: 10.1016/j.future.2015.09.021.

Bradford, N. J. (2005). *Place-based public policy: towards a new urban and community agenda for Canada*. Canadian Policy Research Networks, Work Network. https://www.researchgate.net/publication/237424452_Place-Based_Public_Policy_Towards_a_New_Urban_and_Community_Agenda_for_Canada.

Caputo, A., Marzi, G., & Pellegrini, M. M. (2016). The internet of things in manufacturing innovation processes: development and application of a conceptual framework. *Business Process Management Journal*, 22(2), 383–402. doi: 10.1108/BPMJ-05-2015-0072.

Checchi, R. M., Sevcik, G. R., Loch, K. D., & Straub, D. (2002). *An instrumentation process for measuring ICT policies and culture*. J. Mack Robinson College of Business, Georgia State University. www.cis.gsu.edu/~dstraub/Research/ACITAPIT/Endnote/Publications/Checchi (2002). pdf (accessed 29/03/09).

Cottrill, C. D., Pereira, F. C., Zhao, F., Dias, I. F., Lim, H. B., Ben-Akiva, M. E., & Zegras, P. C. (2013). Future mobility survey: experience in developing a smartphone-based travel survey in Singapore. *Transportation Research Record*, 2354(1), 59–67. doi: 10.3141/235

Davies, R. G., Barbosa, O., Fuller, R. A., Tratalos, J., Burke, N., Lewis, D. . . . & Gaston, K. J. (2008). City-wide relationships between green spaces, urban land use and topography. *Urban Ecosystems*, 11(3), 269–287. doi: 10.1007/s11252-008-0062-y

Edensor, T., & Jayne, M. (Eds.). (2012). *Urban theory beyond the West: a world of cities*. Routledge. doi: 10.4324/9780203802861

Evers, H. D. (2008). *Knowledge hubs and knowledge clusters: designing a knowledge architecture for development* (No. 27). ZEF Working Paper Series. doi: 10.1142/9789814343688_0002

George, J., & Nazeh, M. (2019). Challenges faced by CIOs in cloud and IoT based organizations: a study on IT and business leaders. *JOIV: International Journal on Informatics Visualization*, 3(1), 18–34. doi: 10.30630/joiv.3.1.213.

Gittell, J. H., & Weiss, L. (2004). Coordination networks within and across organizations: a multi-level framework. *Journal of Management Studies*, 41(1), 127–153. doi: 10.1111/j.1467-6486.2004.00424.x.

Goodwin, R. F. (1999). Redeveloping deteriorated urban waterfronts: the effectiveness of US coastal management programs. *Coastal Management*, 27(2–3), 239–269. doi: 10.1080/089207599263857.

Hosseinian, H., Damghani, H., Damghani, L., Nezam, G., & Hosseinian, H. (2019). Home appliances energy management based on the IoT system. *International Journal of Nonlinear Analysis and Applications, 10*(1), 167–175. doi: 10.22075/IJNAA.2019.4061.

Kalimuthu, S. (2021). Sentiment analysis on social media for emotional prediction during COVID-19 pandemic using efficient machine learning approach. In *Computational intelligence and healthcare informatics* (pp. 215–233). doi: 10.1002/9781119818717.ch12.

Kalimuthu, S., Naït-Abdesselam, F., & Jaishankar, B. (2021). Multimedia data protection using hybridized crystal payload algorithm with chicken swarm optimization. In *Multidisciplinary approach to modern digital steganography* (pp. 235–257). IGI Global. doi: 10.4018/978-1-7998-7160-6.ch011

Kataria, A., Rani, S., Bhambri, P., & Khang, A. (2022). Smart city ecosystem-concept, sustainability, design principles and technologies. In *AI-centric smart city ecosystems: technologies, design and implementation* (1st ed.). CRC Press. https://doi.org/10.1201/9781003252542.

Khang, A., Bhambri, P., Rani, S., & Kataria, A. (2022). *Big data, cloud computing and internet of things*. CRC Press. ISBN: 978-1-032-284200. doi: 10.1201/9781032284200.

Khang, A., Chowdhury, S., & Sharma, S. (2022). *The data-driven blockchain ecosystem: fundamentals, applications and emerging technologies*. CRC Press. HB.ISBN: 978-1-032-21624-9 EB.ISBN: 098-1-003-26928-1 PB.ISBN: 098-1-032-21625-6. doi: 10.1201/9781003269281.

Khang, A., Rani, S., Chauhan, M., & Kataria, A. (2021). IoT equipped intelligent distributed framework for smart healthcare systems. *Networking and Internet Architecture*. https://arxiv.org/abs/2110.04997v2, doi: 10.48550/arXiv.2110.04997.

Lewis, J. D. (2002). *Partnerships for profit: structuring and managing strategic alliances*. Simon and Schuster. https://www.simonandschuster.com/books/Partnerships-for-Profit/Jordan-D-Lewis/9780743237635.

Mallo, D., Schoneboom, A., Tardiveau, A., & Vigar, G. (2020). From non-place to place in post-suburbia: city-edge office parks as loci for nature-based micro-interventions. *Journal of Environmental Planning and Management, 63*(13), 2446–2463. doi: 10.1080/09640568.2020.1779675

Mathur, S., & Arora, A. (2020). Internet of things (IoT) and PKI-based security architecture. In *Industrial internet of things and cyber-physical systems: transforming the conventional to digital* (pp. 25–46). IGI Global. doi: 10.4018/978-1-7998-2803-7.ch002.

Mohammed, A. H., Khaleefah, R. M., & Abdulateef, I. A. (2020, June). *A review software defined networking for internet of things*. In 2020 International Congress on Human-Computer Interaction, Optimization and Robotic Applications (HORA) (pp. 1–8). IEEE. doi: 10.1109/HORA49412.2020.9152862

Padhi, P. K., & Charrua-Santos, F. (2021). 6G enabled industrial internet of everything: towards a theoretical framework. *Applied System Innovation, 4*(1), 11. doi: 10.3390/asi4010011

Pawar, A., Kolte, A., & Sangvikar, B. (2021). Techno-managerial implications towards communication in internet of things for smart cities. *International Journal of Pervasive Computing and Communications, 17*(2). doi: 10.1108/IJPCC-08-2020-0117

Ramana, N. V., Nagesh, P., Lanka, S., & Karri, R. R. (2018, November). Big data analytics and IoT gadgets for tech savvy cities. In *International conference on computational intelligence in information system* (pp. 131–144). Springer. doi: 10.1007/978-3-030-03302-6_12

Rana, G., Khang, A., & Khanh, H. H. (2021). *The role of artificial intelligence in blockchain applications*. Reinventing Manufacturing and Business Processes Through Artificial Intelligence. doi: 10.1201/9781003145011.

Rani, S., Bhambri, P., & Chauhan, M. (2021, October). *A machine learning model for kids' behavior analysis from facial emotions using principal component analysis.* In 2021 5th Asian Conference on Artificial Intelligence Technology (ACAIT) (pp. 522–525). IEEE. doi: 10.1109/ACAIT53529.2021.9731203.

Rani, S., Chauhan, M., Kataria, A., & Khang, A. (2021). IoT equipped intelligent distributed framework for smart healthcare systems. *arXiv preprint,* arXiv:2110.04997.

Rani, S., Kataria, A., Chauhan, M., Rattan, P., Kumar, R., & Sivaraman, A. K. (2022). Security and privacy challenges in the deployment of cyber-physical systems in smart city applications: state-of-art work. In *Materials Today: Proceedings.* doi: 10.1016/j.matpr.2022.03.123.

Rani, S., Kataria, A., Sharma, V., Ghosh, S., Karar, V., Lee, K., & Choi, C. (2021). Threats and corrective measures for IoT security with observance of cybercrime: a survey. *Wireless Communications and Mobile Computing.* https://www.researchgate.net/publication/344757378_Threats_and_Corrective_Measures_for_IoT_Security_with_Observance_to_Cybercrime

Rani, S., & Kumar, R. (2022). Bibliometric review of actuators: key automation technology in a smart city framework. In *Materials Today: Proceedings.* Elsevier. doi: 10.1016/j.matpr.2021.12.469.

Rani, S., Mishra, R. K., Usman, M., Kataria, A., Kumar, P., Bhambri, P., & Mishra, A. K. (2021). Amalgamation of advanced technologies for sustainable development of smart city environment: a review. *IEEE Access, 9,* 150060–150087. doi: 10.1109/ACCESS.2021.3125527.

Rath, M., & Pattanayak, B. (2019). Technological improvement in modern health care applications using internet of things (IoT) and proposal of novel health care approach. *International Journal of Human Rights in Healthcare.* doi: 10.1108/IJHRH-01-2018-0007.

Ross, E. A., & Gross, M. (2017). *Social control: a survey of the foundations of order.* Routledge.

Ruiz, A., & Guevara, J. (2020). Sustainable decision-making in road development: analysis of road preservation policies. *Sustainability, 12*(3), 872. doi: 10.4324/9781315129488.

Shaikh, S., & Chitre, V. (2017, May). *Healthcare monitoring system using IoT.* In 2017 International Conference on Trends in Electronics and Informatics (ICEI) (pp. 374–377). IEEE. doi: 10.1109/ICOEI.2017.8300952.

Sharma, R. (2019). Evolution in smart city infrastructure with IOT potential applications. In *Internet of things and big data analytics for smart generation* (pp. 153–183). Springer. doi: 10.1007/978-3-030-04203-5_8.

Singh, N., Hans, A., & Kaur, S. (2016). Layer and RFID based security issues of internet of things. *International Journal of Grid and Distributed Computing, 9*(10), 301–310. doi: 10.14257/ijgdc.2016.9.10.27.

Sta, H. B. (2017). Quality and the efficiency of data in "smart-cities." *Future Generation Computer Systems, 74,* 409–416. doi: 10.1016/j.future.2016.12.021.

Sterling, C., & Kittross, J. M. (2001). *Stay tuned: a history of American broadcasting* (pp. 570–571). Routledge. doi: 10.1093/joc/54.3.570.

Tailor, R. K, Pareek, R., & Khang, A. (Eds.). (2022). Robot process automation in blockchain. *The data-driven blockchain ecosystem: fundamentals, applications, and emerging technologies* (1st ed., pp. 149–164). CRC Press. https://doi.org/10.1201/9781003269281.

Tayyaba, S., Khan, S. A., Ashraf, M. W., & Balas, V. E. (2020). Home automation using IoT. In *Recent trends and advances in artificial intelligence and internet of things* (pp. 343–388). Springer. doi: 10.1007/978-3-030-32644-9.

Tripathi, M. M., Haroon, M., & Ahmad, F. (2022). A survey on multimedia technology and internet of things. In *Multimedia technologies in the internet of things environment*, Volume 2 (pp. 69–87). Springer. doi: 10.1007/978-981-16-3828-2_4.

Vermesan, O., & Friess, P. (Eds.). (2014). *Internet of things-from research and innovation to market deployment*, Volume 29. River Publishers. https://www.riverpublishers.com/pdf/ebook/RP_E9788793102958.pdf.

Zhou, J., Leppanen, T., Harjula, E., Ylianttila, M., Ojala, T., Yu, C. . . . & Yang, L. T. (2013, June). *Cloudthings: a common architecture for integrating the internet of things with cloud computing.* In Proceedings of the 2013 IEEE 17th international conference on computer supported cooperative work in design (CSCWD) (pp. 651–657). IEEE. doi: 10.1109/CSCWD.2013.6581037.

4 Smart City Framework
Technologies and Applications

*Prerna Sharma, Kumar Guarve,
Sumeet Gupta, Kashish Wilson*

CONTENTS

4.1 INTRODUCTION

In a smart city, people, procedures, rules, technology, and other enablers all work together to achieve a set of goals. The smart city is not completely owned by the city. Other value creators participate as well, sometimes in collaboration with one another and sometimes on their own (Albino et al., 2015).

One of our present society's primary challenges is dealing with the difficulties of cities. Overpopulation is one of these challenges, but others include transportation, pollution, sustainability, security, and the establishment of new businesses (Anthopoulos et al., 2009).

DOI: 10.1201/9781003252542-4

We desire higher standards of service delivery as citizens, which necessitate more public resources. We also file a tax reduction claim at the same time. In this circumstance, efficiency is important to achieving the required results. Municipal officials are also contending with a growing demand for active participation in city governance.

As a result, open government initiatives (Ramírez-Alujas and Dassen, 2014) are becoming increasingly popular. According to Jetzek et al. (2013), information reuse could enable not just enhanced efficiency but also avenues for involvement and creativity. The data-driven ecosystems of smart cities are examined in depth in this essay.

The major goal is to determine the role of information reusability in the creation of new services and innovation in the ecosystems around smart cities.

4.2   SMART CITIES—DATASETS AND APPS

A smart city is a vast and varied concept that can be found in a range of organizations, as mentioned in Theresa and Nam (2013).

According to Hall et al. (2000), a smart city is

> a city that monitors and integrates the conditions of all of its critical infrastructures, including roads, bridges, tunnels, rails, subways, airports, seaports, communications, water, power, and major buildings, in order to better optimize resources, plan preventive maintenance activities, and monitor security aspects while maximizing services to its citizens.

According to Giffinger et al. (2000),

> A forward-looking city established on the smart combination of endowments and activities of self-decisive autonomous and aware residents that performs well in the economy, people, government, mobility, environment, and lifestyle.

According to (Washburn et al., 2010), a smart city is

> the application of smart computing technologies to make a city's critical infrastructure components and services, such as city administration, education, healthcare, public safety, real estate, transportation, and utilities, more intelligent, interconnected, and efficient.

A smart city, as defined in this article, is a public-private ecosystem that uses technology to provide services to citizens and their organizations (Rani, Bhambri et al., 2021).

These scholarly definitions differ from those accepted by smart city solution vendors, which are more aligned with governance models, like Dekkers et al. (2006). A smart city can be classified into the following.

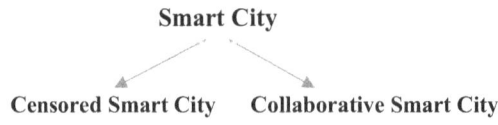

4.2.1 Sensor-Monitored Smart City

Another popular method is the Internet of Things. In this technique of using sensors, the city collects data from thousands of sensors, traffic sensors, air pollution sensors, sound sensors, humidity sensors, cameras, and other devices located across the city.

These sensors would provide vital information for tackling a wide range of issues. Managing the massive amounts of data generated by the sensors, on the other hand, is a feat in and of itself, even more difficult than the condition they are aiming to alleviate or cure (Rani et al., 2022; Rani, Khang et al., 2021).

While maintaining a complicated technical infrastructure, this strategy can provide a detailed grasp of some of the city's most difficult problems.

4.2.2 Collaborative Smart City

A third and increasingly popular approach emphasizes people's daily ability to act on local concerns. To improve municipal management, digital engagement systems, citizen-led active sensing, participatory budgeting, and other ways are being implemented (Rani, Mishra et al., 2021).

Basic technical instruments can be identified in tough implementation jobs. Privacy, engagement, and public procurement regulations all pose significant roadblocks.

4.3 HOLISTIC APPROACH

The most successful way to find the best solution may be to use a holistic technique that integrates these three previous approaches, as well as some other factors. Because resources are limited, this comprehensive approach will have to prioritize which aspects of the smart city should be built first and which demand the most resources (Dhanalakshmi et al., 2022).

As indicated in Figure 4.1, the city serves the surrounding society as a data-driven service provider (which includes individuals, businesses, and organizations). The vast majority of this information is made available as open data, which is a subset of open digital content as specified by the open definition. The smart city may supply these services directly or through subcontracting.

The smart city makes its digital assets available to the ecosystem, allowing it to build on them to offer additional value. The ecosystem organizations contribute value by integrating external digital assets or conducting new processing on data to tailor it to different target groups. In this way, they give societal services.

On the other hand, society, as a consumer of these services, creates new demands on the environment, which may have a number of effects. For instance, there is a demand for greater knowledge of smart cities. It is also likely that some data will need to be rectified or that suggestions for improving the release procedure will be collected.

Cities that are smart are assembling a tremendous amount of data created by their sensors, internal administrative processes, and participatory technologies to create and generate jobs.

The current trend is for these data to be published in new reusable streams and datasets, which stimulates ecosystems to create new and innovative services based on them. In this setting, a range of business models for such services is feasible.

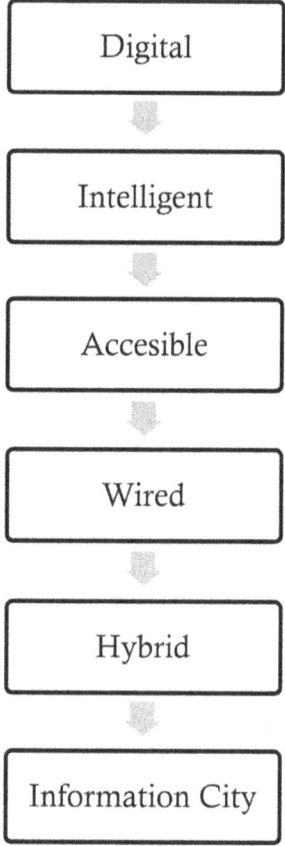

FIGURE 4.1   Representation of technology framework.

Some charge a fee, while others are free to use or operate on a freemium basis. Some of them can also be found through donations, crowdfunding, or even more unorthodox forms of sustainability.

Repurposing public-sector data (PSD) has a big impact on our society. The overall economic impact of PSI reuse has been researched extensively, with widely divergent results. In the EU, statistics range from 0.25% of GDP (With Smart Cities World, 2019) to 1.7% of GDP (Intelligent Cities, 2016) if several market criteria are met.

A major chunk of this potential will likely be provided by smart cities. Although smart cities publish a great lot more data, only information from open data portals maintained by the cities will be analyzed for the purposes of this essay.

4.4   SMART CITY'S DATASET

In 2015, the SHARE-PSI 2.0 Consortium defined reusability as a dataset's ability to be used for reasons other than those for which it was developed. The foundation of a successful smart city data ecosystem of creative services is the reusability of digital assets (i.e., data). The MELODA measure was created by the authors to validate the reusability of any public data source.

Datasets from open data sources are grouped into four areas in the most recent version of the measure (version 3.10), each having a weight attached to it: legal license, accessibility, technological standards, and data model. A global assessment of the dataset's reusability is obtained by integrating these four dimension marks with a formula (Theresa and Nam, 2019).

It is vital to emphasize that, according to WAI standards, accessibility refers to the dataset's ability to be easily accessed by external users, not to the settings that make it easier or possible for disabled people to access information. The term "data source" is always used to refer to a certain sort of data. This item was classified according to the MEPSIR report, which defines seven categories: transportation, geographic, sociological, meteorological, business, legal, and other.

Two other aspects of the datasets that are investigated are timeliness and geolocation. The timeliness of information refers to how frequently it is updated and if it may be considered real-time. True-time data is released as soon as it is generated, with no real consolidation mechanisms in place (or with real-time consolidation). It does not imply that it should be released as soon as it is, but it is extremely usual. It is especially important for data from the city's sensors.

Geolocation refers to datasets that have information linked to them that allows them to be located geographically. As a result, data can be grouped based on its location, such as districts, zip codes, and so on.

4.5   TECHNOLOGY FRAMEWORK

The development of technology is critical in a smart city. Smart city technologies are made up of various combinations of technical infrastructure that combine to generate an array of smart city technologies with differing levels of human-technology interaction.

- **Digital**: A service-oriented infrastructure is necessary to connect people and devices in a smart city. Two examples are innovation services and communication infrastructure. According to G.S. Yovanof and G.N. Hazapis, a digital city is

  > a connected community that combines broadband communications infrastructure, a flexible, service-oriented computing infrastructure based on open industry standards, and innovative services to meet the needs of governments, their employees, citizens, and businesses.

- **Intelligent**: Artificial intelligence and machine learning can be used to educate cognitive technologies to spot patterns in data generated by connected city devices. The efficacy and influence of specific policy measures can be quantified using cognitive systems that analyze the continuous interactions of humans with their urban settings (Campaign Trail, 2019).
- **Accessible**: Citizens can use public services from any connected device in a ubiquitous city. U-city is an extension of the digital city concept because of the ease with which any infrastructure can be accessed.

- **Wired**: Physical components of IT systems are crucial in the early stages of smart city development (Bhambri et al., Cloud & IoT, 2022). Wired infrastructure is required to support IoT and wireless technologies, which are at the heart of more networked living. (With smart cities, 2020) In a linked city, the general public has access to constantly updated digital and physical infrastructure. The Internet of Things, robotics, and other connected technologies can then be leveraged to increase human capital and productivity (Tailor et al., 2022).
- **Hybrid**: A hybrid city is one that blends a physical city with a virtual metropolis that is linked to the physical space. This link can be established by virtual design or the presence of a critical mass of virtual community members in a real metropolis. Hybrid spaces can assist in the implementation of future-state smart city services and integration projects.
- **Information city**: The massive number of interactive devices in a smart city creates a lot of data. The way data is interpreted and stored is critical to smart city growth and security (Mehrotra et al., 2020).

4.5.1 SMART CITIES

Smart grids are a key technology in smart cities. The improved flexibility of the smart grid enables greater penetration of highly variable renewable energy sources, such as solar and wind. Mobile devices are another significant technology that allows people to access smart city services (such as smartphones, tablets, etc.).

Smart cities rely on smart homes (Smart home, 2017) and the technology that they contain. Bicycle-sharing systems are an important part of smart city planning. Smart cities necessitate smart mobility. CCTV and intelligent transportation technology are in the works. Several smart cities have digital libraries as well.

Sensor owners can register and link their devices to send data into an online database for storage, and developers can connect to the database and construct their own applications based on that data.

Supporting technology includes telecommuting, telehealth, and the blockchain, fintech, Internet banking technologies (Khang, Chowdhury et al., Blockchain, 2022).

Electronic cards (also known as smart cards) are another common component in smart city scenarios. These cards have a one-of-a-kind encrypted identification that allows the owner to log into a range of government-provided services (or e-services) without creating multiple accounts.

Governments can use the unique identifier to collect data on citizens and their preferences in order to improve service delivery and identify similar interests among groups (Khang, Bhambri et al., Big Data, Cloud Computing & IoT, 2022). This method has previously been implemented in Southampton.

Retractable bollards can restrict access to city centers (e.g., delivery trucks resupplying outlet stores). An electronic pass is the most common method of opening and closing such impediments, but ANPR cameras coupled to the bollard system can also be employed.

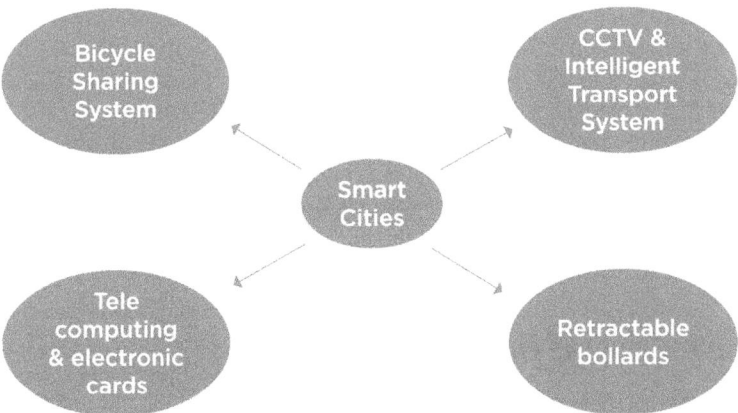

FIGURE 4.2 Representations of various important systems.

4.5.2 Urban Development and Infrastructure

Design teams must work with planners and developers to design and build the optimal solution across the entire project life cycle, from assessing the physical opportunities and restrictions of a site to weighing the profitability of various development options (Silva et al., 2018).

To ensure that all areas of civil engineering and environmental planning are covered, an integrated, multi-disciplinary approach is required for projects ranging from urban rehabilitation to the creation of new cities:

1. Situation on the ground
2. Flood danger exists
3. Energy
4. Infrastructure and strategy for utilities
5. Infrastructure and waste management strategies
6. Transportation and accessibility
7. Environmental impact assessments and how they might be reduced

Many smart systems, such as smart grids linked to infrastructure, now operate in functional silos, each with its own set of hardware and software, all managed by corporations with specialized knowledge in that industry. Each system has its own set of controls and sensor networks (O'Dwyer et al., 2019).

4.5.3 Challenges in Smart Cities

Apart from dealing with financial concerns and cybersecurity, there are a number of other issues to consider while adopting smart city solutions:

1. Infrastructure is the number one concern. Technology infrastructure is at the heart of a smart city, and issues can include power supply, connectivity, and seamless integration. Argentina's capital, Buenos Aires, is a great example

of how to build infrastructure that integrates diverse IT systems and stream-lines information flow. They were able to connect citizen complaints submit-ted through a mobile app to the local government, which then assigned the problem to the nearest vendor to resolve (Vladimir Hahanov et al., 2022).

2. Smart cities must be inclusive. In creating smart cities, all inhabitants' interests must be taken into account, particularly the underserved, includ-ing those who do not have access to the Internet.

3. Residents must be assured that the acquired data remain confidential/pri-vate through clear communication and transparency (Gabrys et al., 2014).

4.6 APPLICATIONS OF SMART CITY

4.6.1 Intelligent Infrastructure

Buildings and urban infrastructures must be planned more efficiently and sustain-ably, and digital technologies are becoming increasingly crucial for cities to have the conditions for continued development. To reduce CO_2 emissions, cities should invest in electric cars and self-propelled vehicles.

In reality, sophisticated technologies are needed to create an infrastructure that is both energy efficient and environmentally friendly. Smart lighting, for example, only turns on when someone passes by it, reducing the demand for electricity; smart lights also allow for the adjustment of brightness settings and the tracking of daily usage (Dameri and Cocchia, 2013).

4.6.2 Management of Air Quality

Smart cities are also putting in place instruments that can record real-time pollution data and forecast emissions. Being able to properly estimate air pollution allows cities to go to the root of their emission issues and create strategies to reduce the amount of pollution they emit.

4.6.3 The Control of Traffic

Finding solutions to optimize traffic is one of the most difficult challenges that large cities face. Finding a remedy, however, is not impossible. Los Angeles, for example, is one of the world's largest cities, and it has developed an intelligent transportation solution to manage traffic flow.

Pavement-integrated sensors transmit real-time traffic flow updates to a central traffic control platform, which analyzes the data and changes traffic signals to the current traffic condition in seconds.

Simultaneously, previous data is used to forecast where traffic will move, and none of these processes require human intervention (Ahvenniemi et al., 2017).

4.6.4 Ingenious Parking

Cities are also implementing sophisticated parking technologies that detect when a car has departed from a parking spot. Sensors embedded in the ground notify the location of available parking spaces to a car via a smartphone app. Others rely on vehicle feedback to establish specific locations.

4.6.5 WASTE MANAGEMENT USING INSIGHT

Waste management systems can improve waste collection efficiency and lower operational costs while also addressing any environmental issues that come with inefficient waste collection.

The garbage container receives a level sensor in these systems, and when a particular threshold is achieved, a truck driver's management platform receives a notification via their smartphone.

The message seems to empty a full container, which helps them avoid drains that are half-empty (Mosannenzadeh and Vettorato, 2014).

4.7 CONCLUSION

Unlike the release of information, which is rather common in the cities investigated, gathering apps created by the reuse ecosystem is currently a pending task for a large number of smart cities.

Those cities have reached a substantial maturity level when it comes to collecting apps in their own portal. The number of applications ranges from two to 110, illustrating the different levels of growth in different cities.

The fact that roughly half of the apps in current portals were created by the city (especially in less populated portals) could indicate that the ecosystems surrounding these cities are still in their infancy.

More research is needed to address the city's return on investment (economic and social) and, as a result, how user ecosystems empower themselves and the city to establish a sustainable business model.

One of the most intriguing and challenging sectors in which to examine the value generation and impact of open data is the release of information by smart cities.

Future studies will focus on how to share information in a way that maximizes the social and economic value offered, as well as how to keep user ecosystems sustainable.

More precise judgments on these concerns are likely to result from more extensive sampling of data combined with continued modeling, blockchain technology, and AI-centric solutions (Rana et al., AI & Blockchain, 2021).

REFERENCES

Ahvenniemi, H., Huovila, A., Pinto-Seppä, I., and Airaksinen, M. What are the differences between sustainable and smart cities? *Cities*. 2017, 60, 234–245. doi:10.1016/j.cities.2016.09.009.

Albino, V., Berardi, U., and Dangelico, R.M. Smart cities: definitions, dimensions, performance, and initiatives. *Journal of Urban Technology*. 2015, 22, 3–21. doi:10.1080/10630732.2014.942092.

Anthopoulos, Leonidas, Fitsilis, Panos, Sideridis, Alexander B., and Patrikakis, Charalampos Z. (Eds.). *Next Generation Society. Technological and Legal Issues*. Lecture Notes of the Institute for Computer Sciences, Social Informatics and Telecommunications

Engineering. Springer Berlin Heidelberg, 23 September 2009, 360–372. doi:10.1007/978-3-642-11631-5.

Bhambri, Pankaj, Rani, Sita, Gupta, Gaurav, and Khang, A. *Cloud and fog computing platforms for internet of things.* CRC Press, 2022. ISBN: 978-1-032-101507, doi:10.1201/9781032101507.

Campaign trail: why uber built a virtual city to promote a product that doesn't exist yet. Marketing Dive. Archived from the original on 26 June 2019. Retrieved 26 June 2019. https://www.marketingdive.com/news/campaign-trail-why-uber-built-a-virtual-city-to-promote-a-product-that-doe/557301/.

Dameri, R.P., and Cocchia, A. *Smart city and digital city: twenty years of terminology evolution.* In Proceedings of the X Conference of the Italian Chapter of AIS, ITAIS 2013, Milan, Italy, 14 December 2013, 1–8. https://www.semanticscholar.org/paper/Smart-City-and-Digital-City%3A-Twenty-Years-of-Dameri-Cocchia/c6b562b4aeb53c6a07c5ac4487d964aad06c8cf9.

Dekkers, M., Polman, F., Te Velde, R., and De Vries, M. MEPSIR. *Final report of study on exploitation of public sector information—benchmarking of EU framework conditions, executive summary and final report.* Part 1 and Part 2, 2006. doi:10.4018/978-1-59904-857-4.ch012.

Dhanalakshmi, R., Anand, Jose, Sivaraman, Arun Kumar, and Rani, Sita. IoT-based water quality monitoring system using cloud for agriculture use. In *Cloud and fog computing platforms for internet of things*, Chapman and Hall/CRC, 2022, 183–196. doi:10.1201_9781003213888.

Gabrys, J. *Programming environments: Environmentality and citizen sensing in the smart city.* Environment and Planning D: Society and Space. 2014, 32, 30–48. doi:10.1068/d16812.

Giffinger, R., Fertner, C., Kramar, H., Kalasek, R., Pichiler-Milanoviu, N., and Meijers, E. *Smart cities: ranking of European medium-sized cities*, Centre of Regional Science (SRF), Vienna University of Technology: Vienna, Austria, 2000. https://www.researchgate.net/publication/261367640_Smart_cities_-_Ranking_of_European_medium-sized_cities.

Hall, R.E. *The vision of a smart city.* In Proceedings of the 2nd International Life Extension Technology Workshop. Paris, France, 28 September 2000. https://www.researchgate.net/publication/241977644_The_vision_of_a_smart_city

Intelligent cities: R&D offshoring, web 2.0 product development and globalization of innovation systems (PDF). Archived (PDF) from the original on 16 May 2018. Retrieved 20 December 2016. https://www.researchgate.net/publication/253386773_Intelligent_Cities_RD_offshoring_web_20_product_development_and_globalization_of_innovation_systems.

Jetzek, T. Avital, M., and Bjorn-Andersen, N. *Generating value from open government data, 2013.* Thirty Fourth International Conference on Information Systems, Milan 2013. http://aisel.aisnet.org/icis2013/proceedings/GeneralISTopics/5/.

Khang, A., Bhambri, Pankaj, Rani, Sita, and Kataria, Aman. *Big data, cloud computing and internet of things*, CRC Press, 2022. ISBN: 978-1-032-284200, doi:10.1201/9781032284200.

Khang, A., Chowdhury, Subrata, and Sharma, Seema. *The data-driven blockchain ecosystem: fundamentals, applications and emerging technologies.* ISBN: 978-1-032-21624, 2022, doi:10.1201/9781003269281. CRC Press.

Mehrotra, Siddharth, and Dhande, Rashi. *Smart cities and smart homes: From realization to reality.* 2015 International Conference on Green Computing and Internet of Things (ICGCIoT), IEEE, 2015, 1236–1239. ISBN: 978-1-4673-7910-6, S2CID: 14156800, doi:10.1109/ICGCIoT.2015.7380652. Archived from the original on 8 June 2020. Retrieved 8 June 2020.

Mosannenzadeh, F., and Vettorato, D. Defining smart city. A conceptual framework based on keyword analysis. *TeMA Journal of Land Use, Mobility and Environment.* 2014, 998. doi:10.6092/1970-9870/2523.

O'Dwyer, E., Pan, I., Acha, S., and Shah, N. Smart energy systems for sustainable smart cities: Current developments, trends and future directions. *Applied Energy.* 2019, 237, 581–591. doi:10.1016/j.apenergy.2019.01.024.

Ramírez-Alujas, A.V., and Dassen, N. Vientos de cambio: *El advance de las políticas de gobiernoabiertoen América Latina y el Caribe.* Technical note. No. IDB-TN-629. Inter-American Development Bank, 2014. https://publications.iadb.org/es/publicacion/16833/vientos-de-cambio-el-avance-de-las-politicas-de-gobierno-abierto-en-america.

Rana, G., Khang, A., and Khanh, H.H. The role of artificial intelligence in blockchain applications. In *Reinventing manufacturing and business processes through artificial intelligence*, CRC Press, 2021, 20, doi:10.1201/9781003145011.

Rani, S., Bhambri, P., and Chauhan, M. *A machine learning model for kids' behavior analysis from facial emotions using principal component analysis.* In 2021 5th Asian Conference on Artificial Intelligence Technology (ACAIT). IEEE, October 2021, 522–525.

Rani, Sita, Kataria, Aman, and Chauhan, Meetali. Fog computing in industry 4.0: applications and challenges—a research roadmap. In *Energy conservation solutions for fog-edge computing paradigms,* Springer: Singapore, 2022, 173–190.

Rani, Sita, Khang, A., Chauhan, Meetali, and Kataria, Aman. IoT equipped intelligent distributed framework for smart healthcare systems. *Networking and Internet Architecture.* 2021, https://arxiv.org/abs/2110.04997v2, doi:10.48550/arXiv.2110.04997.

Rani, Sita, Mishra, Ram Krishn, Usman, Mohammed, Kataria, Aman, Kumar, Pramod, Bhambri, Pankaj, and Mishra, Amit Kumar. Amalgamation of advanced technologies for sustainable development of smart city environment: a review. *IEEE Access.* 2021, 9, 150060–150087.

Silva, B.N., Khan, M., and Han, K. Towards sustainable smart cities: a review of trends, architectures, components, and open challenges in smart cities. *Sustainable Cities and Society.* 2018, 38, 697–713. doi:10.1016/j.scs.2018.01.053.

Smart home, How smart homes can connect to smart cities. 8 September 2017. Archived from the original on 8 June 2020. Retrieved 8 June 2020. https://smartcity.press/how-smart-homes-can-connect-smart-cities/.

Tailor, R.K., Khang, A., and Pareek, R. (Eds.). RPA in blockchain. In *The data-driven blockchain ecosystem: fundamentals, applications, and emerging technologies* (1st ed.). CRC Press, 2022, 149–164. https://doi.org/10.1201/9781003269281.

Theresa A., and Nam, Taewoo. *Conceptualizing smart city with dimensions of technology, people, and institutions* (PDF). Centre for Technology in Government University at Albany, State University of New York, U.S. The Proceedings of the 12th Annual International Conference on Digital Government Research. Archived (PDF) from the original on 7 November 2018. Retrieved 29 June 2019. doi:10.1145/2037556.2037602.

Vladimir Hahanov, Khang A., Abbas, G.L., and Hajimahmud, V.A. Cyber-physical-social system and incident management. In *AI-centric smart city ecosystems: technologies, design and implementation* (1st ed.). CRC Press, 2022. https://doi.org/10.1201/9781003252542

Washburn, D., Sindhu, U., Balaouras, S., Dines, R.A., Hayes, N.M., and Nelson, L.E. *Helping CIOs understand "smart city" initiatives: defining the smart city, its drivers, and the role of the CIO.* Forrester Research, Inc: Cambridge, MA, 2010. http://public.dhe.ibm.com/partnerworld/pub/smb/smarterplan

With smart cities, your every step will be recorded. Archived from the original on 8 June 2020. Retrieved 8 June 2020. https://phys.org/news/2018-04-smart-cities.html

With Smart Cities World, Innovation vs technology. Redefining "smart" in smart-cities. 11
      October 2019. Archived from the original on 9 June 2020. Retrieved 9 June 2020.
Yovanof, Gregory S., and Hazapis, George N. An architectural framework and enabling wire-
      less technologies for digital cities & intelligent urban environments. *Wireless Personal
      Communications.* 19 March 2009, 49(3): 445–463. doi:10.1007/s11277-009-9693-4

5 A Security Approach to Manage a Smart City's Image Data on Cloud

Manish Kumar Mukhija,
Satish Kumar Alaria, Pooja Singh

CONTENTS

5.1   INTRODUCTION

Cloud computing is a significantly versatile and savvy foundation for running HPC and web applications. Regardless, the creating interest of cloud foundation has profoundly extended the energy use of data centers, and it has transformed into a fundamental issue (Alaria et al., 2018).

High energy usage does not simply imply high functional expense, which lessens the net income of cloud providers, but it also prompts high carbon emissions, which are not nature-friendly (Akkaoui et al., 2019).

The commercialization of these advancements is portrayed as cloud computing (Thakare, et al., 2020), where computing is passed on as a utility on a pay-as-you-go basis. By and large, business affiliations used to contribute an enormous proportion

DOI: 10.1201/9781003252542-5

of capital and time in acquiring and backing computational resources (Arunachalam et al., 2021; Banerjee et al., 2022).

The improvement of Cloud computing is rapidly changing the ways of storage by applying based approach to manage enrollment arranged methodology for offering admittance to versatile foundation and organizations on demand. Therefore, customers can store, access, and offer any proportion of data in Cloud.

That is, nearly nothing or medium endeavors/affiliations don't have to worry about purchasing, masterminding, coordinating, and keeping up their own specific computing foundation on premise environment. They can focus on sharpening their middle capacities by abusing different Cloud computing benefits, for instance, on-demand computing resources, faster and more affordable programming improvement limits expecting practically no work or zero exertion.

Furthermore, Cloud computing moreover offers epic proportion of register capacity to affiliations which require treatment of enormous proportion of data created for all intents and purposes reliably. For instance, budgetary as Figure 5.1.

As such, numerous associations not simply see Clouds as a significant on-demand advantage, yet moreover a potential market opportunity.

As demonstrated by an IDC (International Data Corporation) report (Shree et al., 2020), the overall IT cloud organizations' spending is evaluated to increase from $16 billion in 2008 to $42 billion in 2012, which is a compound annual growth rate (CAGR) of 27%.

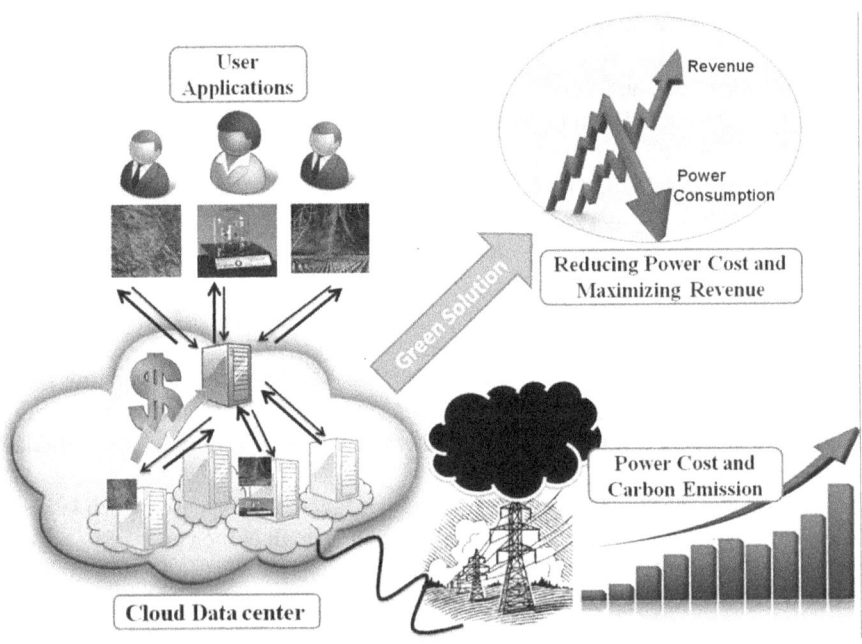

FIGURE 5.1   Cloud and environmental sustainability

5.2 LITERATURE SURVEY

Optimization accepts a huge part in various issues that expect the specific yield (Shree et al., 2020). Security of the data set away in far off specialists forth plainly reliant upon secret key which is used for encryption and interpreting reason.

Various strange key age estimations, for instance, RSA, AES are available to make the key. The key made by such estimations are ought to be smoothed out to give more noteworthy security to your data from unapproved customers similarly as from the outcast auditors (TPA) who will check their data for trustworthiness reason (Kumar et al., 2021).

In this paper a procedure to redesign the baffling key by using cuckoo search computation (CSA) is proposed.

Duties are separated inside a gathering by using the work based induction control (RBAC) in the Azure Internet of Things (IoT) framework, and simply an appropriate level of access is surrendered to customers to perform unequivocal tasks, dependent upon a given situation (Thakare et al., 2020; Long et al., 2019). In any case, a comparative confirmation and endorsement framework is used for "sort of customer," which grows the movement over-trouble on the cloud laborer (Liu et al., 2019).

Also, in view of its RBAC nature, the IoT structure is inefficient in dealing with an amazing situation where various customers request tantamount kinds of resources, by making a couple of repeated positions (Rani, Khang et al., IoT & Healthcare, 2021).

This results in clashing and steadfast execution and the lack of the ability to capably address procedure the board, semantics, reiteration issues in positions, dynamic customer dealing with, work task issues, adaptability, work impact, solitary rights, and security issues in gigantic affiliations.

In this article, essayists arranged and presented a cunning access control model for a basically huge clinical circumstance with capable need based check parts to address the recently referenced issues related with the Azure IoT cloud (Rana et al., 2021).

The proposed model encapsulates the necessity of need based resource access rights across various customers in a huge affiliation, diminishes weakness and ineffectuality, and supports individuals with the consistent execution of approaches (Rao et al., 2019).

Makers evaluated the upsides of the proposed model by differentiating it and existing models and the Azure model, using the clinical benefits use-case situation.

The assessment results show that by melding the need trademark office in the current RBAC model, the proposed model requests the methodology instrument reliant upon need credits and exhibits that the proposed model is prepared for dealing with issues that generally happen while overseeing enormous remarkable circumstances in tremendous affiliations.

In the domain of advanced real structures, the improvement of Smart Living Spaces (SLS) design offers people the opportunity to benefit with better methodologies for living (Belkhiria et al., 2020). Such mechanical example that incorporates a couple of parts of step by step life, allows the occupants of the space to all the more probable modify and control their present situation. SLS stresses, essentially, energy conservation, convenience and comfort similarly as clinical consideration concerns.

They are ceaselessly recognized in utilitarian applications for instance home-grown gadgets and clinical and care systems. The SLS related applications are proposed to chip away at the individual fulfillment for customers by ensuring that their prerequisites are met, subsequently conveying more conservative use of energy in a given living space. Furthermore, the progression of SLS development makes the contraptions around us "clever."

This has brought package of challenges, one of which is the passageway control the board. In a SLS, there should be an energetic access control segment set up to ensure a supported induction to contraptions.

This paper bases on the usage of Role Based Access Control (RBAC) model as a capable method to keep unapproved customers from getting to contraptions in a Smart Building. Then, to guarantee that solitary the supported customers can get to the contraptions, a Smart Building Manager (SBM) is expected to demand that an appraisal engine survey requests (Vladimir Hahanov et al., 2022).

Makers separate the appraisal as an untouchable in the astute design in which makers propose RBAC with regions that stands their game plan model. Makers complete the plan model for the mean to ensure a fine grained induction control to a sharp construction application while progressing from a mono-customer to a multi-customer.

With the happening to web and its associated propels, the affiliations are more stressed in offering security to their resources (Suganthy and Prasanna Venkatesan, 2019). Access control models helps in giving quite far to the customers when the resources are being gotten by them and Role based induction control (RBAC) model is one such access control instrument in which the customers access the resources subject to the positions they obtain in the system.

In the one of a kind environment, the positions that a customer gets in the system is thoroughly depending upon the setting wherein the customer sends a passage interest. This paper explores the thoughts of RBAC models and their increases. The guideline objective of this paper is to take apart the passage control models by recognizing their applications and limitations as a general rule applications.

According to the arrangement considered RBAC model, the SSM framework subject to JAVA language is gotten together with Maven to execute the establishment work module (Mu and Liu, 2019). The front-end advancement, for instance, AJAX and Bootstrap is used to design the WEB page.

The MySQL database is used to store data and Tomcat is used as the specialist. The arrangement of this system gives a good condition to the force the chiefs of the endeavor.

(Liu et al., 2019) to the extent people's lifestyles, direct data can be adequately accumulated and taken apart. Directions to get and utilize these data is a charming issue today. There are various direct affirmation models in insightful world, yet the technique for recognizing conduct are at this point considering standard organizing in industry.

The major clarification is the impact of work circles. Data can reflect individual lead norms, yet work circles disastrously influence rehearses (Bhambri et al., 2022;

Chauhan and Rani, 2021). The working circle here is described all things considered of people working in consistently life. The work circle is related to approvals. Hence, makers familiarize the RBAC model with join the impact of approvals on direct recolonization.

Makers propose a procedure for dynamically creating RBAC models (Kauret al., 2015). Then, makers segment the "min-advantage social events" and join the results of the get-togethers to get their RBAC-outfit model (Rani, Bhambri et al., 2021). In the assessments, makers examine the effects of a couple of models already, then, at that point afterward using "min-advantage social events" (Rani et al., 2022).

The results show that approvals do influence rehearses. Considering "min-advantage social occasions," makers take a gander at the effects of the other three models. The model makers propose achieves the best affirmation (Rani, Kataria et al., 2021).

Cloud computing is an extensive development, which has attracted a lot of thought nowadays (Ghafoorian et al., 2019; Rani and Kumar, 2022). Among the various models that ought to be considered for data storage in the cloud, access control expects an imperative part.

Occupation based permission control (RBAC) is a striking strategy for secure data storage in the cloud. Since the ordinary RBAC models are improper for open and decentralized conditions, lately, a couple of works have composed the trust thought into the RBAC model. Regardless, they have not totally watched out for the essential security estimations of a trust-based system.

Accordingly, in this paper, makers at first present the security goals that should be considered in a powerful trust-based structure. Second, makers propose a sharp trust and reputation based RBAC model that not only can true to form withstand the security risks of trust-based RBAC models, yet moreover is versatile as it has reasonable execution time (Rani, Mishra et al., 2021). Third, makers survey the proposed model using the notable trust association of advogato dataset.

At last, makers contrast the proposed model and actually appropriated ones to the extent mean absolute bungle, execution period of variant trust computation, and gave features. The refined results are illustrative of the need of the proposed model to be used in certified cloud conditions.

Access control expects a huge part in binding the passageway of unmistakable advantages, keeping from assault of unlawful customers or the damages achieved by legal customers' unintentional undertakings (Zou et al., 2019). Regardless, most of approaches for expanding the semantic depiction of RBAC fail to reliably fuse data depiction of RBAC and region express principles together.

In this paper, makers present a semantic standards based RBAC development model for versatile data access and resource conveyance.

Makers first give the regular thought of the development model. Then makers analyze how to depict RBAC approval for the predictable blend in the OWL semantic portrayal and the SWRL semantic rule depiction (Soni and Kumar, 2019).

At long last, makers produce the uniform endorsement sees reliant upon SPARQL question, which join viewpoints supported by both express and evident propensities.

5.3 PROPOSED WORK

The proposed concept is divided into two main segments:

1. Role-based security: According to role-based security, the files are assigned for access for a particular role.
2. TPA-based security: TPA is the third-party auditor that will cross-check the user requests to access the file and generate the OTP and access key.

5.3.1 NEW USER CREATION

In order to access the services related to a cloud-sharing platform, first, the user is required to register using the platform, then the user can access the services.

- Step 1: First, the username, name of the employee, and role of the employee are specified.
- Step 2: A grid is used on the pictures available for selection, and the user has to select a single image from it.
- Step 3: The image is partitioned into small segments, and the image is organized in the second grid.
- Step 4: The size of the image selected is calculated and displayed in bytes.
- Step 5: The user then selects the segments of the image; the selected segment turns gray, and after all the desired image blocks are selected, the generate button is clicked.
- Step 6: The pattern will form on the basis of the selection of the image block, and the basis of the pattern is

  Image (ImageNumber)_part(partnumber1)_sizeofimage_
  Image (ImageNumber)_part(partnumber2)_sizeofimage_
  :::::
  Image (ImageNumber)_part(partnumberN)_sizeofimage_

- Step 7: The pattern is saved with the other details of the user in the database table meminfo, which contains the details of registered users.
- Step 8: End.

5.3.2 EXISTING USER LOGIN

Now, after the registration is done, the registered user can log in using the procedure adopted for registration.

- Step 1: Enter the username.
- Step 2: The grid of the pictures is available for selection, and the user has to select a single image from it.
- Step 3: The image is partitioned into small segments, and the image is organized in the second grid.
- Step 4: The size of the image selected is calculated and displayed in bytes.

- Step 5: The user then selects the segments of the image, the selected segment turns gray, and after all the desired image blocks are selected, the generate button is clicked.
- Step 6: The pattern will form on the basis of the selection of the image block, and the basis of the pattern is

  Image (ImageNumber)_part(partnumber1)_sizeofimage_
  Image (ImageNumber)_part(partnumber2)_sizeofimage_
  ::::
  Image (ImageNumber)_part(partnumberN)_sizeofimage_

- Step 7: If the details are correct:

  Login Successful
  Else
  Login Failed
  [End of If structure.]

- Step 8: End.

5.3.3 FILE UPLOAD

This algorithm is used to upload the file on the server, according to the role.

- Step 1: Access username according to the session variable.
- Step 2: Select the file to share.
- Step 3: Select the role for which the file is to be shared.
- Step 4: Store the details in the database table fuploads.
- Step 5: Stop.

5.3.4 FILE REQUEST

This algorithm is used to request file access from TPA.

- Step 1: Select the name of the file for which access is to be requested.
- Step 2: Save details in the Reqdata, which is the table for user requests.
- Step 3: Auto increment-based request ID is generated.
- Step 4: Stop.

5.3.5 TPA GRANT ACCESS

This algorithm is used to grant the access to the user requesting the file.

- Step 1: Select the request ID.
- Step 2: Fetch the file details and user details.
- Step 3: Generate OTP, which is generated using ten random numbers, and each ranges from 30 to 126, and the character corresponding to these numbers will be combined in order to form the OTP.

- Step 4: Generate hash using SHA-512 algorithm for the file that is requested and the hash of the username who requested the file. Now, extract 20 characters from the hash of the file and 20 characters from the hash of the username to generate the access key.
- Step 5: Save the details in the database table of Reqdata for the requested ID.
- Step 6: End.

5.3.6 USER ACCESSING FILE

This algorithm is used to access the file requested.

- Step 1: Select the request ID.
- Step 2: Fetch the file details.
- Step 3: Enter the OTP and access key.
- Step 4: If the details are correct:
  File Download
  Else
  Invalid Details
  [End of If structure]
- Step 5: End.

5.4 IMPLEMENTATION AND RESULT ANALYSIS

The implementation of the algorithms is graphically created using Visual Studio 2010 and using the database system of Microsoft SQL Server. The system has the general requirement of 2 GB RAM, and the process required for the execution of the implementation is Intel I3 or higher, as shown in Figure 5.2.

Now, to start up the system flow, first, we require the new user to be registered, as shown in Figure 5.3.

In the registration process, we have made the condition of the graphical authentication system, in which we have the picture from the grid, and then the picture segmented and the size of the picture get listed up, as shown in Figure 5.4.

5.4.1 DEVICE 1: RUMKIN TEST

This secret word checker will measure your secret key and give it a score dependent on how great of a secret phrase it is. It will inform you as to whether you picked a typical secret word (don't do that!), and it will likewise consider the likelihood of letters landing near one another, as shown in Table 5.1.

Role and TPA Based Clound Data Sharing

FIGURE 5.2 Loading screen.

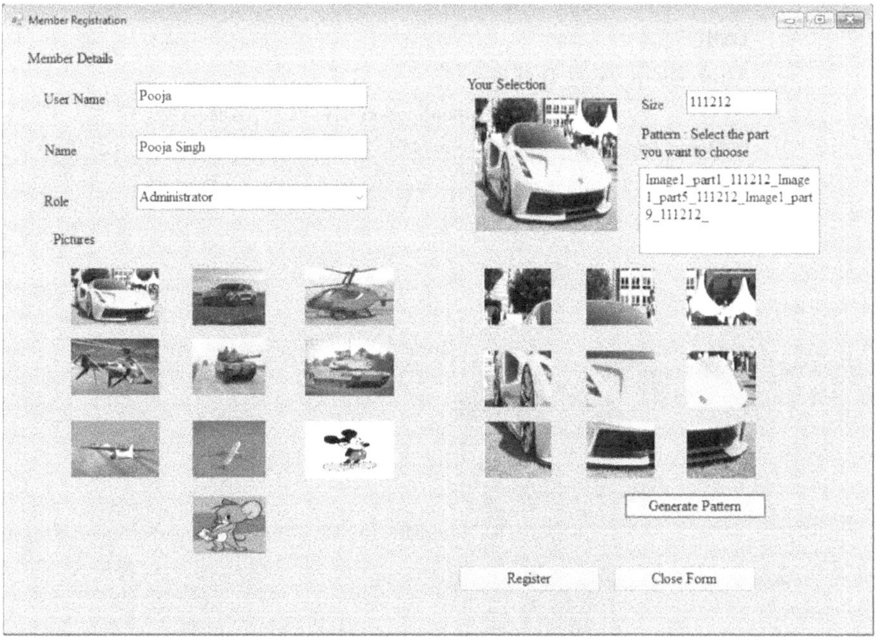

FIGURE 5.3 User registration.

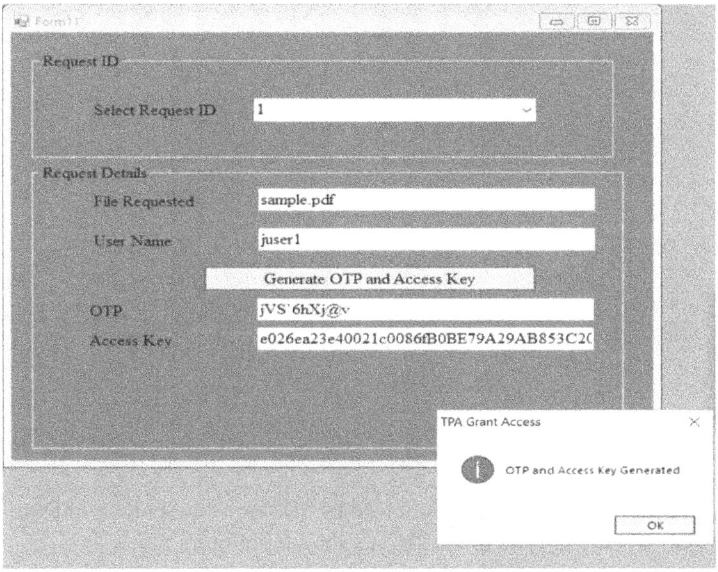

FIGURE 5.4 File access.

TABLE 5.1
Analysis of Keys Test 1

	Authentication Key	Access Key
Proposed Approach	445.1	168.8

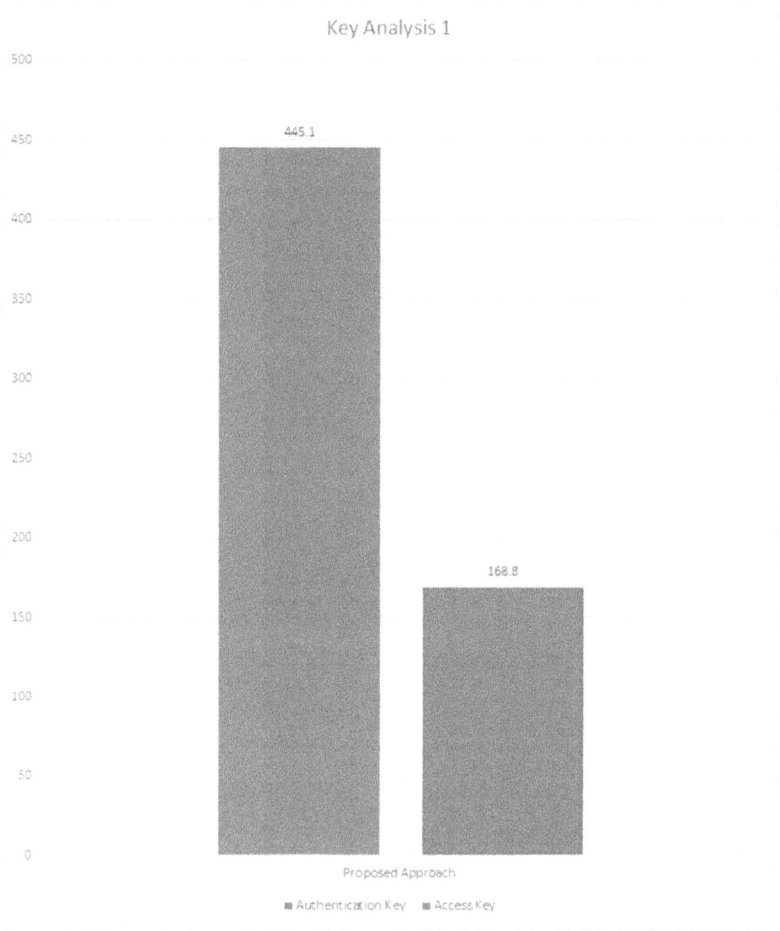

FIGURE 5.5 Graph of key analysis 1.

For example, "Q" is quite often followed by "U," so your secret word's score will not build a lot when you type in the "U," as shown in Figure 5.5.

Authentication Key

Image4_part1_112777_Image4_part4_112777_Image4_part7_112777_Image4_part8_112777_Image4_part9_112777_

Access Key

e026ea23e40021c0086fB0BE79A29AB853C20553

5.4.2 Device 2: Zxcvbn Test

The following is a secret word meter that tests entropy utilizing zxcvbn by Dropbo6. It tests for word reference words, leet-talk, unmistakable examples, and different heuristics to give a reasonable deduction of what the entropy could be.

In case you are gluing passwords from the generator, you will see conflicts. This analyzer is a visually impaired entropy surmise. It does not have the foggiest idea about the arrangement of components your secret phrase is from, nor can it say whether an irregular capacity was utilized, as shown in Figure 5.6.

In this way, the theory might be higher or lower than what you know it to be, as shown in Table 5.2.

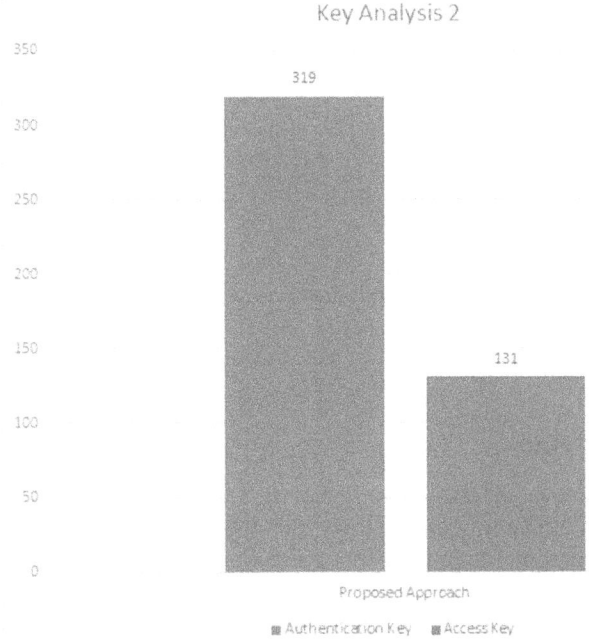

FIGURE 5.6   Graph of key analysis 2.

TABLE 5.2
Analysis of Keys Test 2

	Authentication Key	Access Key
Proposed Approach	319	131

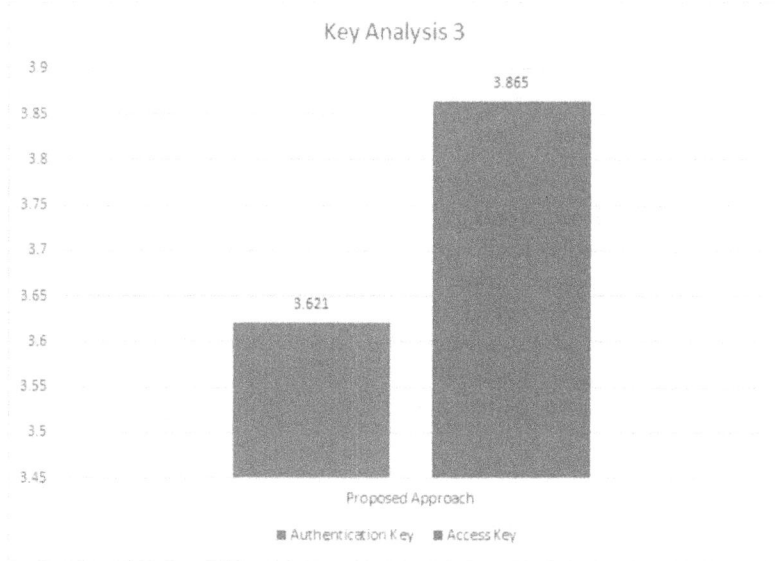

FIGURE 5.7   Graph of key analysis 3.

TABLE 5.3
Analysis of Keys Test 3

	Authentication Key	Access Key
Proposed Approach	3.621	3.865

5.4.3   DEVICE 3: CRYPTOOL

CrypTool is an open-source adventure and implements more than 400 algorithms. In this section, we would like to share with you that the essential result is the free e-getting the hang of programming CrypTool indicating cryptography and crypt-analysis thoughts as in Figure 5.7 and Table 5.3.

As shown by "Hakin9, 2021," the CrypTool project is about making the sometimes daunting subject of cryptography more accessible and easy to understand. Nowadays, this tool is worldwide the most, no matter how you look at it, e-getting the hang of programming in the field of cryptology as Figure 5.8 and Table 5.4.

5.5   CONCLUSION

Two main issues are generally faced when data is shared in a cloud environment are, first, authenticating the user who can access the data and, second, the security of the data itself. Seeing the concern, we proposed the hybrid concept that involves around role-based security and TPA-based security (Khang, Bhambri et al., Big Data, Cloud Computing & IoT, 2022).

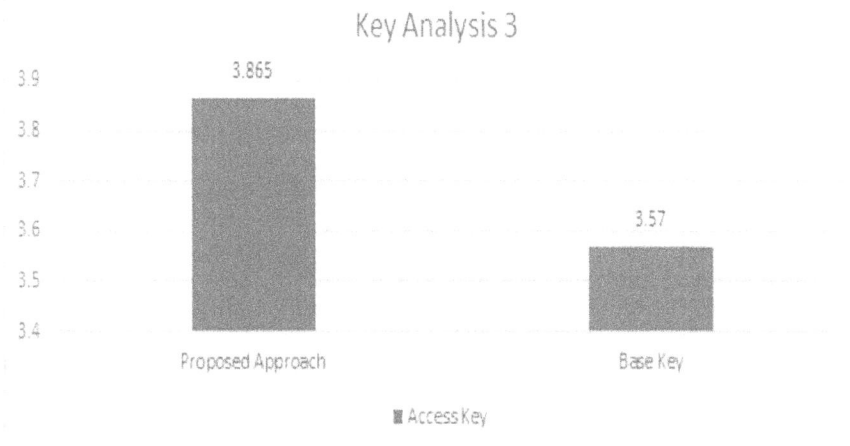

FIGURE 5.8   Graph of analysis 3 base and proposed.

TABLE 5.4
Analysis of Keys Test 2 Base Key

	Access Key
Base Paper	3.57

By making use of role-based security (Khang, Chowdhury et al., Blockchain, 2022), first we authenticate the users using the graphical authentication in which, first, the image must be selected, then the image gets segmented into image blocks, and when selected, then the pattern is formed, which is used for authentication. After that, when the user shares the file, then he/she specifies the role who can access the file (Jadaun et al., 2021).

Once the file is shared, then the second security level is imposed by TPA, which, when it sees the request for the shared file, assigns the OTP and access key for accessing the shared file.

The OTP is a character combination that is an ASCII equivalent to random numbers in the range of 30 to 126, and the access key is formed using the SHA-512 hash for username and file, and the combined extract of 20 characters each of the hash will form the access key.

The authentication pattern and access key are tested over various platforms and tools for testing password strength, and the results are quite satisfactory.

REFERENCES

Akkaoui, R., X. Hei, C. Guo, and W. Cheng, "*RBAC-HDE: On the Design of a Role-based Access Control with Smart Contract for Healthcare Data Exchange,*" 2019 IEEE International Conference on Consumer Electronics—Taiwan (ICCE-TW), Yilan, Taiwan, 2019, pp. 1–2, doi:10.1109/ICCE-TW46550.2019.8991965.

Alaria, S. K., and A. Kumar, "Implementation of New Cryptographic Encryption Approach for Trust as a Service (TAAS) in Cloud Environment," *International Journal of Computer Application* (2250–1797), vol. 4, no. 8, July–August 2018. doi:10.26808/rs.ca.i8v4.03.

Arunachalam, P., N. Janakiraman, A.K. Sivaraman, A. Balasundaram, R. Vincent, S. Rani, B. Dey, A. Muralidhar, and M. Rajesh, "Synovial Sarcoma Classification Technique Using Support Vector Machine and Structure Features," *Intelligent Automation & Soft Computing*, vol. 32, no. 2, pp. 1241–1259, 2021. doi:10.32604/iasc.2022.022573.

Banerjee, K., V. Bali, N. Nawaz, S. Bali, S. Mathur, R. K. Mishra, and S. Rani, "A Machine-Learning Approach for Prediction of Water Contamination Using Latitude, Longitude, and Elevation," *Water*, vol. 14, no. 5, p. 728, 2022. doi:10.3390/w14050728.

Belkhiria, H., F. Fakhfakh, and I. B. Rodriguez, *"Resolving Multi-User Conflicts in a Smart Building using RBAC,"* 2020 IEEE 29th International Conference on Enabling Technologies: Infrastructure for Collaborative Enterprises (WETICE), Bayonne, France, 2020, pp. 181–186. doi:10.1109/WETICE49692.2020.00043.

Bhambri, P., S. Rani, G. Gupta, and Khang, A., *Cloud and Fog Computing Platforms for Internet of Things.* CRC Press, 2022. ISBN: 978-1-032-101507, doi:10.1201/9781032101507.

Chauhan, M., and S. Rani, "Covid-19: A Revolution in the Field of Education in India." In *Learning How to Learn Using Multimedia* (pp. 23–42). Springer, Singapore, 2021. https://pesquisa.bvsalud.org/global-literature-on-novel-coronavirus-2019-ncov/resource/pt/covidwho-1400997

Ghafoorian, M., D. Abbasinezhad-Mood, and H. Shakeri, "A Thorough Trust and Reputation Based RBAC Model for Secure Data Storage in the Cloud," *IEEE Transactions on Parallel and Distributed Systems*, vol. 30, no. 4, pp. 778–788, 1 April 2019. doi:10.1109/TPDS.2018.2870652.

Hakin9, "The CrypTool project is about making the sometimes daunting subject of cryptography more accessible and easy to understand," 2021. https://hakin9.org/download/id-theft-hakin9-032011/

Jadaun, A., S. K. Alaria, and Y. Saini, "Comparative Study and Design Light Weight Data Security System for Secure Data Transmission in Internet of Things," *International Journal on Recent and Innovation Trends in Computing and Communication*, vol. 9, no. 3, pp. 28–32, 2021. https://doi.org/10.17762/ijritcc.v9i3.5476.

Kaur, G., R. Kaur, and S. Rani, "Cloud Computing-A New Trend in IT Era," *International Journal of Science, Technology and Management*, pp. 1–6, 2015. https://www.research-gate.net/journal/SSRN-Electronic-Journal-1556-5068

Khang, A., P. Bhambri, S. Rani, and A. Kataria, *Big Data, Cloud Computing and Internet of Things.* CRC Press, 2022. ISBN: 978-1-032-284200, doi:10.1201/9781032284200.

Khang, A., S. Chowdhury, and S. Sharma, *The Data-Driven Blockchain Ecosystem: Fundamentals, Applications and Emerging Technologies.* CRC Press, 2022. ISBN: 978-1-032-21624, doi:10.1201/9781003269281.

Kumar, M., S. Kumar, and H. Nagar, "Enhanced Text and Image Security Using Combination of DCT Steganography, XOR Embedding and Arnold Transform," *Journal of Design Engineering*, vol. 2021, no. 3, pp. 732–739, 2021. ISSN: 0011-9342. doi:10.1109/ICAIS53314.2022.9742942.

Liu, C., S. Cheng, Y. Luo, and F. Jiang, *"Behavior Recognition Based on RBAC-Ensemble Model,"* 2019 11th International Conference on Knowledge and Smart Technology (KST), Phuket, Thailand, 2019, pp. 29–34.

Long, S., and L. Yan, *"RACAC: An Approach toward RBAC and ABAC Combining Access Control,"* 2019 IEEE 5th International Conference on Computer and Communications (ICCC), Chengdu, China, 2019, pp. 1609–1616.

Mu, Z., and M. Liu, *"Enterprise Rights Management System Based on RBAC Model,"* 2019 International Conference on Robots & Intelligent System (ICRIS), Haikou, China, 2019, pp. 234–237. doi:10.1109/ICRIS.2019.00068.

Rani, S., P. Bhambri, and M. Chauhan, *"A Machine Learning Model for Kids' Behavior Analysis from Facial Emotions Using Principal Component Analysis,"* In 2021 5th Asian Conference on Artificial Intelligence Technology (ACAIT). IEEE, October 2021, pp. 522–525. doi:10.1109/ACAIT53529.2021.9731203.

Rani, S., A. Kataria, M. Chauhan, P. Rattan, R. Kumar, and A. K. Sivaraman, "Security and Privacy Challenges in the Deployment of Cyber-Physical Systems in Smart City Applications: State-of-Art Work." In *Materials Today: Proceedings.* Elsevier, 2022. doi:10.1016/j.matpr.2022.03.123.

Rani, S., A. Kataria, V. Sharma, S. Ghosh, V. Karar, K. Lee, and C. Choi, "Threats and CORRECTIVE Measures for IoT Security with Observance of Cybercrime: A Survey," *Wireless Communications and Mobile Computing,* 28 April 2021. Hindawi.

Rani, S., Khang, A., M. Chauhan, and A. Kataria, *"IoT Equipped Intelligent Distributed Framework for Smart Healthcare Systems,"* Networking and Internet Architecture, 2021, https://arxiv.org/abs/2110.04997v2, doi:10.48550/arXiv.2110.04997.

Rana, G., Khang, A., R. Sharma, A. K. Goel, and A. K. Dubey, *"The Role of Artificial Intelligence in Blockchain Applications,"* Reinventing Manufacturing and Business Processes Through Artificial Intelligence, 2021, doi:10.1201/9781003145011.

Rani, S., and R. Kumar, "Bibliometric Review of Actuators: Key Automation Technology in a Smart City Framework." In *Materials Today: Proceedings.* Elsevier, 2022.

Rani, S., R. K. Mishra, M. Usman, A. Kataria, P. Kumar, P. Bhambri, and A.K. Mishra, "Amalgamation of Advanced Technologies for Sustainable Development of Smart City Environment: A Review," *IEEE Access,* vol. 9, pp. 150060–150087, 2021. doi:10.1109/ACCESS.2021.3125527.

Rao, K. R., I. G. Ray, W. Asif, A. Nayak, and M. Rajarajan, "R-PEKS: RBAC Enabled PEKS for Secure Access of Cloud Data," *IEEE Access,* vol. 7, pp. 133274–133289, 2019. doi:10.1109/ACCESS.2019.2941560.

Shree, S. R., A. Chilambu Chelvan, and M. Rajesh, *"Optimization of Secret Key Using Cuckoo Search Algorithm for Ensuring Data Integrity in TPA,"* 2020 International Conference on Computer Communication and Informatics (ICCCI), Coimbatore, India, 2020, pp. 1–5. doi:10.1109/ICCCI48352.2020.9104102.

Soni, K., and S. Kumar, *"Comparison of RBAC and ABAC Security Models for Private Cloud,"* 2019 International Conference on Machine Learning, Big Data, Cloud and Parallel Computing (COMITCon), Faridabad, India, 2019, pp. 584–587. doi:10.1109/COMITCon.2019.8862220.

Suganthy, A., and V. Prasanna Venkatesan, *"An Introspective Study on Dynamic Role-Centric RBAC Models,"* 2019 IEEE International Conference on System, Computation, Automation and Networking (ICSCAN), Pondicherry, India, 2019, pp. 1–6. doi:10.1109/ICSCAN.2019.8878827.

Thakare, A., E. Lee, A. Kumar, V. B. Nikam, and Y. Kim, "PARBAC: Priority-Attribute-Based RBAC Model for Azure IoT Cloud," *IEEE Internet of Things Journal,* vol. 7, no. 4, pp. 2890–2900, April 2020. doi: 10.1109/JIOT.2019.2963794.

Vladimir Hahanov, Khang, A., Gardashova Latafat Abbas, Vugar Abdullayev Hajimahmud, "Cyber-Physical-Social System and İncident Management." In *AI-Centric Smart City Ecosystems: Technologies, Design and Implementation* (1st ed.). CRC Press, 2022. https://doi.org/10.1201/9781003252542.

Zou, Y., J. Deng, C. Xu, X. Liang, and X. Chen, *"Semantic Rule Based RBAC Extension Model for Flexible Resource Allocation,"* 2019 12th International Symposium on Computational Intelligence and Design (ISCID), Hangzhou, China, 2019, pp. 221–224. doi:10.1109/ISCID.2019.10134.

6 The Role of Nanoelectronic Devices in a Smart City Ecosystem

Mukesh Patidar, Namit Gupta,
Ankit Jain, Nilesh Patidar

CONTENTS

6.1 INTRODUCTION

Smart cities will eventually increase to accommodate larger populations, and many activities within cities will become increasingly automated if current breakthroughs in society and science are any indication.

A smart city is an infrastructure framework that addresses growing urbanization challenges by integrating integrated and automated modern information and communication technologies to help cities operate more efficiently, achieve sustainability goals, and improve people's quality of life.

In general, smart city development demands the development of more efficient and less polluting transportation networks, more comfortable residential buildings that consume less energy, and effectively managed public services of all types (waste disposal, smart devices, gas, electricity, water) (Rani, S., Kataria, A., et al. 2022).

Many facets of city life are already being influenced by nanotechnology, and this trend is expected to continue in the future years. Nanotechnology utilizes the power of ultra-small science to enable exceptionally small sensors to handle large quantities of information across different platforms (Khang, A., Bhambri, P., et al. 2022).

Nanotechnology is so small that it is invisible to the naked human eye, generating large quantities of information. It can also introduce additional nanomaterials for

DOI: 10.1201/9781003252542-6

smart living that can be used to harness our surroundings for electricity, safety, sustainable food production, medicine, entertainment, communications, digital finance, data science, transportation, and water management, among other things (Rani, S., Khang, A., et al., IoT & Healthcare 2021).

The use of nanoelectronics increases the capabilities of electronic devices with less weight, reduces the size of circuits, and lowers latency. The following is a summary of the contributions to society:

1. Nanoscale circuits are possible thanks to the QCA (quantum-dot cellular automata) circuit's high device density. It aids in the reduction of device area and manufacturing at the nanoscale.
2. Low power dissipation aids low-power networks, which are necessary for developing a low-polluting smart city.
3. Both the environment and the people will be in good health.

In recent years, the development of QCA nanotechnology has become critical attention for different issues such as low power consumption, faster operation speed and minimum time, low circuit delay/latency, and reduced complexity of design for QCA circuits and structures (Rani, S., Kataria, A., Sharma, V., et al. 2021; Rani, S., Kaur, S. 2012).

The smart emerging QCA nanotechnology is a candidate for replacing traditional CMOS (Nadim, M.R., Navimipour, N.J. 2018; Mohammadi, Z., Mohammadi, M. 2014; Lent, C.S., Tougaw, P.D. 1997) MOSFETs and μ-VHDL (VHSIC Hardware Description Language) by Moore's law. The QCA (quantum-dot cellular automata), SET (single-electron technology), CNT-FET, and nano-wire transistor are alternative nanotechnologies for resolving these drawbacks of CMOS technology (Khosroshahy, M.B., et al. 2017).

The quantum-dot technology plays an important role in the development of smart cities because its ability to overcome the scaling issue and provide computational performance (Mahalakshmi, K.S., et al. 2016), high device density (Khosroshahy, M.B., et al. 2017; Mahalakshmi, K.S., et al. 2016; Das, J.C., De, D. 2015), minimal size of devices and circuits, high-frequency operation (THz), ultra-low power consumption (Khosroshahy, M.B., et al. 2017; Das, J.C., De, D. 2015), fast operation (Khosroshahy, M.B., et al. 2017; Das, J.C., De, D. 2015), and heat dissipation reduction (Mahalakshmi, K.S., et al. 2016; Das, J.C., De, D. 2015) as compared to traditional transistor technology.

On the basis of its smart city proposal, Indore, Madhya Pradesh (India), won the first phase of the competition for the implementation of smart solutions/smart city proposal (SCP) (Rani, S., Khang, A., 2021; Rani, S., Kumar, R. 2022).

This smart city proposal included both a PAN-City and an area-based development (ABD) solution (www.smartcityindore.org/smart-city-indore/), as shown Figure 6.1.

The ABD area's main components are the conservation of heritage structure, IT connectivity and IT-enabled government services, smart parking, solar power, security system by CCTV cameras, and transportation and walkability, among others (Rana, G., Khang, A., et al. AI & Blockchain 2021).

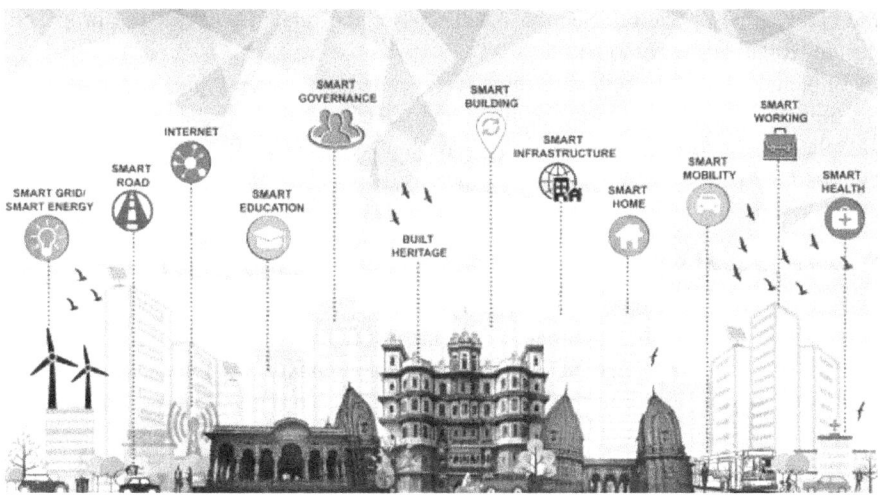

FIGURE 6.1 The Indore smart city project called Indore Smart City Development Limited.

QCA nanotechnology is a transistor-less nanotechnology where the Columbia relation between QCA cells and QCA wire creates the path for the propagation of information in the forward direction (Rani, S., Mishra, R.K., et al. 2021).

The quantum dots have several areas and applications, such as quantum computing, creating smart digital devices, smart building construction, quantum-dot light-emitting diodes (QD-LED), nano-sensors (Nazari, A. 2020), and nanoscale photo-detectors (Nazari, A. 2020).

The main contributions of our proposed work in this chapter are as follows:

1. We presented a smart circuit design of a hybrid half adder-subtractor (HHAS) in QCA on a three-input majority gate using synchronized "clocking schemes."
2. We will present a new schematic design of the signal propagation direction for single-layer HHAS (SHHAS) and multilayer HHAS (MHHAS) using a minimum-wire-length "routing algorithm" used in a controllable inverter in QCA.
3. The presented structure has an ultra-efficient design area, minimum quantum cells, low latency, and minimized complexity circuit, which plays a more effective role in smart devices at room temperature.
4. We have optimized energy dissipation by using QCADesigner-E (Energy) version 2.2 tool with a coherence vector engine operating at high frequencies (THz).

The rest of the presented chapter work is organized as follows: In Section 6.2, we recall the motivation of related work. Section 6.3 introduces the fundamentals of the QCA design structure (majority gate, inverter, and wire) and clocking scheme with the clock cycle operation. Section 6.4 discusses the related work of the proposed design (Rana, G., et al. 2021).

Implementation of coplanar single-layer and multilayer design is explained in Section 6.5. Energy dissipation analysis and polarization effects on output cells are calculated in Section 6.6 and Section 6.7, respectively. The results and discussion are explored in Section 6.8. Finally, Section 6.9 is the conclusion.

6.2 MOTIVATION OF WORK

Nanotechnology developments have enabled the development of nanoscale devices with unique and valuable features that are critical to many creative smart city applications, such as intelligent drug delivery, nanoscale sensing, and industrial and environmental monitoring systems (Khan, T., et al. 2020). The smart city concept is built on the effective use of existing resources in a coordinated and technologically savvy manner.

The widespread usage of intelligent sensors or nano-sensors contributes significantly to the overall design of a smart city's infrastructure or intelligent building (Nazari, A. 2020; Lu, Y., et al. 2019).

In a smart city, real-time sensing (RTS) and quick monitoring technologies are critical for protecting human and environmental resources from dangerous causes. It has been documented in the existing literature for a variety of smart nano-sensor device applications such as water or waste-water quality monitoring (Nazari, A. 2020; Lu, Y., et al. 2019), soil quality monitoring (including environmental monitoring) (Nazari, A. 2020; Lu, Y., et al. 2019), air quality monitoring (Nazari, A. 2020; Lu, Y., et al. 2019), and smart security applications (Nazari, A. 2020; Lu, Y., et al. 2019).

The primary purpose is to implement QCA SHHAS and MHHAS design and analyze different QCADesigner-E-based parameters using a synchronized clocking scheme with 90-degree QCA cells and avoid various QCA-cell defects.

6.3 PROPOSED TECHNOLOGY FOR SMART DESIGN

In order to address the ever-increasing need for quick, green communication and secure, reliable, and intelligent systems, the demand for systems with high computing capabilities has developed at a fast rate (Nawaz, S.J., et al. 2019).

The intrinsic parallelism provided by quantum mechanics' core notions, as well as the prospects presented by recent quantum computing technology accomplishments clearly show that QC technology has the ability to outperform traditional computing systems (Nawaz, S.J., et al. 2019).

The QCA technology operates on very low power for nanoscale circuits and networks in the carbon emission reduction for low polluted smart cities.

The fundamental element of QCA technology is quantum-dot (QD) cells. The QD cells force columbic repulsion between neighboring cells, and no voltage or current is used in QCA (Lent, C.S., Tougaw, P.D. 1997). In the QCA cell, two polarizations are used ($P = -1$, and $P = +1$) (Lent, C.S., Tougaw, P.D. 1997) with 90-degree and 45-degree cell orientations.

The essential element of QCA cells is wire, inverter, and majority gate (Walter, M., et al. 2018; Pidaparthi, S.S., Lent, C.S. 2021). The majority gate is the dominant component of the QCA design. It is the actual output based on the majority of inputs; if the majority gate consists of three input cells, known as a three-input majority gate (MV3). The logic equation is represented in Equation 6.1.

$$F_{in}(A, B, C) = Y_{out} = A.B + A.C + B.C \qquad (6.1)$$

The two inputs (AND logic gate and OR logic gate) can be implemented by three input majority gates for fixed one input polarity, −1.00 and +1.00, respectively, for the AND OR logic gates. The majority-AND and majority-OR logic gate expirations are shown in Eq. 6.2 and Eq. 6.3, respectively.

$$F_{in}(A, B, 0) = Y_{out} = A.B + A.0 + B.0 = A.B \qquad (6.2)$$

$$F_{in}(A, B, 1) = Y_{out} = A.B + A.C + B.C = A + B \qquad (6.3)$$

The QCA design circuits or devices require QCA clocks for synchronization and control information. The four phases of the synchronization clocking scheme concept design are shown in Figure 6.2, and the polarization state of the QCA cells is shown in Table 6.1 (Walter, M., et al. 2018). The clocking scheme is the most important part of QCA nanotechnology.

Each phase has a 90-degree phase difference, and one complete zone has four phases with a total of 360 degrees. The four phases of the zone are namely (P1-Switch, P2-Hold, P3-Release, and P4-Relax) (Khan, T., et al. 2020; Walter, M., et al. 2018).

It is needed for all sequential, combinational circuits and ALU. The QCA clocking scheme was used for designing the different nanoelectronics quantum-dot circuits, such as XOR gate (Haotian, C., et al. 2019; Patidar, M., Gupta, N. 2017; Gupta, A., et al. 2020), digital logic gates, combinational logic gates, sequential logic gates, and reversible logic gates.

Table 6.1 shows the operation of QCA clock phases (Walter, M., et al. 2018).

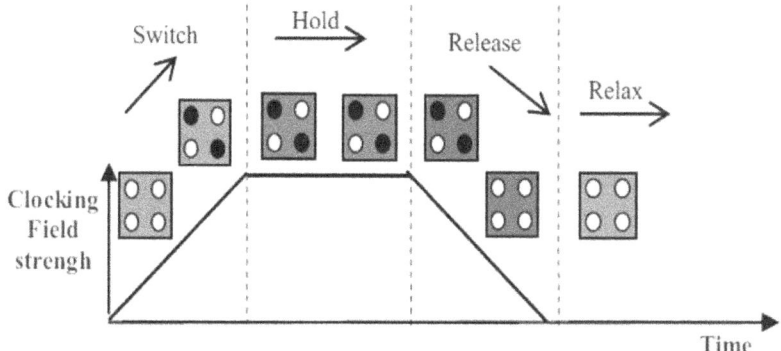

FIGURE 6.2 QCA synchronized clocking scheme.

TABLE 6.1

Operation of QCA Dataset

Phase (P)	Clock Cycle	Inter-Dot Barrier	Polarization Condition
P1	Switch phase	Low to high	Polarizes
P2	Hold phase	Held high	Polarizes
P3	Release phase	Lowered	Depolarized
P4	Relax phase	Low	Depolarized

6.4 RELATED WORKS

The concept of making cities smart is critical in light of the growing population in metropolitan areas and the need to meet demand efficiently. Nanotechnology is used in a variety of disciplines, including manufacturing, agriculture, biomedicine, and military equipment (Srikanth, N.G., Sharan, P.C. 2016).

In this section, we have discussed the related work of adder-subtractor and exclusive-OR logic. A number of research publications dealing with the design of adder-subtractor employing QCA technology can be found in the literature.

The exclusive-OR is an essential part of all adder and subtractor designs. The presented work and existing work design by the QCADesigner tool used two different methods: bi-stable approximation and coherence vector energy options simulation engine.

The author Roy, S.S. (2018) proposed an adjoined adder-subtractor using 32 QCA cells, occupied design area 0.03 μm^2 on the single-layer structure is shown in Figure 6.3.

Ahmad, P.Z. et al. (2017) proposed a controlled half adder-subtractor QCA design using 45 quantum cells, utilizing 0.05 μm^2 total design area, and the delay observed at 0.75 clock cycle using bi-stable simulation engine is shown in Figure 6.4.

Bahar, A.N. et al. (2018) have proposed an ultra-efficient XOR gate designed using coplanar single-layer. This ultra-efficient XOR structure occupies a 0.01 μm^2 cell area, 14 cells, and the clock cycle delay count is 0.5.

Laajimi, R. et al. (2017) have proposed a QCA coplanar design of 8-input, 4-input, and 2-input quantum-dot exclusive-OR gate using occupied QCA cells, such as 90, 35, and 10, respectively, and utilized a total design area of 0.114 μm^2, 0.036 μm^2, and 0.008 μm^2, respectively.

Dallas, H., and Mehran, M. (2015) have designed half-subtractor design by QCADesigner tool. It occupied a cell design area of 0.054 μm^2 and utilized 55 quantum cells with three clock cycle phases.

Ramachandran, S.S., and Kumar, K.J.J. (2017) have proposed a modified half-subtractor QCA structure by using XOR logic occupying a design area is 0.06 μm^2, which is better than the conventional half subtractor QCA design occupying an area of 0.13 μm^2.

Poorhosseini, M., and Hejazi, A.R. (2018) have proposed an efficient XOR structure and half-adder design. The proposed single-layer half-adder design has cell

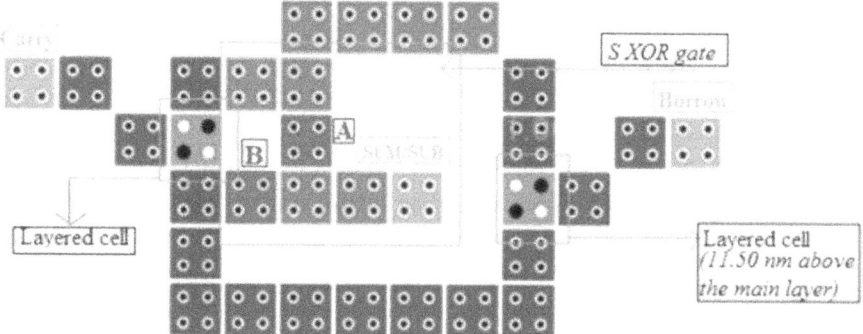

FIGURE 6.3 Diagram of an adjoined adder-subtractor.

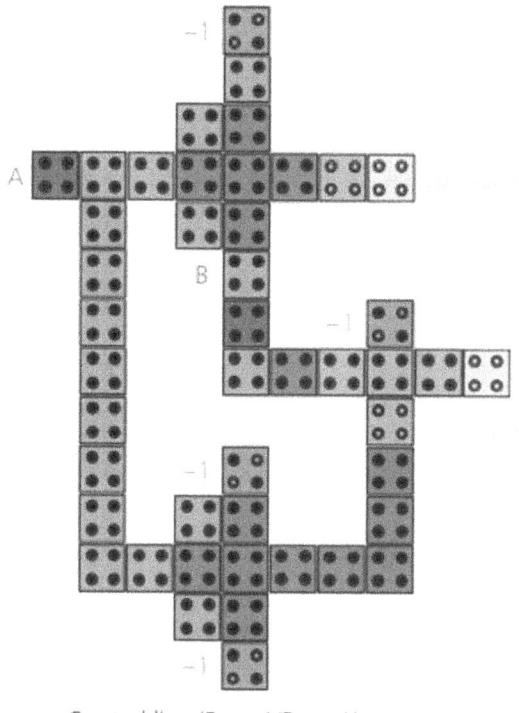

Control line (P = +1/P = −1)

FIGURE 6.4 Controlled half adder-subtractor design.

complexity, total design area, and the cell area is 44 QCA cells, 0.05 μm², and the cell area 0.014 μm², respectively.

Moustafa, A. (2019) has proposed a building block of half-adder and its QCA design. The proposed half-adder Moustafa, A. (2019) utilized the 48 QCA cells, and the occupied design area is 0.052 μm² with a measured latency of 0.5 clock cycle.

TABLE 6.2
Truth Table Dataset

Inputs		Outputs		
		Sum (S)/		
A	B	Difference (D)	Carry (C)	Borrow (B)
Operations		A′B + AB′ or XOR	AB or AND	AB′
0	0	0	0	0
0	1	1	0	1
1	0	1	0	0
1	1	0	1	0

In the literature work, we have observed that most of the designs utilized an excess number of QCA cells that region the operated design area is increased. Most of the work used five input majority voters, a long QCA wire, and not an adequately synchronized clock from inputs to outputs; because of this reason, the delay is increased.

The half-adder and half-subtractor are performed the digital logic operations between two digits. The logical process of add and sub circuits is simplified by the K-map (SOP) method. The mathematical expirations of sum/difference, carry, and borrow are represented in Eq. 6.4, Eq. 6.5, and Eq. 6.6, respectively.

The truth table is shown in Table 6.2. These logic operations are used in computers and processors in the ALU.

$$\text{Sum}(S)/\text{Difference}(D)\text{ bit}: S/D = A'B + AB' = A \oplus B \tag{6.4}$$

$$\text{Carry}(C)\text{ output bit}: \text{Carry}(C_0) = AB \tag{6.5}$$

$$\text{Borrow}(B)\text{ output bit}: \text{Borrow}(B_0) = A.B' \tag{6.6}$$

Table 6.2 presents the truth table for half adder-subtractor.

6.5 IMPLEMENTATION OF THE PROPOSED WORK

Smart devices should be able to communicate with one another and collaborate with other smart devices in their vicinity to achieve their objectives. Various levels of danger are associated with these devices and the physical systems with which they are interconnected, depending on their nature.

This section explains the implementation of the proposed design of the coplanar single-layer and coplanar crossover MHHAS QCA design.

For this purpose, we have a need for exclusive-OR, which is extracted from the design (Laajimi, R., et al. 2017). The two input exclusive-OR gate output is obtained by Eq. 6.7, where A and B are two binary inputs.

$$\text{Exclusive-OR}(A, B) = A \oplus B = A'B + AB' \tag{6.7}$$

The HHAS design's proposed layout is based on QCA nanotechnology by using quantum-dot cells. Each cell consists of four quantum dots. In this work, we have

TABLE 6.3

Parameters of Coherence Vector Eenergy

Parameter	Value	Simulation Type/Method
Number of samples	12,800	Design tool: QCADesigner-E tool
Default temperature (K)	1.000	V.2.2
Time step(s)	1.0000e-016	
Total simulation time(s)	5.000e-011	Simulation types: Coherence vector
Clock slop (RAMP, GAUSS)	1.000e-012 s	energy options
Convergence tolerance	0.001000	
Relative permittivity	12.9000	Methods:
Input and clock period(s)	4.000000e-012	(a) Euler method
Clock high (J)	9.800000e-022	(b) Runge Kutta
Clock low (J)	3.800000e-023	
Clock shift	0.000000e+000	Layer types:
Default clock	Clock-0	(a) Coplanar single-layer
Clock amplitude factor	2.0000	(b) Coplanar multi-layer
Layer separation (nm)	11.5000	
Cell Size	18×18 nm^2	The radius of effect: 80.00 nm

considered each quantum-dot cell size as 18×18 nm^2; the default clock is clock-0, the layer separation between two-layer is 11.500 nm, the effect of the radius is 80 nm, and the maximum iterations/samples is 100. The used simulation parameters with values for the proposed design are initialized, as shown in Table 6.3.

Both presented transistor-less coplanar SHHAS and MHHAS QCA circuits design are constructed by two 3-input majority gates (Maj-3), one inverter, one ultra-efficient exclusive-OR logic gate, and QCA wire with an applied synchronization clock method.

The simplified majority gate expression for the proposed design is represented in Eq. 6.8, Eq. 6.9, and Eq. 10 for sum/difference, carry, and borrow, respectively.

$$Sum / Diff = M\left(M\left(A, B', 0 \right), M\left(A', B, 0 \right), 1 \right) \qquad (6.8)$$

$$Carry = M\left(A, B, 0 \right) \qquad (6.9)$$

$$Borrow = M\left(\bar{A}, B, 0 \right) \qquad (6.10)$$

Table 6.3 presents the design parameters of coherence vector energy (Pidaparthi, S.S., Lent, C.S. 2021; Gupta, A., et al. 2020; Roy, S.S. 2018).

The single-layer coplanar typical wire crossing is one of the essential features in QCA nanotechnology. The coplanar plane allows for the physical intersection of vertical and horizontal (rotating or regular) QCA wires on the same plane.

The novel coplanar HHAS design is a single-layer QCA design constructed at nano-level-based nanoelectronic design, which uses 32 QCA cells, minimum design area of 0.0361 µm^2, and latency of 0.5 clock cycles. The presented design is implemented by one ultra-efficient-XOR logic, which is designed by using 12 QCA cells.

The SHHAS schematic design is shown in Figure 6.5. The implemented proposed QCA layout design is shown in Figure 6.6, and the simulation results are in Figure 6.7.

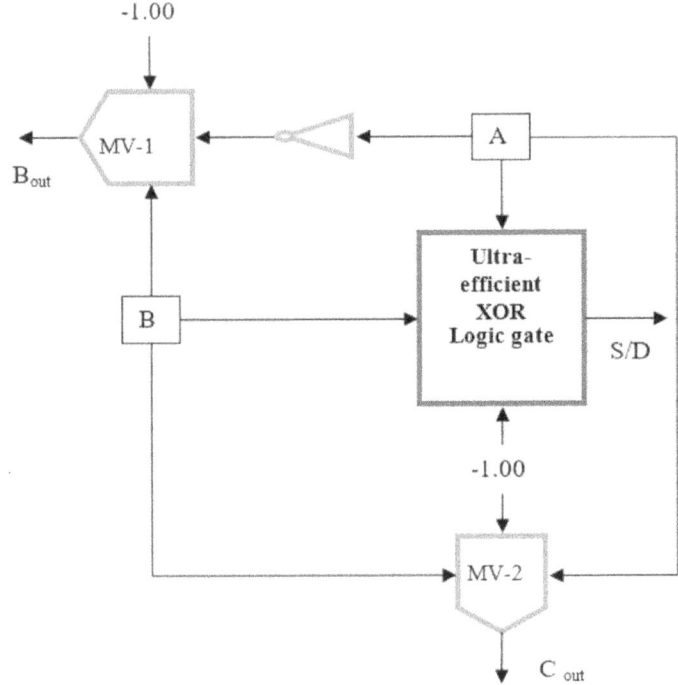

FIGURE 6.5 SHHAS—signal flow diagram.

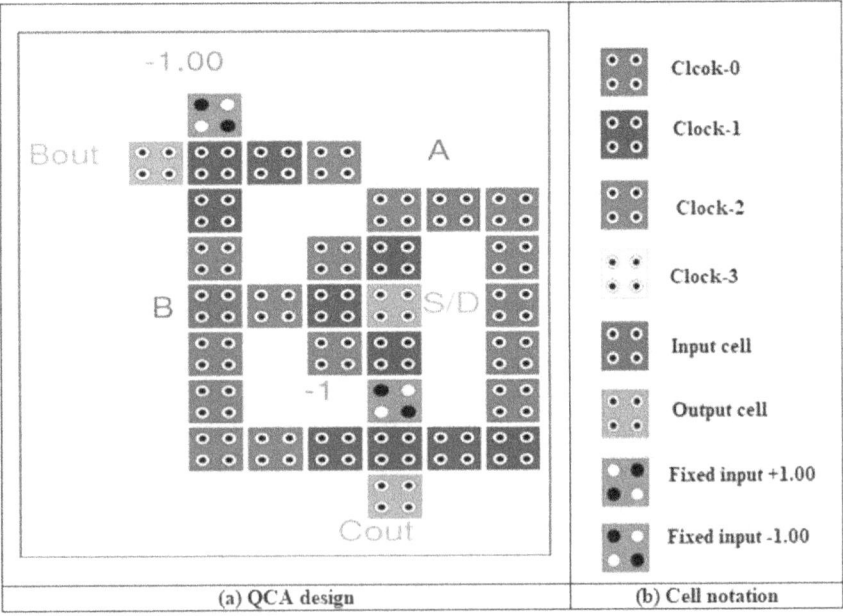

FIGURE 6.6 The implemented proposed QCA layout design.

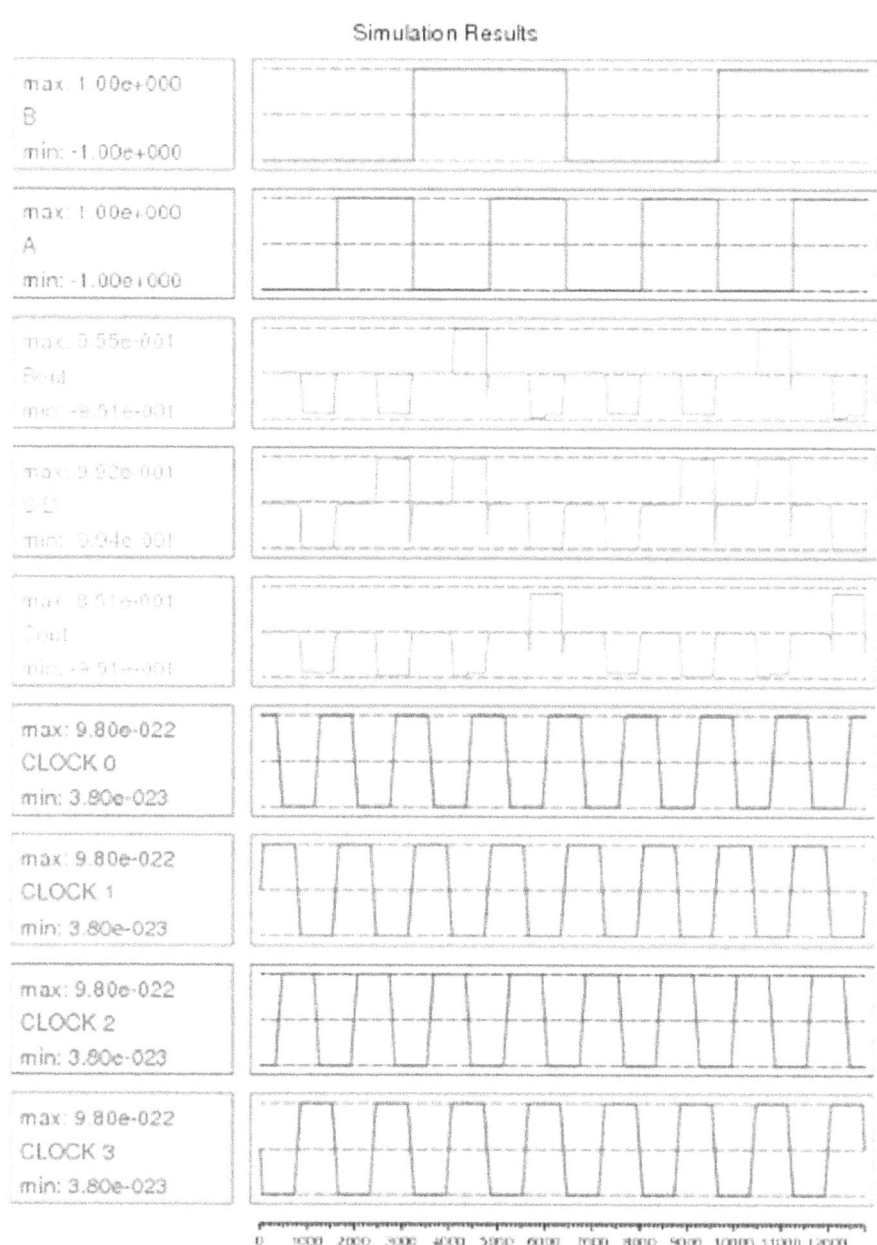

FIGURE 6.7　Results for coplanar hybrid half-adder-subtractor.

TABLE 6.4
Data for Multilayer HHAS

Inputs		NOT-A	-1.00	S/D	Cout	Bout
A	B	Ā	P = –1	XOR	Maj3 (A, B, -1)	Maj3 (Ā, B, -1)
0	0	1	0	0	0	0
0	1	1	0	1	0	1
1	0	0	0	1	0	0
1	1	0	0	0	1	0

The presented design results are verified in Table 6.2 and used simulation parameters, as shown in Table 6.3.

The QCA layout design signal flows from the input source (A and B) to destination output sources (sum/difference, borrow, and carry) by using a four-phase, synchronized QCA clock (P0-Switch → P1-Hold → P2-Release → P2-Relax) signal in a forwarding direction. In Table 6.4, we have explained the operations of SHHAS and MHHAS.

The majority voter (MV-1) performed the Cout procedure (Maj3 (A, B, -1)), the majority voter (MV-2) performed the Bout operation (Maj3 ($\bar{A}$, B, -1)) and sum/difference. They performed the XOR operation $(A \oplus B = A'B + AB')$ between two inputs.

The coplanar multilayer QCA wire crossing is designed by the multilayer placement of quantum-dot cells. The multilayer wire crossing is either 90-degree or 45-degree cells. One wire of the intersection is transferred to another layer, and after passing through the crossing, the wire is returned to the original layer. The QCADesigner tool implements the novel MHHAS design by using a coplanar crossover plane.

The presented multilayer design uses 41 QCA cells, the design area is 0.03 μm^2, and the latency is 0.5 clock cycle (CC) with coherence simulation energy vector simulation engine parameters used are represented in Table 6.3. The total table area is the total rectangular area required for QCA design. The circuit using each cell size is considered 18 nm × 18 nm, and the distance between two quantum cells is 2 nm.

The schematic design, QCA layout design, and operation of MHHAS are shown in Figure 6.8, Figure 6.9, and Table 6.4, respectively. The presented design has used a cell size of 18 nm × 18 nm with a dot diameter of 5 nm.

The presented MHHAS uses different types of QCA cells, such as 32-normal regular 90- degree QCA cells, two vertical (via) cells, and seven crossover cells, as shown in Figure 6.10 at coplanar crossover plane by applied synchronization clock scheme. The simulation result is shown in Figure 6.11.

Table 6.4 shows the coplanar single-layer and multilayer operations.

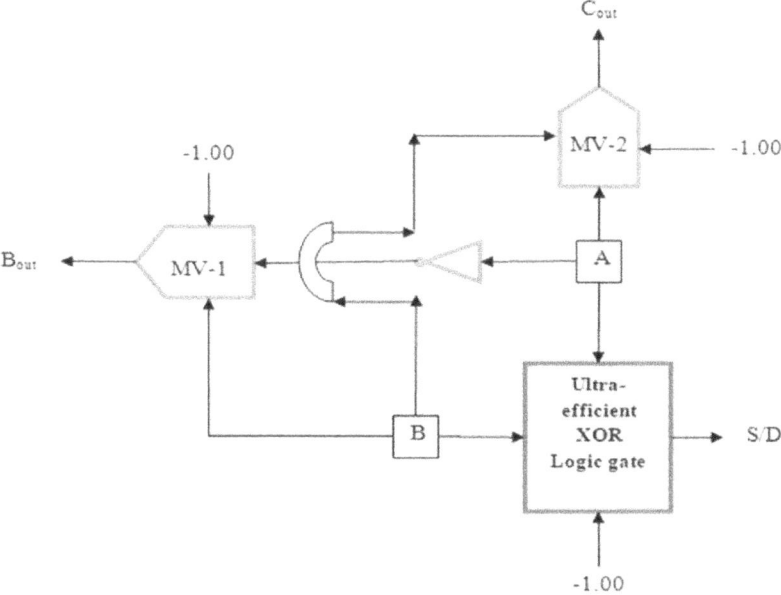

FIGURE 6.8 Multilayer HHAS.

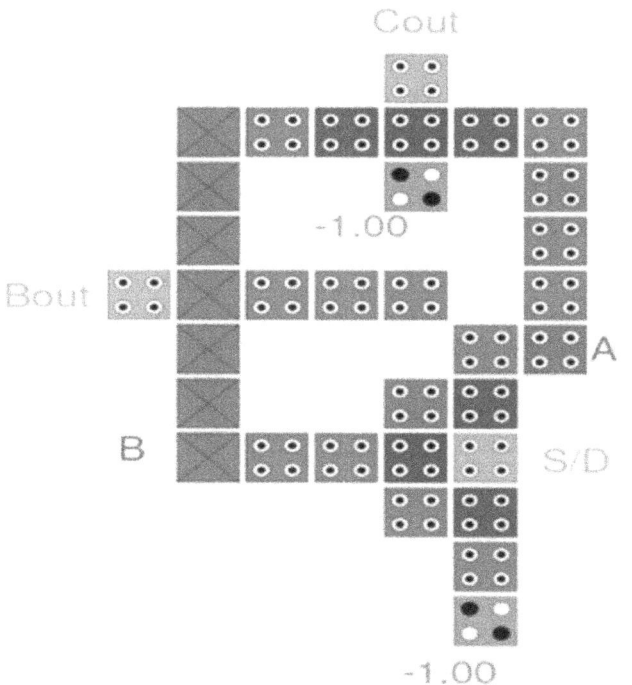

FIGURE 6.9 QCA circuit.

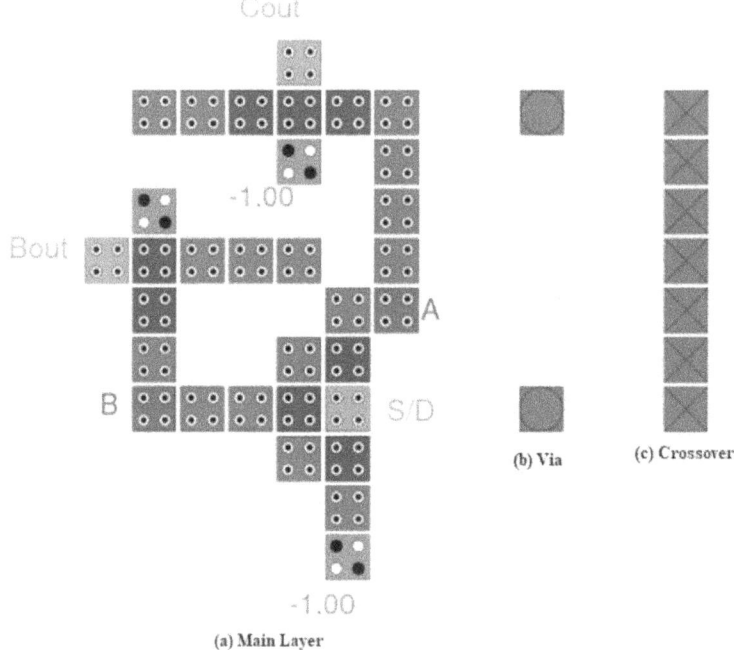

FIGURE 6.10   QCA circuit design.

6.6   ENERGY DISSIPATION ESTIMATION USING QD-E TOOL

Nano-construction materials are sophisticated in their design to accomplish energy efficiency and comfort in the surroundings (Rehan, N.M. 2021). The power/energy calculated by using the Hamiltonian matrix is represented in Eq. 6.11 (Wang, L., Xie, G. 2019).

The heat dissipated occurs during the digital binary transaction (from 0→1 or 0→1) at room temperature (T = 300 K) is 0.017 eV, is taken out to be physical irreversibility (Zhang, Y., et al. 2018).

The electron transfer by tunnelling from one-quantum-dot to another-quantum-dot enables switching the device state (Nawaz, S.J., et al. 2019). Power dissipation can be extraordinarily low (Pidaparthi, S.S., Lent, C.S. 2021).

$$H = \begin{bmatrix} -\dfrac{1}{2}\sum_{i\neq j} P_j E_k^{ij} & -\gamma \\[2mm] -\gamma & \dfrac{1}{2}\sum_{i\neq j} P_j E_k^{ij} \end{bmatrix} \tag{6.11}$$

The energy dissipation has been calculated by using QD-E tool by CVSE (w/Energy) (Bilal, B., et al. 2018). The parameters of coherence vector on the saturation energy of clock signal is $\gamma_H = 9.8e-22$ J and $\gamma_L = 3.8e-23$ J.

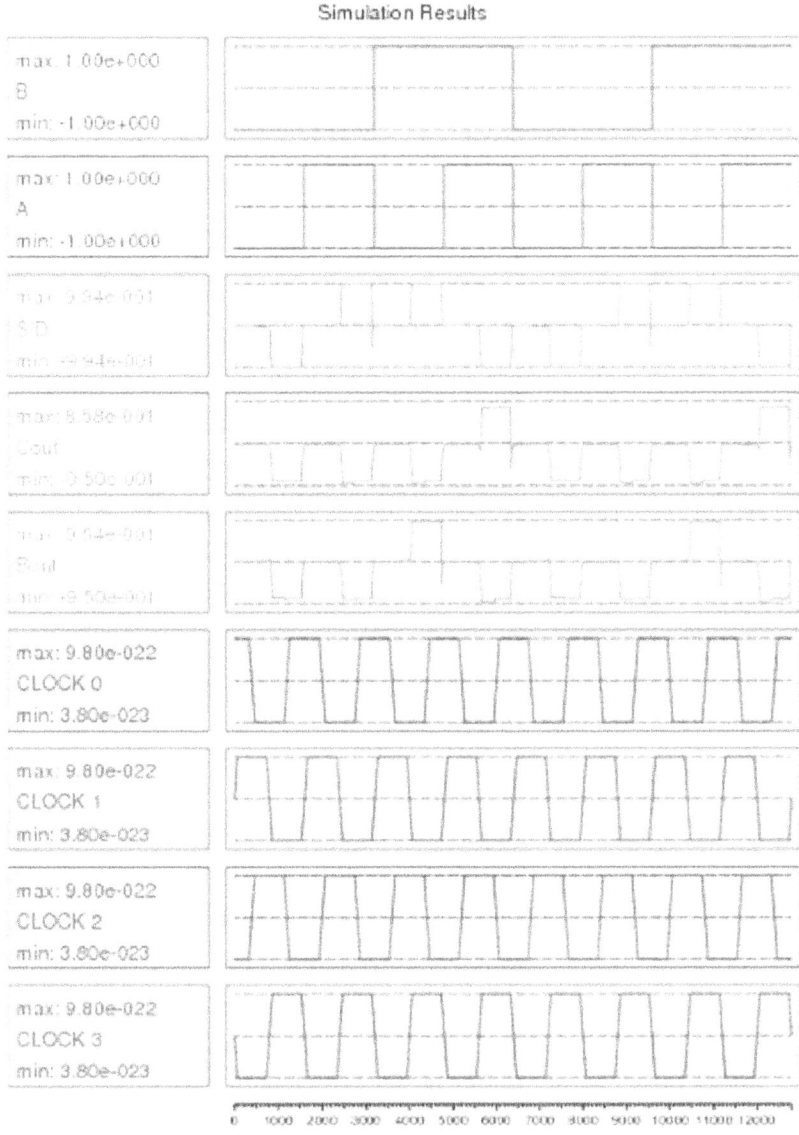

FIGURE 6.11   Simulation result

The relaxation time and relative permittivity are $1e-15$ s, and 12.90 respectively (Torres, F.S., et al. 2018; Patidar, M., Gupta, N. 2021b). The kink energy ($E_{i,j}$) can be expressed as in Eq. 6.12 (Patidar, M., Gupta, N. 2020)

$$E_{i,j} = \frac{1}{4\pi\varepsilon_r\varepsilon_0} \sum_n \sum_m \frac{q_{i,n}q_{j,m}}{\left|r_{i,n} - r_{i,j}\right|} \qquad (6.12)$$

TABLE 6.5
Energy dissipation analysis results

E_{btx}	Sum_bath	E_{ctx}	E_{Etx}	Avg_Ebath				
3.0449e-003		2.3278e-004	-3.2997e-004					
3.7114e-003		-6.5874e-005	-3.8117e-004					
2.4385e-003		4.5376e-005	-2.6513e-004					
3.1471e-003	3.39e-002 eV (Error: ± -3.56e-003 eV)	-2.9208e-004	-3.1649e-004	3.08e-003 eV (Error: ± -3.24e-004 eV)	Total simulation time: 40 s	Default temperature: 1K	simulation with 500000 samples	Coherence vector energy
3.0449e-003		2.3278e-004	-3.2997e-004					
3.7114e-003		-6.5874e-005	-3.8117e-004					
2.4385e-003		4.5376e-005	-2.6513e-004					
3.1471e-003		-2.9208e-004	-3.1649e-004					
3.0449e-003		2.3278e-004	-3.2997e-004					
3.7114e-003		-6.5874e-005	-3.8117e-004					
2.4385e-003		4.5376e-005	-2.6513e-004					

Table 6.5 shows the Energy dissipation analysis results of coplanar single-layer QCA design.

TABLE 6.6
Energy dissipation analysis results

E_{btx}	Sum_bath	13 pt	E_{Etx}	Avg_Ebath				
3.0741e-003		6.4284e-004	-3.2742e-004					
3.3804e-003		6.0889e-004	-3.3661e-004					
3.1053e-003		5.9647e-004	-3.3207e-004					
4.8034e-003	3.83e-002 eV (Error: ± -3.98e-003 eV)	-1.1066e-003	-4.9342e-004	3.48e-003 eV (Error: ± -3.61e-004 eV)	Total simulation time: 41 s	Default temperature: 1K	simulation with 500000 samples	Coherence vector energy
3.0741e-003		6.4284e-004	-3.2742e-004					
3.3804e-003		6.0889e-004	-3.3661e-004					
3.1053e-003		5.9647e-004	-3.3207e-004					
4.8034e-003		-1.1066e-003	-4.9342e-004					
3.0741e-003		6.4284e-004	-3.2742e-004					
3.3804e-003		6.0889e-004	-3.3661e-004					
3.1053e-003		5.9647e-004	-3.3207e-004					

Table 6.6 shows the Energy dissipation analysis results of coplanar crossover multilayer design.

For every QCA clock cycle (QCC), the expectation value of QCA energy is expressed as in Eq. 6.13 (Patidar, M., Gupta, N. 2020)

$$E = \langle H \rangle = \frac{\hbar}{2} \times \lambda \times \Gamma \tag{6.13}$$

Where in Eq. 6.13 Γ is the energy environment of the cell, λ is the coherence vector, and $\hbar$ is the reduced plank constant. The energy dissipation (E.D.) calculation of the proposed QCA coplanar SHHAS and crossover MHHAS is shown in Table 6.5 and Table 6.6, respectively. Where E_{btx} = E_bath_total, S_b = Sum_bath, E_{ctx} = E_clk_total, E_{Etx} = E_Error_total, and A_b = Avg_bath.

In this analysis, we observed the no-energy dissipation is observed in (Sum_E_{bath}) and (Avg_E_{bath}) for all the entire fixed input cells such as (+1.00 & -1.00) and inputs cells (A & B) of circuits.

The Sum_bath (S_b) value for coplanar SHHAS and crossover-multi-layer HHAS observes by QCADesigner-E tool is $3.39e-002$ eV $\left(\text{Error}: \pm -3.56e-003 \text{ eV}\right)$ and $3.83e-002$ eV $\left(\text{Error}: \pm -3.98e-003 \text{ eV}\right)$ respectively.

6.7 AOP CALCULATION USING RUNGA-KUTTE AND EULER

The average output polarization (AOP) of the coplanar SHHAS and MHHAS steadily decreased up to a temperature of 6K. The presented design efficiently worked and got desired outcomes in the temperature range of 1–6K. After this temperature range from T = 7K, the circuit started to break down.

Temperature effect (TE) for AOP calculated by the equation is represented in Eq. 6.14 (Torres, F.S., et al. 2018; Patidar, M., Gupta, N. 2021b; Khan, A., Arya, R. 2021; Abdullah-Al-Shafi, M., et al. 2018; Abdullah-Al-Shafi, M., Bahar, A.N. 2017).

$$\text{AOP} = \left\{ \frac{P_{max} - P_{min}}{2} \right\} \tag{6.14}$$

P_{max} is the maximum polarization value and P_{min} is the minimum polarization value. For example, at T = 1K, the P_{max} and P_{min} of output (B_o) quantum-dot cell of SHHAS using Runge Kutta method (RKM) 9.55e-001 and -9.51e-001, respectively. Therefore, the AOP calculation for the output quantum-dot cell is gating B_0=3.5058 in Eq. 6.15.

The relative assessment of AOP by the proposed coplanar SHHAS and MHHAS are organized in Table 6.7 and Table 6.8, respectively. For this analysis, we have observed that the coherence Runge Kutta method takes maximum or total simulation time (TST) for simulation as compared to the Euler method (EM) in both proposed designs.

For example, the Runge Kutta and Euler methods took simulation times of 66 s and 22 s at the temperature of 1K for a coplanar single-layer structure. Similarly, we observed that all the nano quantum computing circuits work efficiently within 1–5K (Khan, A., Arya, R. 2021).

Eq. 6.15 to Eq. 6.20 represent the AOP calculation for the coplanar single-layer design.

Temperature effect (TE) on output polarization for Borrow (B_o) output at temperature (T = 1K):

$$\text{AOP for } B_o = \left\{ \frac{(9.55e-001) - (-9.51e-001)}{2} \right\} \cong 3.5058 \text{ for RKM} \left(\text{TST} = 66s\right) \tag{6.15}$$

$$\text{AOP for } B_o = \left\{ \frac{(9.55e-001) - (-9.51e-001)}{2} \right\} \cong 3.5058 \text{ for EM} \left(\text{TST} = 22s\right) \tag{6.16}$$

TE on output polarization for sum/difference (S/D) output at temperature (T = 1K):

$$\text{AOP for S / D} = \left\{ \frac{(9.92e-001)-(-9.94e-001)}{2} \right\} \cong 3.6530 \text{ for RKM} \left(\text{TST} = 66s \right) \quad (6.17)$$

$$\text{AOP for S / D} = \left\{ \frac{(9.92e-001)-(-9.94e-001)}{2} \right\} \cong \text{for } 3.6530 \text{ EM} \left(\text{TST} = 22s \right) \quad (6.18)$$

Table 6.7 shows the temperature effect on AOP for coplanar single-layer QCA design using the coherence vector engine.

TE on output polarization for carrying (C_o) output at temperature (T = 1K):

$$\text{AOP for } \left(C_o \right) = \left\{ \frac{(8.51e-001)-(-9.51e-001)}{2} \right\} \cong \text{for } 3.3145 \text{ RKM} \left(\text{TST} = 66s \right) \quad (6.19)$$

TABLE 6.7
AOP Calculation for Coplanar Single-Layer QCA Design

AOP Calculation for Coplanar Single-Layer QCA Design

Temp. in (K)	Output	Runge Kutta Method (RKM)				Euler Method (EM)			
		Ma7.	Min.	Avg.	TST	Ma7.	Min.	Avg.	TST
	B_0	9.55e-001	-9.51e-001	3.5058		9.55e-001	-9.51e-001	3.5058	
1K	S/D	9.92e-001	-9.94e-001	3.6530	66s	9.92e-001	-9.94e-001	3.6530	22s
	C_0	8.51e-001	-9.51e-001	3.3145		8.51e-001	-9.51e-001	3.3145	
	B_0	9.55e-001	-9.51e-001	3.5058		9.55e-001	-9.51e-001	3.5058	
2K	S/D	9.92e-001	-9.94e-001	3.6530	70s	9.92e-001	-9.94e-001	3.6530	23s
	C_0	8.51e-001	-9.51e-001	3.3145		8.51e-001	-9.51e-001	3.3145	
	B_0	9.55e-001	-9.51e-001	3.5058		9.55e-001	-9.51e-001	3.5058	
3K	S/D	9.92e-001	-9.94e-001	3.6530	68s	9.92e-001	-9.94e-001	3.6530	21s
	C_0	8.50e-001	-9.51e-001	3.3127		8.50e-001	-9.51e-001	3.3127	
	B_0	9.55e-001	-9.51e-001	3.5058		9.55e-001	-9.51e-001	3.5058	
4K	S/D	9.92e-001	-9.94e-001	3.6530	69s	9.92e-001	-9.94e-001	3.6530	23s
	C_0	8.42e-001	-9.50e-001	3.2961		8.42e-001	-9.50e-001	3.2961	
	B_0	9.54e-001	-9.49e-001	3.5003		9.54e-001	-9.49e-001	3.5003	
5K	S/D	9.92e-001	-9.94e-001	3.6530	71s	9.92e-001	-9.94e-001	3.6530	24s
	C_0	8.26e-001	-9.47e-001	3.2612		8.26e-001	-9.47e-001	3.2612	
	B_0	9.51e-001	-9.45e-001	3.4874		9.51e-001	-9.45e-001	3.4874	
6K	S/D	9.92e-001	-9.94e-001	3.6530	67s	9.92e-001	-9.94e-001	3.6530	24s
	C_0	8.01e-001	-9.42e-001	3.2060		8.01e-001	-9.42e-001	3.2060	
	B_0	9.46e-001	-9.36e-001	3.4617		9.46e-001	-9.36e-001	3.4617	
7K	S/D	9.92e-001	-9.94e-001	3.6530	72s	9.92e-001	-9.94e-001	3.6530	23s
	C_0	7.69e-001	-9.33e-001	3.1306		7.69e-001	-9.33e-001	3.1306	

TABLE 6.8
AOP Calculation for Coplanar Crossover Multilayer QCA Design

AOP Calculation for Coplanar Crossover Multilayer QCA Design

Temp. in (K)	Output	Runge Kutta Method (RKM)				Euler Method (EM)			
		Ma7.	Min.	Avg.	TST	Ma7.	Min.	Avg.	TST
1K	B_0	9.54e-001	-9.50e-001	3.5022		9.54e-001	-9.50e-001	3.5022	
	C_0	8.58e-001	-9.50e-001	3.3256	84s	8.58e-001	-9.50e-001	3.3256	28s
	S/D	9.94e-001	-9.94e-001	3.6567		9.94e-001	-9.94e-001	3.6567	
2K	B_0	9.54e-001	-9.50e-001	3.5022		9.54e-001	-9.50e-001	3.5022	
	C_0	8.58e-001	-9.50e-001	3.3256	88s	8.58e-001	-9.50e-001	3.3256	29s
	S/D	9.94e-001	-9.94e-001	3.6567		9.94e-001	-9.94e-001	3.6567	
3K	B_0	9.54e-001	-9.50e-001	3.5022		9.54e-001	-9.50e-001	3.5022	
	C_0	8.58e-001	-9.50e-001	3.3256	85s	8.56e-001	-9.50e-001	3.3219	29s
	S/D	9.94e-001	-9.94e-001	3.6567		9.94e-001	-9.94e-001	3.6567	
4K	B_0	9.54e-001	-9.50e-001	3.5022		9.54e-001	-9.50e-001	3.5022	
	C_0	8.58e-001	-9.49e-001	3.3237	90s	8.50e-001	-9.49e-001	3.3090	30s
	S/D	9.94e-001	-9.94e-001	3.6567		9.94e-001	-9.94e-001	3.6567	
5K	B_0	9.53e-001	-9.47e-001	3.4948		9.53e-001	-9.47e-001	3.4948	
	C_0	8.34e-001	-9.47e-001	3.2759	90s	8.34e-001	-9.47e-001	3.2759	31s
	S/D	9.94e-001	-9.94e-001	3.6567		9.94e-001	-9.94e-001	3.6567	
6K	B_0	9.50e-001	-9.42e-001	3.4801		9.50e-001	-9.42e-001	3.4801	
	C_0	8.11e-001	-9.42e-001	3.2244	89s	8.11e-001	-9.42e-001	3.2244	30s
	S/D	9.93e-001	-9.94e-001	3.6548		9.93e-001	-9.94e-001	3.6548	
7K	B_0	9.44e-001	-9.33e-001	3.4525		9.44e-001	-9.33e-001	3.4525	
	C_0	7.81e-001	-9.33e-001	3.1527	87s	7.81e-001	-9.33e-001	3.1527	30s
	S/D	9.93e-001	-9.94e-001	3.6548		9.93e-001	-9.94e-001	3.6548	

$$\text{AOP for } (C_0) = \left\{ \frac{(8.51e-001)-(-9.51e-001)}{2} \right\} \cong \text{for } 3.3145 \text{ EM} \left(\text{TST} = 22s \right) \quad (6.20)$$

Table 6.8 shows the temperature effect on AOP for coplanar crossover multilayer QCA design using a coherence vector engine.

AOP calculation for multilayer design from Eq. 6.21 to Eq. 6.26.

TE on output polarization for Borrow (B_0) output at temperature (T = 1K):

$$\text{AOP for } B_0 = \left\{ \frac{(9.54e-001)-(-9.50e-001)}{2} \right\} \cong 3.5022 \text{ for RKM} \left(\text{TST} = 84s \right) \quad (6.21)$$

$$\text{AOP for } B_0 = \left\{ \frac{(9.54e-001)-(-9.50e-001)}{2} \right\} \cong 3.5022 \text{ for EM} \left(\text{TST} = 28s \right) \quad (6.22)$$

TE on output polarization for Carry (C_o) output at temperature (T = 1K):

$$\text{AOP for } (C_o) = \left\{ \frac{(8.58e-001)-(-9.50e-001)}{2} \right\} \cong \text{ for } 3.3256 \text{ RKM} (\text{TST} = 84s) \quad (6.23)$$

$$\text{AOP for } (C_o) = \left\{ \frac{(8.58e-001)-(-9.50e-001)}{2} \right\} \cong \text{ for } 3.3256 \text{ EM} (\text{TST} = 28s) \quad (6.24)$$

TE on output polarization for Sum/Difference (S/D) output at temperature (T = 1K):

$$\text{AOP for } \frac{S}{D} = \left\{ \frac{(9.94e-001)-(-9.94e-001)}{2} \right\} \cong 3.6530 \text{ for RKM} (\text{TST} = 84s) \quad (6.25)$$

$$\text{AOP for } \frac{S}{D} = \left\{ \frac{(9.94e-001)-(-9.94e-001)}{2} \right\} \cong 3.6530 \text{ for EM} (\text{TST} = 28s) \quad (6.26)$$

6.8   RESULTS AND DISCUSSION

This section discusses the implementation of QCA design for considering various important QCA design parameters using the QCADesigner-E tool. It presents an efficient method of coplanar SHHAS and MHHAS using a regular clocking scheme with minimal majority voter using the minimal number of QCA cells (Tailor, R K., et al. 2022).

The presented design of SHHAS and MHHAS occupied total design areas of 0.0248 μm² and 0.0300 μm², having quantum cell 32 and 41 designs, respectively.

The presented design has considered default factors: square cell size of 18 nm × 18 nm, the radius of effect of 80 nm, quantum-dot dimension of 5 nm, relative permittivity of 12.900, default clock factor of 2.00, and convergence tolerance of 0.001 (Abdullah-Al-Shafi, M., Bahar, A.N. 2018; Seyedi, S., Navimipour, N.J. 2018; Patidar, M., Gupta, N. 2021a; Mosleh, M.A. 2019). Both designs occupied a minimum clock delay of 0.50 clock cycle using the synchronization clocking method without rotated cells.

The proposed design results are compared with the existing adjoined adder-subtractor (Roy, S.S. 2018) and controlled half adder-subtractor (Ahmad, P.Z., et al. 2017). The comparative analysis is shown in Table 6.9 and graphically represented in Figure 6.12.

Based on the output results, it is observed that the proposed coherence single-layer and MHHAS are efficient in terms of QCA cell complexity, reduced design area, less latency count, and less design complexity over existing designs.

The coplanar single-layer HHAS, having improvement in the cell area occupied design area, has a latency of 28.94%, 50.40%, and 25.00%, respectively, compared to the controlled half adder-subtractor (Ahmad, P.Z., et al. 2017) on coplanar single-layer design energy dissipation calculated in novel QCADesigner-E (Energy)

TABLE 6.9
The Coplanar Single-Layer and Multilayer Design

The Existing and Proposed Design	Adjoined Adder-Subtractor (Roy, S.S. 2018)	Controlled Half Adder-Subtractor (Ahmad, P.Z., et al. 2017)	Coplanar Single-Layer Nanoscale HHAS	Coplanar Multilayer Nanoscale HHAS
QCA cells	32	45	32	41
Occupied design area (μm^2)	0.0300	0.0500	0.0248	0.0300
Cell area (μm^2)	0.01036	0.01458	0.01036	0.01328
% Area usage	34.53	29.16	41.77	44.26
Latency (clock cycle)	0.50	0.75	0.50	0.50
Cell density	1,066.6	900.00	1,290.32	1,366.66
Quantum cost	0.0075	0.0281	0.0620	0.0075
Rotated cell (45-deg.)	No	No	No	No
Layer	Single-layer	Single-layer	Single-layer	Multilayer

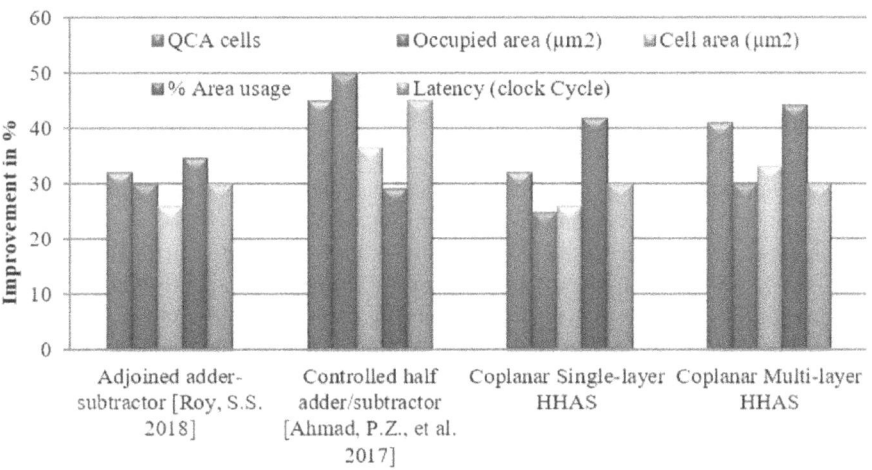

Existing and proposed QCA designs

FIGURE 6.12 QCA designs.

tool version 2.0.3 using the Hamiltonian matri7. The quantum cost is computed by Eq. 6.27 (Patidar, M., Gupta, N. 2021a).

$$\text{Quantum Cost} = \left(\text{Occupied design area}\right) \times \left(\text{Delay}\right)^2 \qquad (6.27)$$

Table 6.9 shows the design results analysis of the coplanar single-layer and multilayer design.

The AOP calculation analysis asserts that the proposed coplanar MHHAS and MHHAS circuit output such as Bout (B_o), S/D, and Cout (C_o) is stable within the temperature range of 1K to 7K in both coherence vector Runge Kutta method (RKM) and Euler (EM) method.

The maximum polarization strength of single-layer QCA design outputs, such as B_o, S/D, and C_o, are 9.55e-001, 9.92e-001, and 8.51e-001, respectively, and the minimum polarization strength of single-layer QCA design outputs, such as B_o, S/D, and C_o, are -9.51e-001, -9.94e-001, and -9.51e-001, using RKM and EM, at the temperature of 1K.

In this analysis, we have also observed similarly the maximum and minimum polarization values in both design analysis processes, such as RKM and EM potential, and the application for many sectors in human life is increasing constantly, even though it is still in a hybrid pattern as an emerging or middle-stage technology.

Through techniques such as machine learning and neural networks, global tech companies are investing in developing machines to think and behave more like humans.

6.9 CONCLUSION

A smart city is an infrastructure system that handles expanding urbanization concerns by integrating integrated and automated information and communication technologies to greatly enhance city operations, fulfill environmental targets, and improve people's life quality.

This chapter has implemented the novel coplanar single and multilayer QCA design, optimized energy consumption, and calculated the temperature effects on output polarization cells using the QCADesigner-E tool.

In the coplanar SHHAS, the improved cell area, occupied design area, and latency are 28.94%, 50.40%, and 25.00%, respectively, compared to controlled half adder-subtractor (Ahmad, P.Z., et al. 2017) on coplanar single-layer design.

In this energy optimization, we have obtained that the total energy dissipation Sum bath (S_b) value for coplanar SHHAS and crossover MHHAS, as observed by QD-E is $3.39e-002$ eV $\left(\text{Error}: \pm -3.56e-003 \text{ eV}\right)$ and $3.83e-002$ eV $\left(\text{Error}: \pm -3.98e-003 \text{ eV}\right)$, respectively.

In this AOP calculation, the proposed design results in fast output in coherence EM compared to RKM. The coherence EM is better than other RKMs for polarization output in the coplanar single-layer and multilayer QCA design at temperatures 1–7K. The future of the smart city goal is nano-enabled smart living.

REFERENCES

Abdullah-Al-Shafi, M., Bahar, A.N. 2017. A novel binary to grey and grey to binary code converter in majority voter-based QCA nano computing. *Journal of Computational and Theoretical Nanoscience* 14, pp. 2416–2421. https://doi.org/10.1166/jctn.2017.6842.

Abdullah-Al-Shafi, M., Bahar, A.N. 2018. An architecture of 2-dimensional 4-dot 2-electron QCA full adder and subtractor with energy dissipation study. *Active and Passive Electronic Components* 2018(5062960), pp. 1–10. https://doi.org/10.1155/2018/5062960.

Abdullah-Al-Shafi, M., et al. 2018. Average output polarization dataset for signifying the temperature influence for QCA designed reversible logic circuits. *Data in Brief* 19, pp. 42–48. https://doi.org/10.1016/j.dib.2018.05.009.

Ahmad, P.Z., et al. 2017. Design of novel QCA-based half/full subtractors. *Nanomaterials and Energy* 6(2), pp. 59–66. https://doi.org/10.1680/jnaen.15.00020.

Bahar, A.N., Waheed, S., Hossain, N., Asaduzzaman, M. 2018. A novel 3-input XOR function implementation in quantum dot-cellular automata with energy dissipation analysis. *Alexandria Engineering Journal* 57(2), pp. 729–738. https://doi.org/10.1016/j.aej.2017.01.022.

Bilal, B., Ahmed, S., Kakkar, V. 2018. Modular adder designs using optimal reversible and fault-tolerant gates in field-coupled QCA nanocomputing. *International Journal of Theoretical Physics* 57, pp. 1356–1375. https://doi.org/10.1007/s10773-018-3664-z.

Dallas, H., Mehran, M. 2015. Novel subtractor design based on Quantum-Dot Cellular Automata (QCA) nanotechnology. *International Journal of Nanoscience and Nanotechnology* 11(4), pp. 257–262. http://www.ijnnonline.net/article_17011.html.

Das, J.C., De, D. 2015. Reversible comparator design using quantum dot-cellular automata. *IETE Journal of Research* 62(3), pp. 323–330. https://doi.org/10.1080/03772063.2015.1088407.

Gupta, A., Kumar, V., Ahmed, S., Gautam, S. 2020. Impact of nanotechnology in the development of smart cities. In: Ahmed, S., Abbas, S., Zia, H. *Smart Cities—Opportunities and Challenges*. Lecture Notes in Civil Engineering, 58. Springer, Singapore, pp. 845–857. https://doi.org/10.1007/978-981-15-2545-2_68.

Haotian, C., Hongjun, L., Zhang, Z., Xin, C., Guangjun, X. 2019. Design and analysis of a novel low-power exclusive-or gate based on Quantum-Dot Cellular Automata. *Journal of Circuits, Systems and Computers* 28(8), p. 1950141. https://doi.org/10.1142/S021812661950141X.

Khan, A., Arya, R. 2021. Optimal demultiplexer unit design and energy estimation using quantum-dot cellular automata. *Journal of Supercomputing* 77, pp. 1714–1738. https://doi.org/10.1007/s11227-020-03320-z.

Khan, T., et al. 2020. Nanosensors for smart cities nanosensor networks for smart health care. *Nanosensors for Smart Cities*, pp. 387–403. https://doi.org/10.1016/B978-0-12-819870-4.00022-0.

Khang, A., Bhambri, P., Rani, S., Kataria, A. 2022. *Big Data, Cloud Computing and Internet of Things*. CRC Press. ISBN: 978-1-032-284200. https://doi.org/10.1201/9781032284200.

Khang, A., Chowdhury, S., Sharma, S. 2022. *The Data-Driven Blockchain Ecosystem: Fundamentals, Applications and Emerging Technologies*. CRC Press. HB.ISBN: 978-1-032-21624-9 EB.ISBN: 098-1-003-26928-1 PB.ISBN: 098-1-032-21625-6. https://doi.org/10.1201/9781003269281.

Khosroshahy, M.B., et al. 2017. Quantum-Dot Cellular Automata circuits with reduced external fixed inputs. *Microprocessors and Microsystems* 50, pp. 154–163. https://doi.org/10.1016/j.micpro.2017.03.009

Laajimi, R., Touil, L., Ajimi, A., Bahar, A.N. 2017. A novel design for XOR gate used for Quantum-Dot Cellular Automata (QCA) to create a revolution in nanotechnology structure. *International Journal of Advanced Computer Science and Applications* 8(10), pp. 279–287. https://doi.org/10.14569/IJACSA.2017.081036

Lent, C.S., Tougaw, P.D. 1997. A device architecture for computing with quantum dots. *Proceedings of the IEEE* 85, pp. 541–557. https://doi.org/10.1109/5.573740

Lu, Y., et al. 2019. Multifunctional nanocomposite sensors for environmental monitoring. In Song, K., Liu, C., J.Z. Guo, J.Z. (Eds.), *Polymer Based Multifunctional Nanocomposites and Their Application*, pp. 157–174. Elsevier. https://doi.org/10.1016/B978-0-12-815067-2.00006-8.

Mahalakshmi, K.S., et al. 2016. Performance estimation of conventional and reversible logic circuits using QCA implementation platform. In *2016 International Conference on Circuit, Power and Computing Technologies (ICCPCT), Nagercoil*, pp. 1–9. IEEE. https://doi.org/10.1109/ICCPCT.2016.7530135.

Mohammadi, Z., Mohammadi, M. 2014. Implementing a one-bit reversible full adder using quantum-dot cellular automata. *Quantum Info Process* 13, pp. 2127–2147. https://doi.org/10.1007/s11128-014-0782-2.

Mosleh, M.A. 2019. Novel full adder/subtractor in Quantum-Dot Cellular Automata. *International Journal of Theoretical Physics* 58, pp. 221–246. https://doi.org/10.1007/s10773-018-3925-x

Moustafa, A. 2019. Efficient quantum-dot cellular automata for half adder using building block. *Quantum Information Review* 7(1), 1–6. https://doi.org/10.18576/qir/070101.

Nadim, M.R., Navimipour, N.J. 2018. A new three-level fault tolerance arithmetic and logic unit based on quantum-dot cellular automata. *Microsystem Technologies* 24(2), pp. 1295–1305. https://doi.org/10.1007/s00542-017-3502-x

Nawaz, S.J., et al. 2019. Quantum machine learning for 6G communication networks: State-of-the-art and vision for the future. *IEEE Access* 7, pp. 46317–46350. https://doi.org/10.1109/ACCESS.2019.2909490

Nazari, A. 2020. Nanosensors for smart cities Nanosensors for smart cities: an introduction. *Nanosensors for Smart Cities* 3(8), pp. 3–8. https://doi.org/10.1016/B978-0-12-819870-4.00001-3.

Patidar, M., Gupta, N. 2017. Efficient design and simulation of novel Exclusive-OR gate based on nanoelectronics using quantum-dot cellular automata. In *Proceeding of the Second International Conference on Microelectronics, Computing & Communication Systems (MCCS 2017)*. Lecture Notes in Electrical Engineering, 476, Springer, Singapore. https://doi.org/10.1007/978-981-10-8234-4_48

Patidar, M., Gupta, N. 2020. An efficient design of edge-triggered synchronous memory element using quantum-dot cellular automata with optimized energy dissipation. *Journal of Computational Electronics* 19, pp. 529–542. https://doi.org/10.1007/s10825-020-01457-x

Patidar, M., Gupta, N. 2021a. An ultra-efficient design and optimized energy dissipation of reversible computing circuits in QCA technology using zone partitioning method. *International Journal of Information Technology*. https://doi.org/10.1007/s41870-021-00775-y

Patidar, M., Gupta, N. 2021b. Efficient design and implementation of a robust coplanar crossover and multilayer hybrid full adder—subtractor using QCA technology. *Journal of Supercomputing* 77, pp. 7893–7915. https://doi.org/10.1007/s11227-020-03592-5

Pidaparthi, S.S., Lent, C.S. 2021. Energy dissipation during two-state switching for quantum-dot cellular automata. *Journal of Applied Physics* 129, p. 024304. https://doi.org/10.1063/5.0033633

Poorhosseini, M., Hejazi, A.R. 2018. A fault-tolerant and efficient XOR structure for modular design of complex QCA circuits. *Journal of Circuits, Systems and Computers* 27(7), pp. 1–9. https://doi.org/10.1142/s0218126618501153.

Ramachandran, S.S., Kumar, K.J.J. 2017. Design of a 1-bit half and full subtractor using a quantum-dot cellular automata (QCA). *2017 IEEE International Conference on Power, Control, Signals and Instrumentation Engineering (ICPCSI), Chennai*, pp. 2324–2327. https://doi.org/10.1109/ICPCSI.2017.8392132

Rana, G., Khang, A., Khanh, H.H. 2021. The role of artificial intelligence in blockchain applications. In *Reinventing Manufacturing and Business Processes Through Artificial Intelligence*. CRC Press. https://doi.org/10.1201/9781003145011.

Rani, S., Kataria, A., Chauhan, M. 2022. Fog computing in industry 4.0: Applications and challenges—a research roadmap. In *Energy Conservation Solutions for Fog-Edge Computing Paradigms*, pp. 173–190. Springer, Singapore.

Rani, S., Kataria, A., Sharma, V., Ghosh, S., Karar, V., Lee, K., Choi, C. 2021. Threats and corrective measures for IoT security with observance of cybercrime: a survey. *Wireless Communications and Mobile Computing* 2021, pp. 1–30.

Rani, S., Kaur, S. 2012. Cluster analysis method for multiple sequence alignment. *International Journal of Computer Applications*, 43(14), pp. 19–25. https://doi.org/10.5120/6171-8595

Rani, S., Khang, A., Chauhan, M., Kataria, A. 2021. IoT equipped intelligent distributed framework for smart healthcare systems. *Networking and Internet Architecture*, https://arxiv.org/abs/2110.04997v2. https://doi.org/10.48550/arXiv.2110.04997

Rani, S., Kumar, R. 2022. Bibliometric review of actuators: key automation technology in a smart city framework. *Materials Today: Proceedings*. https://doi.org/10.1016/j.matpr.2021.12.469.

Rani, S., Mishra, R.K., Usman, M., Kataria, A., Kumar, P., Bhambri, P., Mishra, A.K. 2021. Amalgamation of advanced technologies for sustainable development of smart city environment: A review. *IEEE Access* 9, pp. 150060–150087. https://doi.org/10.1109/ACCESS.2021.3125527.

Rehan, N.M. 2021. Nanotechnology as a sustainable approach for achieving sustainable future. *World Journal of Engineering and Technology* 9, pp. 877–890. https://doi.org/10.4236/wjet.2021.94060.

Roy, S.S. 2018. Implementation of novel binary logic gates with temperature stability factor analysis in Quantum-dot Cellular Automata. *Journal of Photonic Materials and Technology* 4(1), pp. 8–14. https://doi.org/10.11648/j.jmpt.20180401.12.

Seyedi, S., Navimipour, N.J. 2018. An optimized three-level design of decoder based on nanoscale Quantum-Dot Cellular Automata. *International Journal of Theoretical Physics* 57, pp. 2022–2033. https://doi.org/10.1007/s10773-018-3728-0.

Srikanth, N.G., Sharan, P.C. 2016. A novel quantum-dot cellular automata for 4-bit code converters. *Optik-Optics* 127(10), pp. 4246–4249. https://doi.org/10.1016/j.ijleo.2015.12.119.

Tailor, R.K., Khang, A., Pareek, R. (Eds.). 2022. RPA in Blockchain. In *The Data-Driven Blockchain Ecosystem: Fundamentals, Applications, and Emerging Technologies*, 1st ed., pp. 149–164. CRC Press. https://doi.org/10.1201/9781003269281.

Torres, F.S., et al. 2018. An energy-aware model for the logic synthesis of quantum-dot cellular automata. *IEEE Trans CAD* 37(12), pp. 3031–3041. https://doi.org/10.1109/TCAD.2018.2789782.

Walter, M., et al. 2018. An exact method for design exploration of quantum-dot cellular automata. In *IEEE 2018 Design, Automation & Test in Europe Conference & Exhibition (DATE)-Dresden, Germany, 2018 Design, Automation & Test in Europe Conference & Exhibition*, pp. 503–508. https://doi.org/10.23919/DATE.2018.8342060.

Wang, L., Xie, G. 2019. A power-efficient single layer full adder design in field-coupled QCA nanocomputing. *International Journal of Theoretical Physics* 58, pp. 2303–2319. https://doi.org/10.1007/s10773-019-04121-8.

Zhang, Y., Xie, G., Sun, M. 2018. An efficient module for full adders in Quantum-dot Cellular Automata. *International Journal of Theoretical Physics* 57, pp. 3005–3025. https://doi.org/10.1007/s10773-018-3820-5.

7 Autonomous Robots for a Smart City
Closer to Augmented Humanity

*Vugar Abdullayev Hajimahmud, Alex Khang,
Vladimir Hahanov, Eugenia Litvinova, Svetlana
Chumachenko, Abuzarova Vusala Alyar*

CONTENTS

7.1 INTRODUCTION

One of the most popular technologies in recent times is robots. The first robots have existed since ancient times. For example, automatic water clocks, found in Egyptian and Greek cultures, are cited as examples of the first robots (Intel, 2020).

The first person to propose the meaning of the word *robot*, or rather its modern meaning, was the Czech writer Josef Čapek. He first used the word in *Rossum's Universal Robots*. The Czech word for *robot* means *worker*. Today, in the simplest case, the definition of a robot could be a machine that works instead of a human.

This is because robots are now being used more and more to reduce the workload of human workers (Autonomous Robots for Smart City, 2022).

Demand for this technology has grown steadily since its inception, especially in areas where human labor needs to be replaced.

Robots have undergone a great evolution from the past to the present and have evolved according to different fields. Today, robots are actively used not only in manufacturing but also in medicine, education, and the military (Rani and Gupta, 2017; Rani et al., 2022).

Large, medium, small, and even micro robots have been created and used. Human-derived and, in a sense, autonomous forms of robots have also been created. Such robots are known as autonomous robots.

It should be noted that the most advanced robots today are artificial-intelligence-based robots, which are a form of autonomous robots (Rani, Kataria et al, 2021).

The future of robots continues to be predicted by many experts. One of the most worrying problems for humans is that robots will completely replace human workers in the future.

7.2   THE CONCEPT OF AN AUTONOMOUS ROBOT

Autonomous robots belong to the category of developing devices, including unmanned aerial vehicles (aerial robots) that can be programmed to perform tasks without human intervention or interaction (Khang et al., AI & Blockchain, 2021).

Like a human, a robot will move instead of a human. Although this sounds a bit different, the idea that robots will completely replace human workers in the near future is still relevant (Rani, Kataria, Chauhan, and Khan, 2021).

However, if we look at the current era, such robots are expected to be useful in places where people cannot enter a disaster or overcome labor shortages. If robots try to perform even simple and unconscious actions for humans, they will be a very advanced technology (Alfrianta, 2019).

Another feature of autonomous robots is that they can think and cooperate on their own. For example, for large loads that a robot cannot carry, more than one robot must work together to move the load and move in the same direction. Such a situation requires learning to "cooperate" in order to communicate with other robots (Kobayashi, 2020).

Autonomous robots are expected to play an active role in various areas, such as the service industry, healthcare, and education (Rani, Khang et al., IoT & Healthcare, 2021).

7.3   AUTONOMOUS ROBOT OR SEMI-AUTOMATIC ROBOT?

In general, in a sense, robots can be classified as autonomous, controlled, and self-contained robots.

First, let us look at the difference between autonomous robots and self-contained robots. These two are somewhat different (Singh et al., 2017). In this regard, we can look at the following description.

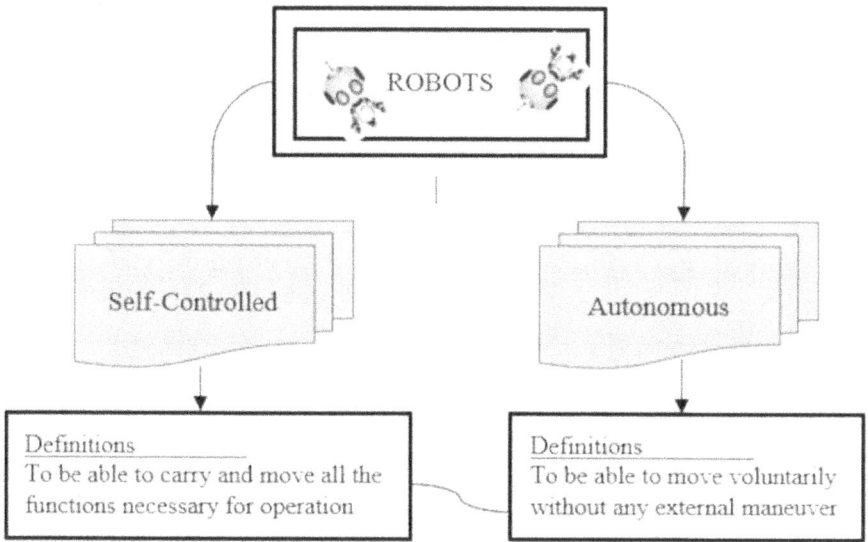

FIGURE 7.1   Self-contained and autonomous robot concepts.

7.3.1 AUTONOMOUS ROBOT

A robot that can move without receiving energy from the outside, as well as a robot that thinks and makes decisions on its own, is called an autonomous robot.

There are a number of advantages to using autonomous robots:

1. Cost reduction
2. Improving security
3. Increase productivity
4. Addressing staff shortages
5. Execution of daily (routine) work
6. Increase efficiency

Another feature of autonomous robots is that they can think and cooperate on their own. For example, for large loads that a robot cannot carry, more than one robot must work together to move the load and move in the same direction. Such a situation requires learning to cooperate in order to communicate with other robots.

7.3.2 MANUAL ROBOT

Another concept is the manual robot. Manual robots are generally dependent on operators. That is, they do not have the ability to make their own decisions like autonomous robots do.

Another related concept is the remote-controlled robot. In most cases, a robot can be operated by any remote control device (remote operation of a system). These robots are called remote-controlled robots. There is a direct interaction between man

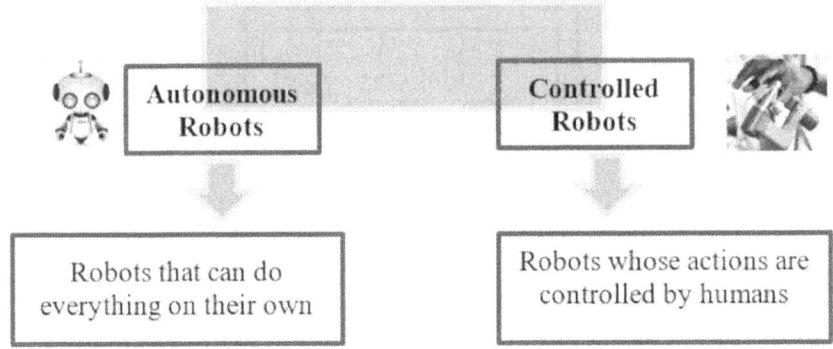

FIGURE 7.2   Comparison of controlled and autonomous robots.

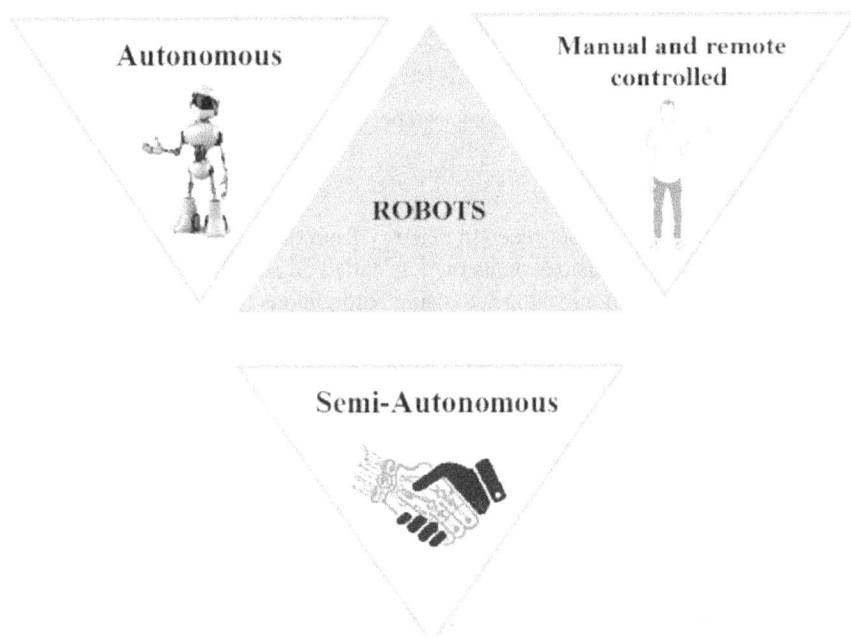

FIGURE 7.3   Classification of robots.

and robot. Moreover, basically, humans have complete control over the robot's movements (Padhan, 2021).

In short, while autonomous robots are less human-based, manual and remote-controlled robots interact directly with humans. Remote-controlled and manual robots can be considered similar concepts.

7.3.3 SEMI-AUTONOMOUS ROBOT

Another concept is semi-autonomous robots. As the name suggests, these robots are neither completely autonomous nor completely dependent on human workers. Such robots both receive instructions from human workers and can analyze the information they collect through their sensors.

The main component of such robots is a microchip embedded inside them, also known as the robot's brain.

7.4 TOWARD AUGMENTED HUMANITY BY AUTONOMOUS ROBOTS

The concept of augmented humanity is one of the new terms in the dictionary of technology. This term is reminiscent of the term *augmented reality*.

Separately, augmented reality can also be confused with the concept of virtual reality. However, these two concepts are different. Augmented reality (or AR) is defined as a virtual environment that simulates the real world through sound, graphics, GPS data, and so on.

That is, it is a computer-altered reality. Virtual reality offers a world designed and animated in the virtual world instead of the real, physical world.

Simply put, augmented reality is connected to the physical world; that is, while the physical world is grounded, virtual reality is completely virtual, so the cyber world is grounded (Vladimir Hahanov et al., 2022).

The main purpose of augmented humanity is to create a relationship between the virtual and physical world, to act as a bridge. In other words, the target is for technology and people to live together in the future. At this point, augmented humanity can also see as the next level of humanity.

There are various technologies for the realization of augmented humanity. One of them is autonomous robots.

The augmented humanity market is projected to exceed $70 billion by the end of 2021 and will continue to grow in 2025 and beyond as new market needs increase the demand for new and next-generation applications (Sachin, 2019).

7.5 WORK PROCESS OF AUTONOMOUS ROBOTS

First, it should be noted that the system of autonomous robots consists of four parts. This also carries out its work process.

1. Visual system
2. Decision-making system
3. Management (sub)system
4. Communication system

7.5.1   Visual System

The visual system is the part that is responsible for understanding the situation in the environment. As its name suggests, it obtains an image from the environment, analyzes it, generally recognizes the "target," and then sends the results to the decision-making system.

7.5.2   Decision-Making System

A decision-making system is one that, after receiving the results, analyzes them and makes a "reasonable" decision. It then sends the decision to the management system.

7.5.3   Management System

The management system receives commands (decisions) from the main system and transmits them to another system. In addition, it manages the system and monitors the correct operations of parts, controls, and so on.

7.5.4   Communication System

The communication system communicates with robots in the field and computers outside the field via a wireless network to perform operations from the system.

In general, the work process of an autonomous robot is carried out in stages with the help of these four systems.

7.6   ROBOT SENSORS AND ALGORITHM

Sensors are one of the most important parts of a robot, especially autonomous robots. Thus, the robot can monitor the environment through sensors and select the target.

When combined with sensor software algorithms, it allows robots to understand and navigate the environment, detect and prevent collisions with objects, and discover information about locations. With the help of the information obtained through sensors, the robot performs the necessary operations (Rani, Khang et al., IoT & Healthcare, 2021).

Sensors can operate in two ways: active and passive. The main sensors for autonomous robots are location (GPS), navigation, and sound sensors.

More advanced autonomous robots use stereo vision to see the world around them. It uses two cameras, and the image recognition program allows them to find and classify different objects. Robots can also use microphones and odor sensors to analyze the world around them (Harris, 2020).

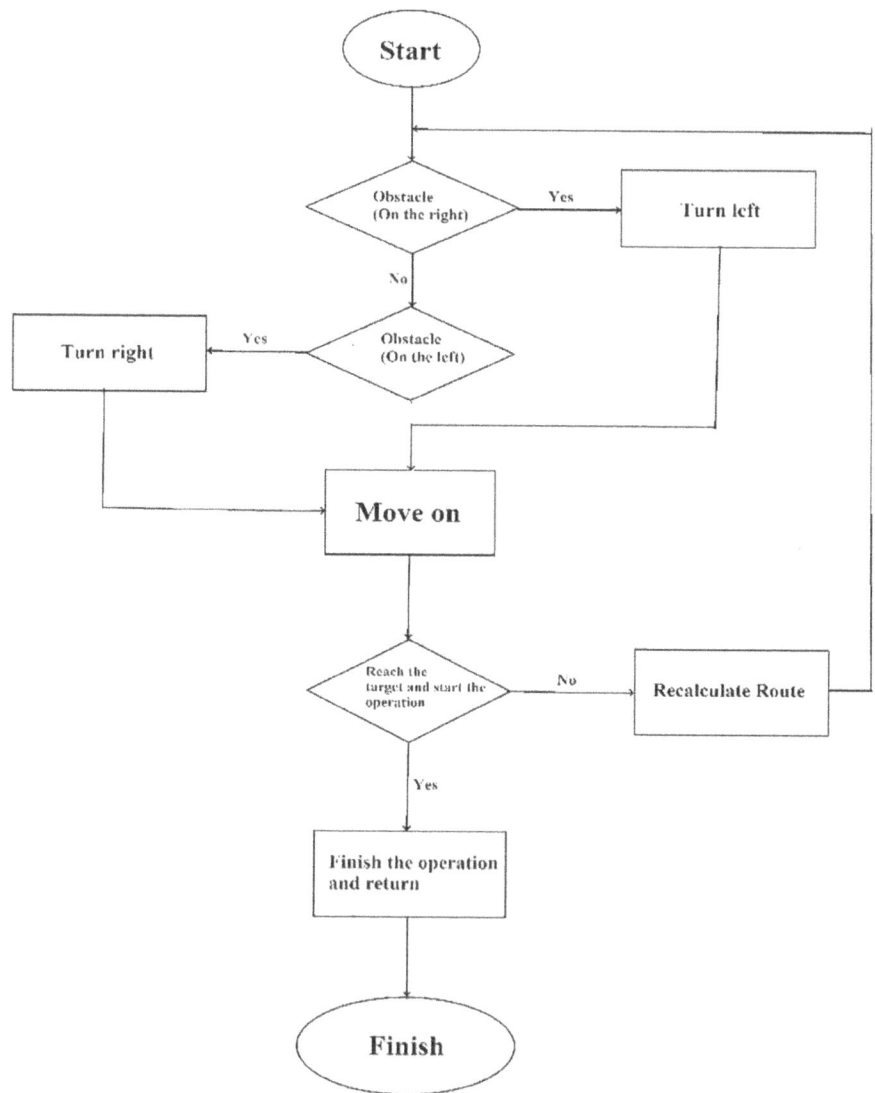

FIGURE 7.4 Simple motion algorithm of the cleaning robot.

Some autonomous robots can only work in a familiar, limited environment. Mowing robots, for example, depend on buried boundary markers to determine the boundaries of their yards. An office-cleaning robot may need a map of the building to maneuver from one point to another.

In simple terms, the following algorithm can define one of the daily operations performed by a cleaning robot, as shown in Figure 7.4.

For the same example, a simple trajectory on a coordinate system can be as follows:

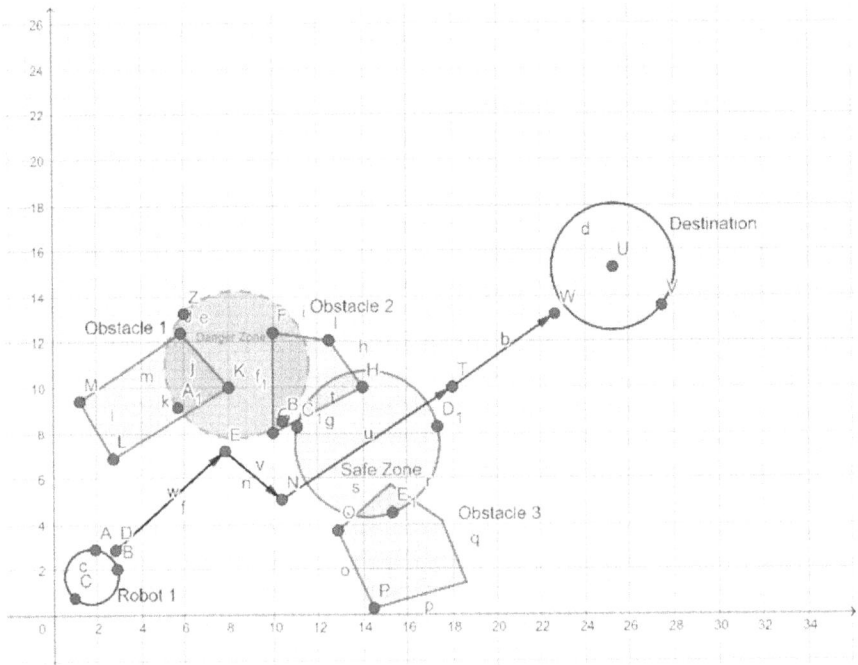

FIGURE 7.5   Simple trajectory of the cleaning robot.

In this way, obstacles 1 and 2, to the left and in front, create a dangerous zone on the way to the robot's target—that is, it cannot pass.

On the other hand, it creates a safe zone for the robot because there is enough distance between obstacle 3 and obstacle 2. Moreover, using this way, the robot can reach its target.

In part, unlike the previous figure, obstacles 1 and 2, in front and to the right, create a danger zone for the robot to reach its target and create a safe zone for the robot, as there is enough distance between obstacle 3 and obstacle 1. And using this way, the robot can reach its target.

7.7   MAIN FIELDS OF ROBOT APPLICATION

Like many technologies, robots have a wide range of applications. Industry, medicine, entertainment, daily home, and so on refer to the applications of autonomous robots.

Autonomous robots can be used for various purposes in medicine. For example, doctors can accompany patients when they visit. This helps doctors in many ways. Any patient's personal information can be displayed on the screen of the robot accompanying the doctor. In this way, the doctor can deal with issues, such as carrying extra money with him or having paramedics accompany him on visits.

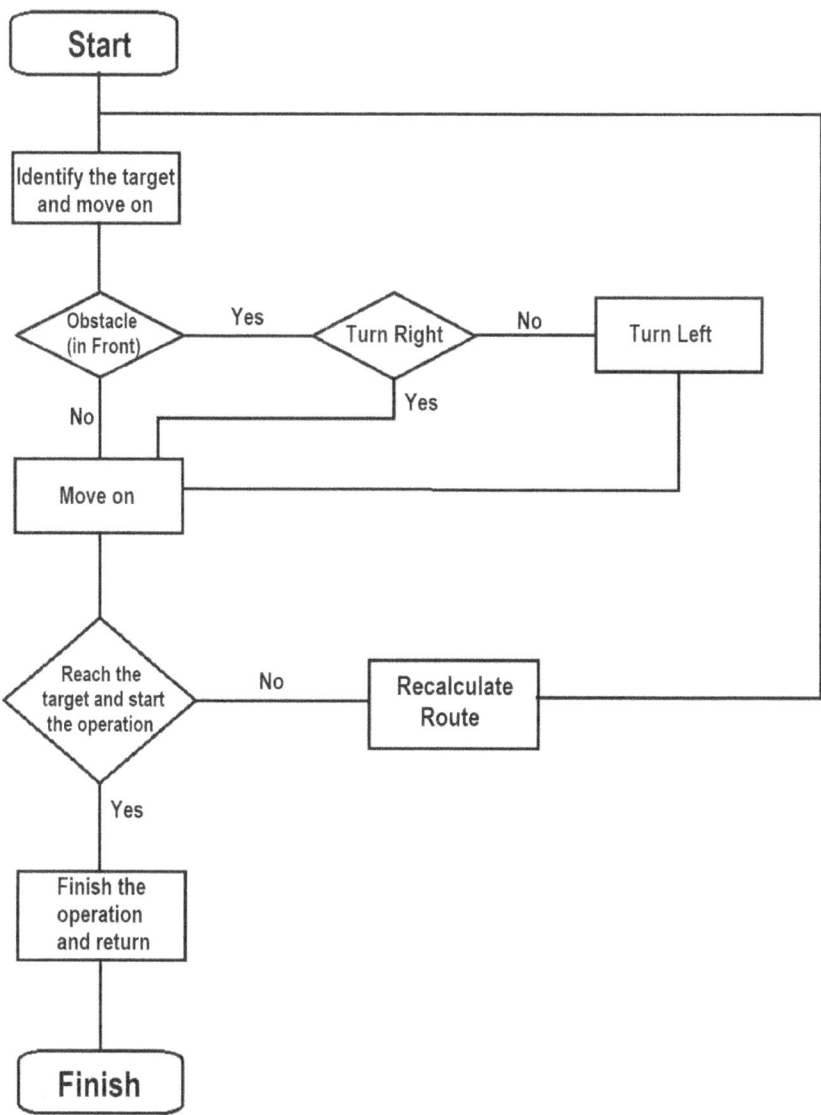

FIGURE 7.6 Simple motion algorithm of the cleaning robot.

In addition, such robots can be used in wards inside the hospital. They can also clean places. This not only simplifies human work but also shows that no human labor is needed. That is, in some cases, robots instead of human workers can do all this work.

In addition, such robots can measure their own energy values and can return to their charging (power) stations if necessary.

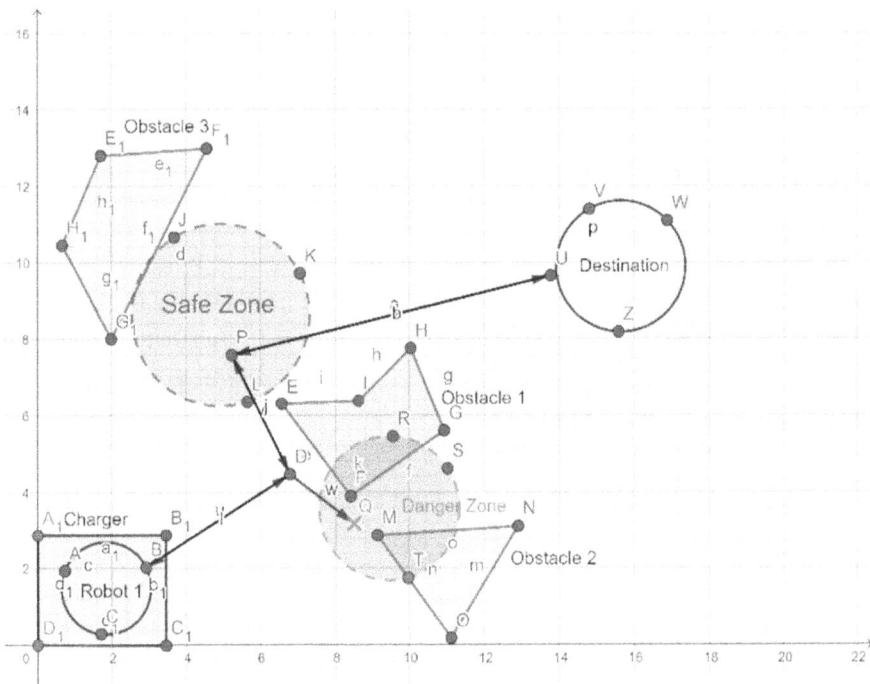

FIGURE 7.7 Simple trajectory of the cleaning robot.

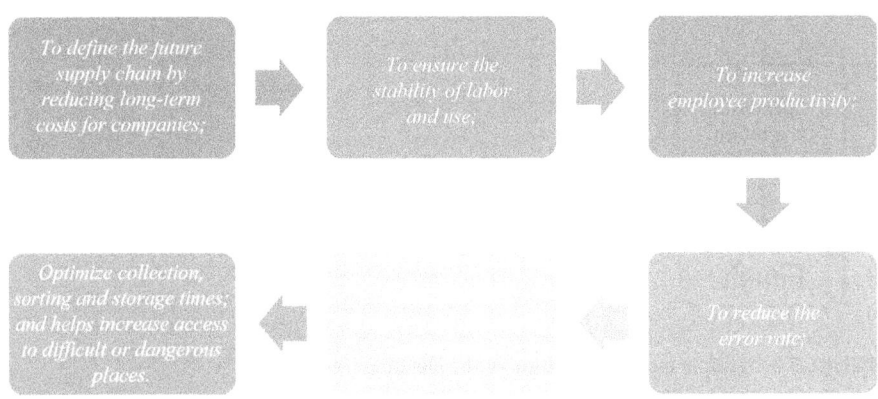

FIGURE 7.8 Autonomous robots used in companies

Autonomous robots can also be used for various purposes in the field of production. Autonomous robots are shown in Figure 7.8.

In addition, such robots are expected to be useful in places where people cannot enter in the event of a disaster or to overcome labor shortages. If robots try to perform even simple and unconscious actions for humans, they will become very advanced technologies.

7.8 CONCLUSION

The number of AI-based robots is increasing, and the field of application is expanding and will continue to expand (Bhambri et al., Cloud & IoT, 2022).

Autonomous robots are expected to evolve over the next five years, especially within supply chain operations, involving lower-cost, potentially hazardous, or high-risk tasks.

In other words, AI-centric applications, autonomous robots, which are expected to develop further in the future, are expected to help such people in many areas, even in everyday life (Khang et al., AI & Blockchain, 2021). In addition, humans now need robots in their daily lives. In this regard, more robots that are autonomous are expected to be used at home in the coming years.

Therefore, if we show cleaning robots in the simplest form now, more may come to our homes in the future. In addition, other robots will program many robots.

REFERENCES

Alfrianta, Albert, "Simple Avoidance Algorithm, Implemented on e-puck Robot and Simulated on Webots Robot Simulator," October 22, 2019, https://medium.datadriven-investor.com/simple-avoidance-algorithm-implemented-on-e-puck-robot-and-simu-lated-on-webots-robot-simulator-3143c096d285

Autonomous Robots for Smart City, Other Resources, "Autonomous Robots and Their Types with Applications," 2022, www.elprocus.com/different-types-of-autonomous-robots-and-real-time-applications/

Bhambri, Pankaj, Sita Rani, Gaurav Gupta, Khang, A., *Cloud and Fog Computing Platforms for Internet of Things.* CRC Press, 2022, ISBN: 978-1-032-101507, doi:10.1201/9781032101507.

Harris, Tom, 2020, "How Robots Work: Autonomous Robots," https://science.howstuff-works.com/robot4.htm.

Intel, 2020, "Robotics in Healthcare to Improve Patient Outcomes", www.intel.com/content/www/us/en/healthcare-it/robotics-in-healthcare.html

Khang, A., Pankaj Bhambri, Sita Rani, Aman Kataria, *Big Data, Cloud Computing and Internet of Things.* CRC Press, 2022, ISBN: 978-1-032-284200, doi:10.1201/9781032284200.

Khang, A., Subrata Chowdhury, Seema Sharma, *The Data-Driven Blockchain Ecosystem: Fundamentals, Applications and Emerging Technologies.* CRC Press, 2022, ISBN: 978-1-032-21624, doi:10.1201/9781003269281.

Khang, A., Geeta Rana, Ravindra Sharma, Alok Kumar Goel, Ashok Kumar Dubey, "The Role of Artificial Intelligence in Blockchain Applications," *Reinventing Manufacturing and Business Processes Through Artificial Intelligence.* CRC Press, 2021, doi:10.1201/9781003145011.

Kobayashi, Kunikazu, "Autonomous Robots that Think and Act for Themselves are Coming to Home!," Lecture No. 06723, 2020, https://yumenavi.info/lecture.aspx?GNKCD=g006723

Padhan, Asmita, "What are Remote Controlled Robots? How to Build a Remote Controlled Robot?," May 4, 2021, www.skyfilabs.com/blog/what-are-remote-controlled-robots-and-how-to-build

Rani, Sita, and O. P. Gupta. "CLUS_GPU-BLASTP: Accelerated Protein Sequence Alignment Using GPU-Enabled Cluster," *The Journal of Supercomputing* 73, no. 10 (2017): 4580–4595. doi:10.1007/s11227-017-2036-4.

Rani, Sita, Aman Kataria, and Meetali Chauhan. "Fog Computing in Industry 4.0: Applications and Challenges—A Research Roadmap," in *Energy Conservation Solutions for Fog-Edge Computing Paradigms*, pp. 173–190. Springer, Singapore, 2022.

Rani, Sita, Aman Kataria, Vishal Sharma, Smarajit Ghosh, Vinod Karar, Kyungroul Lee, and Chang Choi. "Threats and Corrective Measures for IoT Security with Observance of Cybercrime: A Survey," *Wireless Communications and Mobile Computing* 2021, (2021): 1–30. doi:10.1155/2021/5579148.

Rani, Sita, Khang, A., Meetali Chauhan, Aman Kataria, "IoT Equipped Intelligent Distributed Framework for Smart Healthcare Systems," *Networking and Internet Architecture*, 2021, https://arxiv.org/abs/2110.04997v2, doi:10.48550/arXiv.2110.04997.

Sachin, Kurlekar, "Sensors in Autonomous Mobile Robots for Localization and Navigation," February 8, 2019, https://internetofthingsagenda.techtarget.com/blog/IoT-Agenda/Sensors-in-autonomous-mobile-robots-for-localization-and-navigation

Singh, Preetkamal, O. P. Gupta, and Sita Saini. "A Brief Research Study of Wireless Sensor Network," *Advances in Computational Sciences and Technology* 10, no. 5 (2017): 733–739. Research India Publications, https://ripublication.com/acst17/acstv10n5_07.pdf

Vladimir Hahanov, Khang, A., Gardashova Latafat Abbas, Vugar Abdullayev Hajimahmud, "Cyber-Physical-Social System and İncident Management," in *AI-Centric Smart City Ecosystems: Technologies, Design and Implementation* (1st ed.). CRC Press, 2022. https://doi.org/10.1201/9781003252542.

8 Drone Networks and Monitoring Systems in Smart Cities

B. Prabu, R. Malathy, M.N.A.
Gulshan Taj, N. Madhan

CONTENTS

8.1 INTRODUCTION

Fast and precise sensor investigation has numerous applications in multiple domains pertinent to society today. These incorporate the detection and recognizable proof of chemical leaks, gas releases, forest fires, surveillance issues, and other security-related challenges. There are other structural applications in the form of building health monitoring, construction, and environmental monitoring.

Commercial unmanned aerial vehicles (UAVs), which are more popularly termed as robots, have seen a gigantic expansion in the most recent couple of years, making these gadgets exceptionally available to the public. Portable drones, like UAVs, can be utilized for surveillance, monitoring, observation, checking information about structures frameworks, and health-related issues inside buildings.

AI algorithms process huge amounts of data to find patterns and detect features. Hence, the integration of UAV and AI algorithms has been studied and used in today's world.

Hence, it is perceived that flexible, self-ruling, and dynamic mobile drones can do surveillance jobs in remote locations and found major applications in smart city projects. Over time, many algorithms and prediction methods were formulated and

DOI: 10.1201/9781003252542-8

implemented by many researchers worldwide (Rani, Kataria et al., 2022; Rani, Kataria, Chauhan et al., 2022).

The AI execution has decreased quantities of difficulties to UAV configurations other than upgrading the abilities and making way for the application of smart city surveillance.

Machine learning is a superior algorithm that is characterized as a framework's capacity to accurately decode outside information, and the information database may contain information about the population, number of events, and number of vehicles, to name a few.

8.2 SMART CITY

8.2.1 Definition of a Smart City

There are many definitions of a smart city. In simple words, a smart city is a city that uses technology to optimize the functions of the city. This technology promotes the economy and sustainability of resources. The concept comprises ICT, which is used to improve the lives of people (Rani, Kataria, et al., 2021).

The characteristic of a smart city includes a smart economy, smart environment, smart mobility, smart governance, smart living, and smart people. The three aspects of city growth are sustainability (improving the city's environment), smartness, and inclusiveness (Rani, Kataria, Chauhan, and Khang, 2021).

There are two main functioning features: sensing and automation. Components like sensors, cameras, road sensors, wireless connectivity, and GPS can sense and transmit information. The automation parts include Arduino, Raspberry Pi, and embedded systems.

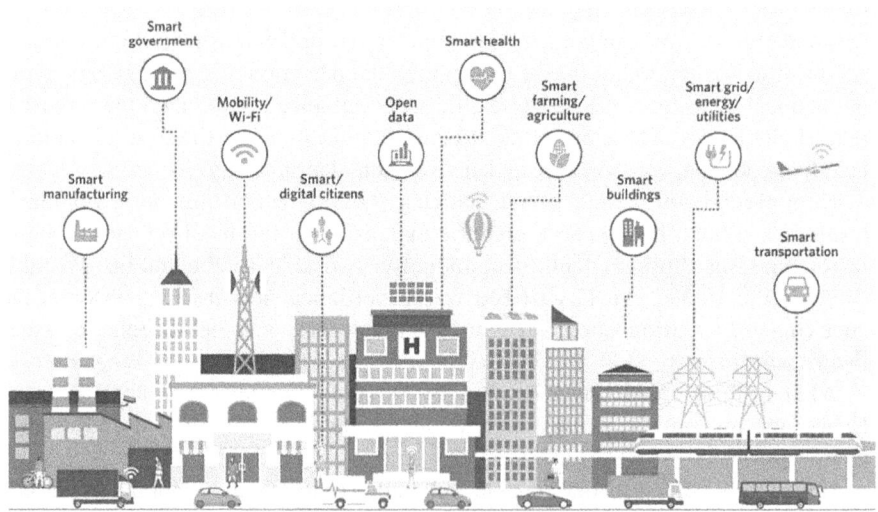

FIGURE 8.1   Smart city application sectors.

8.2.2 Technology in a Smart City

Technologies are the driving fuel to establishing a smart city. The following are important technologies used in a smart city: (1) APIS, (2) artificial intelligence (AI), (3) Internet of Things, (4) machine learning (ML), (5) cloud computing, (6) dashboards, (7) machine-to-machine communication, and (8) mesh networking. Of these, the Internet of Things (IoT) is the most widely used method whose framework involves the usage of sensors to sense the required data. Data from such sensors are processed through the cloud for the output (Tailor et al., 2022).

Combined with machine learning, IoT is used for various applications, from vehicles to home appliances and street-sensing applications (Rani, Khang et al., IoT & Healthcare, 2021).

IoT's interest in smart technology is increasing daily as the world pursues instant response and there is a gradual increase in population. Thus, a smart solution is the need of the hour. ICT also helps to improve decision-making.

8.2.2.1 APIs

API (application programming interface) simplifies product-to-product communication and helps save time and money. The application ranges from simple things like marketing to locating a satellite in space. They allow data to work seamlessly.

When it comes to the application of API in cities, it is crucial as working with APIs enables users to connect data they need without surfing through generated data, as shown in Figure 8.3.

FIGURE 8.2 Smart city diagram.

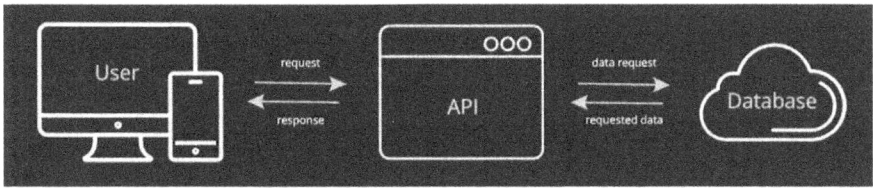

FIGURE 8.3 API working principle.

8.2.2.2 AI and ML

Artificial intelligence (AI) is the concept of science and technology concerned with making machines learn, act, and make decisions like humans. Approaches include making machines think and act like humans do.

The best examples of AI are smart assistants like Siri, Alexa, and other website recommendation portals.

Machine learning (ML) is one of the fastest growing and most important branches of AI. It improves performance and its ability through algorithms, input, and past data (Khang et al., AI & Blockchain, 2021).

These programs can access input and utilize them to learn. Representation, evaluation, and optimization are the three components of ML. In the traditional method

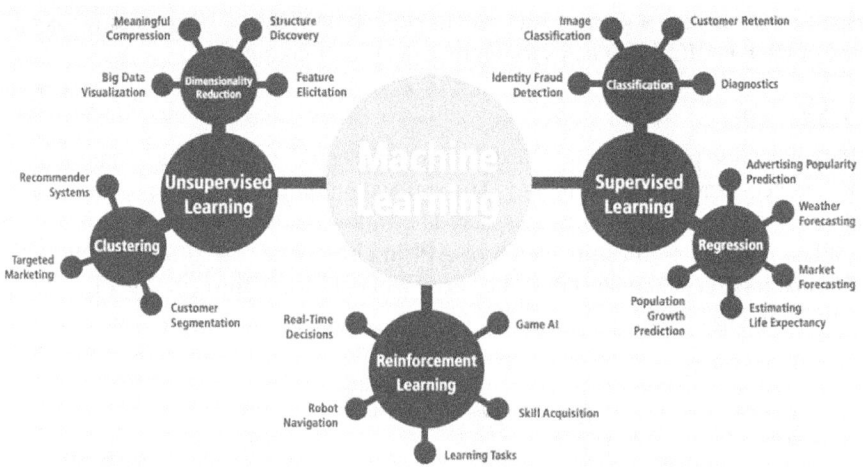

FIGURE 8.4 Various branches of artificial intelligence.

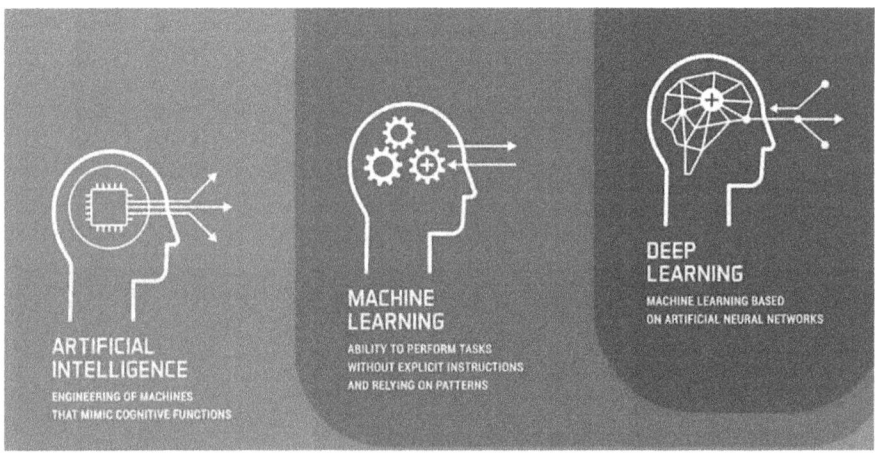

FIGURE 8.5 Relationship between AI, ML, and DL.

of programming, data and programs are used to fetch output. But in ML, data and output are processed to get a program that can be accessed and used for traditional programming.

8.3 DRONES

Drones are an upcoming area of interest in the world of modernization. Earlier, they were of limited use and employed in air deployments in military bases. An UAV is built with composite material that is light to reduce weight and for easy maneuverability. It has technologies and systems, including a global positioning system, cameras to capture objects, sensors, and other such things for various purposes.

8.3.1 CONTROL SYSTEMS IN DRONES

The technology behind drones has two main parts: the drone itself and the ground control unit. The ground control unit is nothing but a device, like a mobile, tablet, or laptop, to operate and initiate action when needed. Drones have sensors mounted on their faces. These sensors are the most important features and act as eyes for the drone.

Nowadays, drones are embedded with dual systems, like GPS. When a drone is initiated, it will search for the GNSS and detect a satellite constellation, which is a cluster of satellites, for coordinating positions. Figure 8.6 shows drones in fieldwork.

The ground controller will reveal how many satellites are connected. After connecting with satellites and GNSS, UAVs are ready for flight.

At first, the current location of the drone will appear in the ground controller. Then, the home location will be fixed with respect to the ground controlling unit. They fly with the help of a battery. If battery power is low, drones will return to the

FIGURE 8.6 Drones in fieldwork.

concerned home location. We can also make the drone return home by pressing the home button on the controller, as shown in Figure 8.6.

8.3.2 COMPONENTS OF DRONES

Drones are embedded with multiple sensors, cameras, batteries, propellers, and so on, with obstruction detection being vital in drones, as shown in Figure 8.7.

The following one or more sensors are fused for the detection of obstacles: (1) vision sensors, (2) infrared sensors, (3) ultrasonic sensors, (4) lidar, and (5) monocular vision sensors. These sense hurdles in the drone's way.

The propulsion system is the most important part of drone combination (known as drone technology). This system consists of motors, propellers, and electronic specifications. In quadcopters, there are two pairs of propellers (i.e., four propellers, a clockwise pair and an anticlockwise pair). The other components are (1) standard prop, (2) pusher prop, (3) brushless motors, (4) motor mount, (5) landing gears, (6) retractable landing gears, (7), ESC, (8) GPS module, and (9) gimbal.

8.3.3 DRONES IN A SMART CITY

We can use drones for various applications in smart cities. When we use stable and fixed devices like cameras and sensors, their range is quite low and, at some point, not feasible. Thus, this difficulty can be countered using aerial and mobile devices like UAVs.

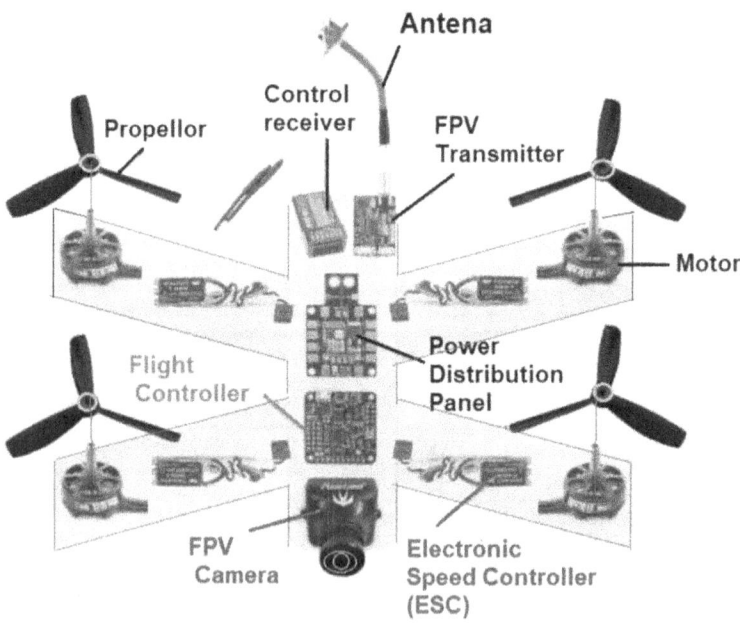

FIGURE 8.7 Components of drones.

The main advantages of UAVs are that they are cost-effective, can access remote places inaccessible to humans, and provide immediate response and real-time communication. The interesting places where UAVs could be most advantageous are traffic management, crowd monitoring, pollution control, disaster management, precision agriculture, surveillance, and surveying.

8.3.3.1 Drones in a Traffic Monitoring System

Traffic management is a major problem in the modern city, and it is a mission when it comes to the management of vehicular congestion. Surveillance methods used now are stable, but their range of vision is limited.

IoT-based sensors fixed on sidewalks (traffic signals, road junctions, and intersections) have challenges like inaccurate details, vehicle-to-vehicle interactions, and their efficiency during congestion and identification, as shown in Figure 8.8.

The main objectives of employing UAVs in STM (smart traffic monitoring) are real-time data collection and transmission, faster site clearance, emergency response, efficient monitoring, immediate access to the congestion site, and so on. Thus, UAVs can be used to reduce the duration of responses.

Fixed-wing drones are adequate and suitable for a wide range of sensing and for surveying congestion-prone zones. On the other hand, rotary-wing UAVs can be used for monitoring and accident site investigation.

In general, the key functions of UAVs in smart traffic management are identification and control of vehicles, tracking, detection of conflicts, and facilitating the exchange of captured data in real time, as shown in Figure 8.9.

8.3.3.1.1 Devices Required for a Traffic Monitoring System

Devices required for establishing a UAV system of traffic monitoring are an aircraft unit, camera and gimbal, HD transmitter and receiver, ground control unit, and suitable software for post-processing techniques.

FIGURE 8.8 Applications of drones in traffic management.

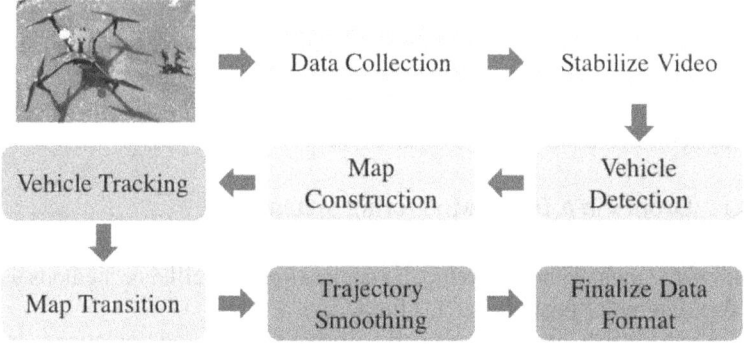

FIGURE 8.9   Vehicle detection using drones.

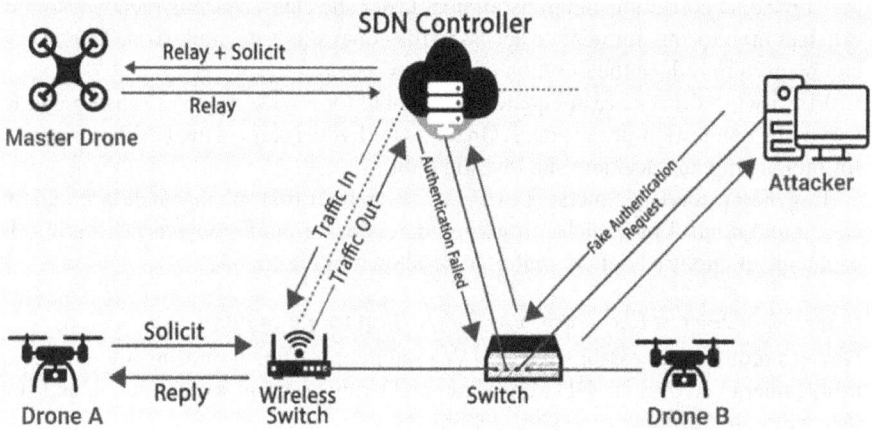

FIGURE 8.10   Demo of a vehicle tracking system.

UAVs could perform unmanned flights (automatic flights) using a global positioning system and software that enables us to plot their surveillance path.

There are two modes of flight: auto mode, as mentioned before, and piloting mode, which can be done manually by using a controller. This enables us to switch drones if they are not working properly or for emergency response.

They are equipped with an MHz transceiver so that they can share collected information in real time with the ground controller unit. The camera and gimbal sections consist of electronic stabilizers to achieve a stable camera configuration.

Similarly, they can be operated manually or by sensors for fixed capturing. Wireless communication through camera and gimbal enables us to zoom, record, and shoot images. These are controlled by GHZ controllers, as shown in Figure 8.10.

The ground control unit consists of a transceiver, a monitoring unit like a laptop or a tablet, UAV and camera controller, and a database for storing captured data. Suitable software is installed to plot the route path and to be used for post-processing, extraction of data, and output generation.

FIGURE 8.11   Monitoring System in Drones

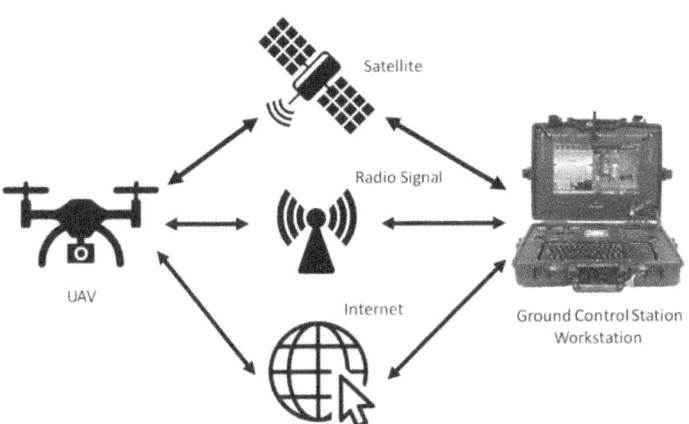

FIGURE 8.12   Transfer of data from UAV to ground control station.

The methodology in smart traffic monitoring includes the flight of UAVs in a zone, capturing details, transmitting them through wireless communication, extracting data, and processing to get the output, as shown in Figure 8.11.

8.3.3.1.2   Methodology in a Traffic Monitoring System

Primarily, the methodology includes three modules: establishing the aircraft unit in the plotted zone, sensing to assist monitoring, and using input data to identify vehicle and traffic congestion.

Two types of scenarios are monitored by drones: first, collecting congestion data, like queue length, total delay time, and traffic count at intersections; second, real-time incident reports. In this segment, broadcasting can be done by using a 4G/LITE signal, as shown in Figure 8.12.

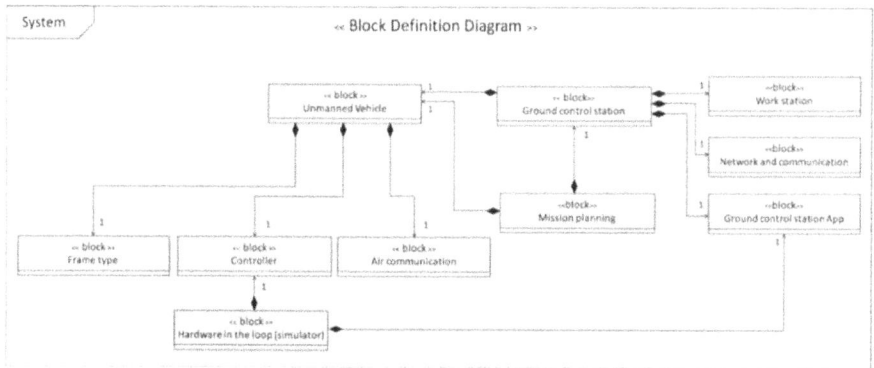

FIGURE 8.13 Block diagram showing the overall system-level design of the UAV, HTL simulator, and ground control station.

MODULE 1

The first step of the system is to assemble an aircraft unit. It includes unfolding the UAV, battery attachment, fitting of propellers, and the attachment of power cables if it is to be in a stable position for monitoring a particular junction. Then, take drones to the plotted area.

The route plot can be achieved by using specific software that uses GPS and satellite connections for efficient performance, as shown in Figure 8.13. Thus, remote-sensing technologies help us get necessary details like coordinates, location of congestions, and so on.

MODULE 2

A major part of a smart traffic system is data collection. In this section, drones that are plotted with route paths can do an automatic flight. They collect data in a circular.

For example, in incidents like a traffic jam/congestion, the deployed drones take photographs of the congestion and transmit them to the traffic control system. Information like queue length, the flow of vehicles, duration of traffic congestion, and behavior of vehicles is received by the ground control unit through wireless networks.

The data is processed and analyzed by the person in charge. Necessary steps can be taken using the transmitted data. In monitoring highways or road intersections, e-drones could be established in a way that it concentrates on the area to be covered.

It is powered via cables for continuous surveillance. Vehicle flow, velocity, speed, and acceleration are noted. Vehicle tracking and coordinating its location can be done if needed for investigation. In accidents, drones send immediate responses to the control unit so that adequate measures are taken.

Investigation and tracking are faster than in a manual operation. UAVs use HD video and image-capturing techniques and features to take several observations to identify the impact, tire marking locations, source vehicle, and scope of clearance.

MODULE 3

The last part of the module is processing. GPS coordinates are essential for geo-tagging. Assuming that drones are deployed and their camera gimbals are controlled on their own, the coordinates at that time of capturing are recorded in the log. Based on the coordinates of the accident site, drones are sent to the site and capture the scenario at defined plot points.

The transmitted images are recorded in the laptop/system using a capturing card for processing. The images are processed by using various applications. They are processed to get an independent map of the captured images and to generate a 3D model of the accident site based on the point cloud. This model helps measure the required detailing like tire mark length, impact on the vehicle, and damage to surroundings.

8.3.3.2   Drones in Surveillance and Disaster Management

8.3.3.2.1   Rescue Drones

A drone is well suited to provide a hawk's eye view of the incident for investigating on-site and reconnaissance. Thus, they can be used as vital tools in surveillance, safety, and search-and-rescue missions. Initially, there were several technologies for such rescue systems. The main drawback of such systems is their poor coverage or being unable to reach many regions for analyses.

To optimize information, it is recommended that UAVs integrate detection techniques for rescuing people or vehicles in danger, as shown in Figure 8.14. For example, a mission of detecting people can be done by heat-related features.

FIGURE 8.14   Drones in rescue operations.

8.3.3.2.2   Mechanism behind Rescue Drones

The technique and method are mainly concerned with integrating necessary GIS and sensing sensors for search-and-rescue missions. Search and detection algorithms were created to ensure proper results.

The developed algorithms can sense victims with global positioning coordinates in real time. After flight, a map of the investigated zone can provide the rescue crew with decisions and responses.

It is launched by letting the drone fly over the required area. The application of drones in search and rescue can be useful in disasters like earthquakes, floods, landslides, storms, and hurricanes.

UAVs can transmit sensed data from remote areas faster and without risk to lives. Above all, in natural disasters, data from the craft can assist the crew in understanding the condition and identify the person in danger who requires help and support.

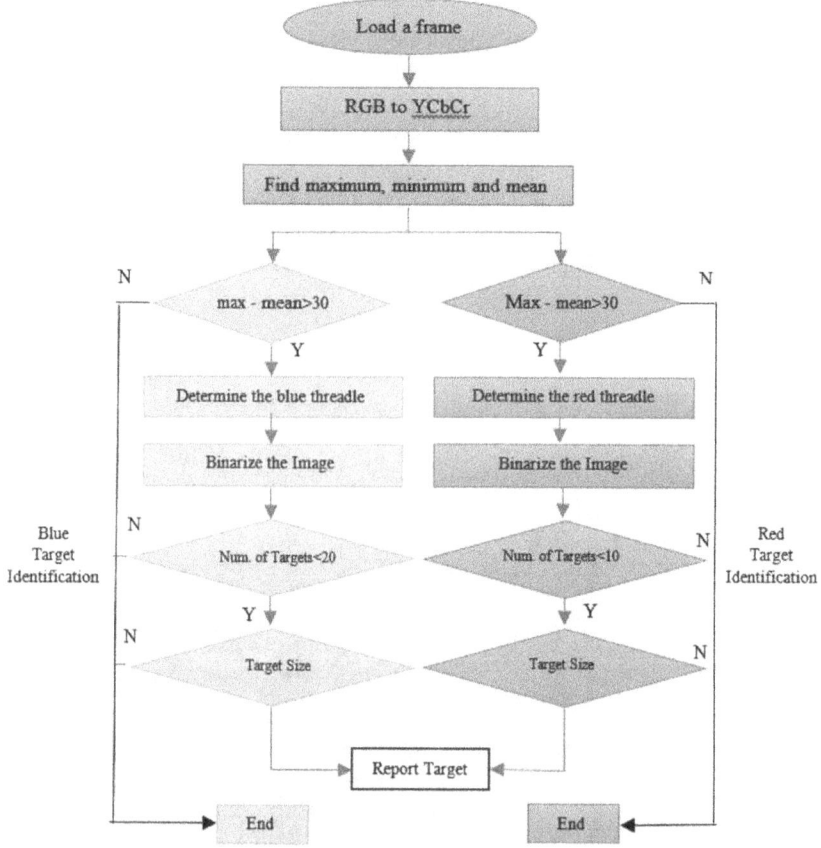

FIGURE 8.15   Algorithm of drones for rescue operations.

YCBCR = rgb2ycbcr(RGB) converts the red, green, and blue values of an RGB image to luminance (Y) and chrominance (Cb and Cr) values of a YCbCr image.

A step ahead, drones can be used to provide essentials like water and other vital products to persons who are injured and need to be rescued. In wildfires, drones can be deployed to assist firefighters. The area of usage is obtained by calculating the spread of fire, how far it is spreading, conditions of wildlife, trees, temperature, air quality, and any danger to rescuers.

By using the wireless communication system, captured images are passed on to control systems. On the basis of the processed image, a decision is made. The generated map shows the presence of the target object/person with their GPS coordinates, as shown in Figure 8.16.

Post-processing highlights the targeted objective and transfers the required information. Rescuers can then do the necessary inspection without entering the site.

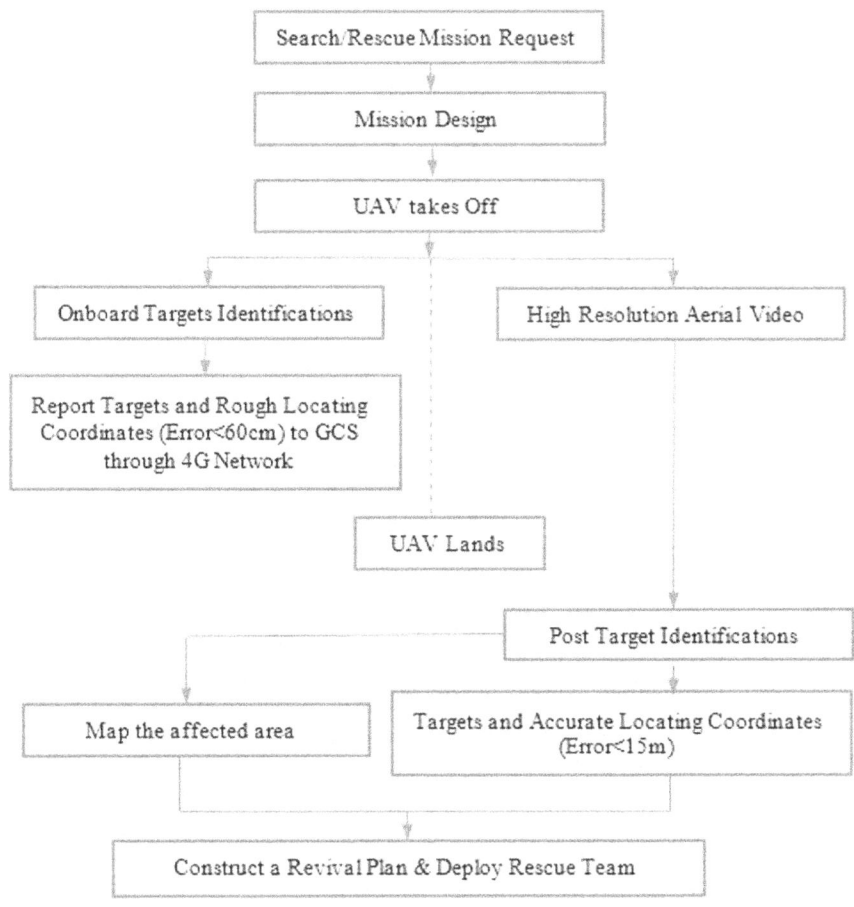

FIGURE 8.16 Working principle of rescue drones.

8.3.3.3   Drones in Crowd Monitoring

Another exciting application of drones in search-and-rescue applications is crowd monitoring. As in the previous cases, the working technique involves using real-time images by drones; these images are analyzed, and crowd density is determined using segmentation.

Data is used to identify and map density under various gimbal orientations. The system has deployed crowd monitoring techniques using real-time images taken by UAVs; collected data is investigated, and crowd density is estimated using image segmentation procedures, as shown in Figure 8.17.

A color-based segmentation method is used to detect, identify, and map crowd density under different camera positions and orientations. The craft is to be equipped with sensors, like an inertial navigation system and cameras with infrared and thermal features, for the identification of people in exact locations.

The control unit processes the sensed data and assists field staff in emergency situations. The field in charge acts quickly. In recent times, drones have been used for the analysis and management of crowds at mass gatherings as they are low-cost and a more easily operable, as shown in Figure 8.18.

The sensed images are processed to analyze the situation and its safety. The number of persons in the crowd, their behavior, and identification of missing persons are some key areas where crowd monitoring by drones can be implemented. Image segmentation has been concluded as the tool to detect and differentiate a person or target from the surrounding atmosphere.

The accuracy of the final output will be higher than that of the conventional method of monitoring crowds with the help of surveillance cameras. This image shows how crowd monitoring is done by drones. Both the earlier figures demonstrate

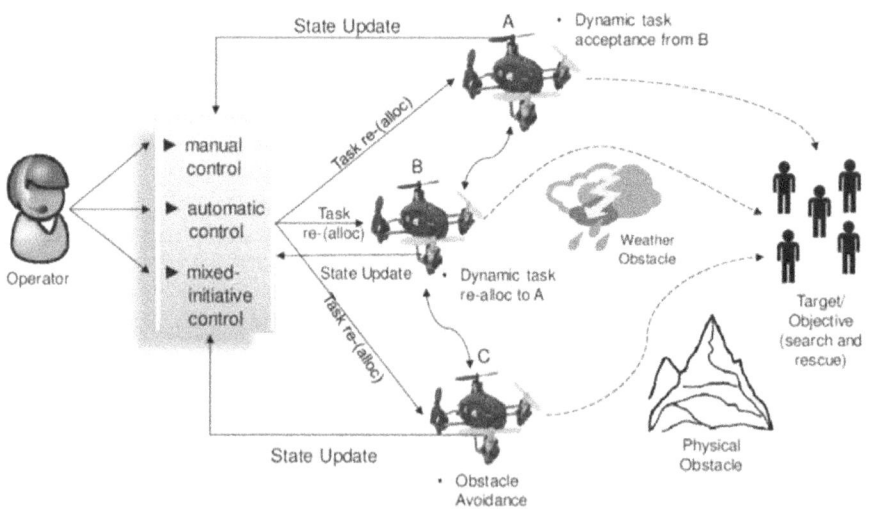

FIGURE 8.17   Operating principle in crowd monitoring.

FIGURE 8.18 Crowd detection and counting using drones.

how drones would be able to detect the presence of humans by monitoring rescue events.

The following are advantages of using drones in gatherings or inspecting a crowd: (1) easy monitoring, (2) precise crowd monitoring, (3) time saving, (4) immediate response, (5) quality aerial imaging, (6) wide vision coverage, and (7) medical assistance, if necessary.

8.3.3.4 Pollution Monitoring in Smart City via Drones

Pollution can be defined as the alteration of the environment and atmosphere. In general, the addition of harmful and foreign particles, which changes the condition of the environment, is known as pollution.

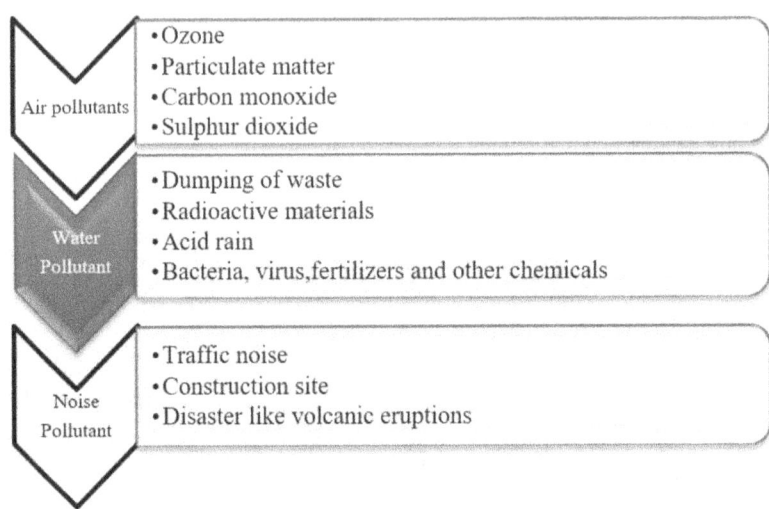

FIGURE 8.19 Major pollutants in a smart city.

When we talk about the revolution of cities and urbanization, the increase in pollution should be addressed. The intensity of pollution is increasing daily. Thus, a smart solution is needed to counter and monitor such pollution effectively. As we already know, the major types of pollution are air, water, and noise pollution. The pollutants are containments that alter the stability of the environment.

Resources can be of two types: natural and man-made. Figure 8.19 show the major pollutants.

UAVs can be an excellent tool for monitoring the environment. This process is done by fitting some sensors embedded with IoT. Generally, drones are used for deploying, and default configuration can be undertaken to reach a target autonomously, capture data, share it in real time, and so on. In addition, when embedded with necessary sensors, drones' purposes can be expanded.

This can be achieved by two important phases: design and configuration. It includes the design and modification of existing drones and UAVs, per the required application.

The computer and engineering phase includes sharing data sensed by pollution-detecting sensors and accessing them through wireless networks. E-drones, or environmental drones, are modified to monitor environmental conditions.

Air pollution is caused by the presence of harmful gases and particulate matter in the atmosphere. In smart cities or developing countries, the intensity of air pollutants is increasing each day.

Emissions from transport vehicles and the burning of plastic, wood, and tires release harmful gases and oxides, which pollute the air. Inspection of air quality can be done by using drones. Major air pollutants are (1) sulfur dioxide, (2) carbon monoxide, (3) particulate matter, (4) leads, (5) carbon dioxide, and (6) nitrogen oxide.

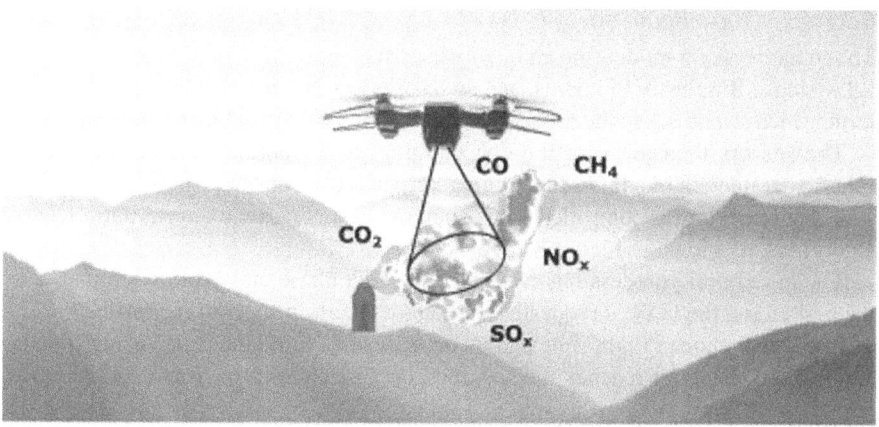

FIGURE 8.20 Air quality monitoring using drones.

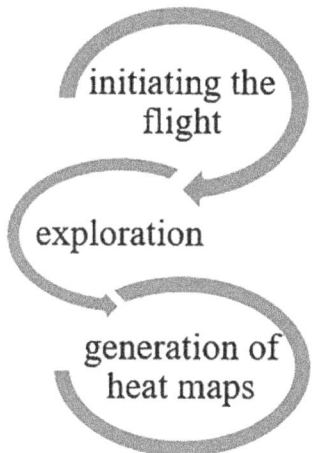

FIGURE 8.21 Systematic approach in heat detection.

8.3.3.4.1 Methodology

As in previous cases, the methodology involves three stages. The first stage is plotting the target and readying the drone for the mission. Second, the exploration phase includes the rapid and looping movement of drones to sense the required data. The last is the generation of heat maps after post-processing the gathered input.

To achieve the proper and desired output, suitable algorithms are developed. These algorithms play a vital role in the mechanism of quality monitoring by drones. These are the driving fuels as the performance and response of the aircraft depend on them.

Before creating an algorithm, the threshold values of individual air pollutants are noted. These values are used as benchmarks during monitoring, as shown in Figure 8.21.

8.3.3.4.2   Tools Required

The components for monitoring air quality consist of the aircraft unit, adequate sensors, IoT systems, Raspberry Pi boards, and so on. Sensors are installed on a separate platform, which is attached to the drone so that its stability and aerodynamics are unaltered.

The sensors then measure the concentration of air pollutants and other atmospheric parameters like humidity, temperature, oxygen concentrations, and so on.

The aircraft unit is controlled using chemotaxis and particle swarm optimization algorithms, as shown in Figure 8.22.

The algorithmic process has two halves: search phase and exploration phase. In the first phase, the UAV is requested to monitor a certain specific region.

Initially, the mean flight time and its path are coded into the drone using specific applications. When it reaches the specified target location, the sensor in the drone

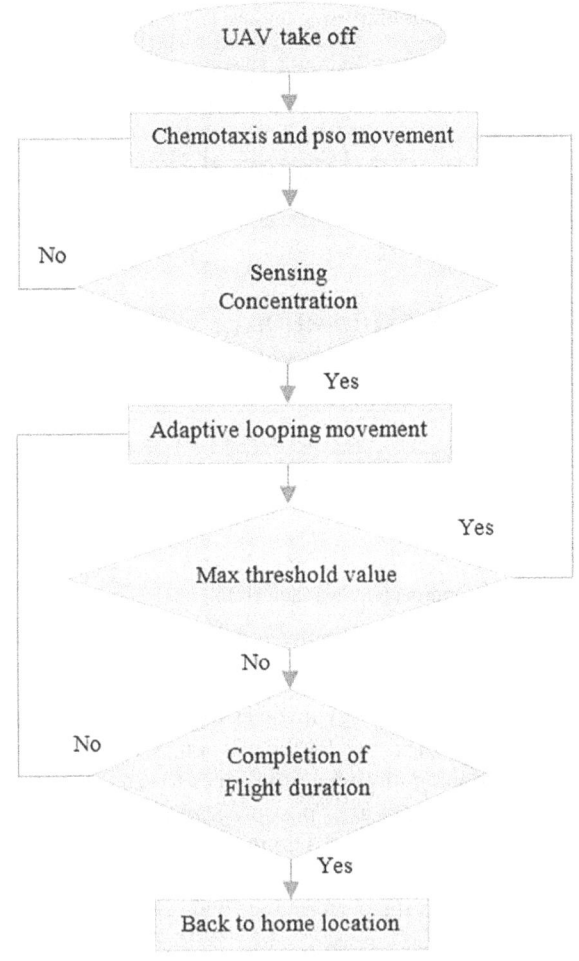

FIGURE 8.22   Algorithm of drones

initiates its work. It starts operating in a spiral movement until the required conditions are satisfied. The conditions are (1) whether the targeted zone is monitored completely or (2) completion of mean flight time.

The sensors compare sensed data with the threshold value of pollutant concentration. When the data sensed is more compared to previously sensed data, it starts monitoring the surrounding areas. Thus, with the help of data gathered by heat maps, the identification of concentration of gases in air is completed. Data from each GPS coordinate are noted and tabulated to form an air quality index. A cluster of e-drones can be used to monitor a huge area in a minimum time.

8.3.3.5 Management of Water Pollution

Water is the main primary source of life on earth. Without water, the sustainability of life is impossible. Currently, industries and water transportation needs have increased, affecting the quality of water greatly, as shown in Figure 8.23.

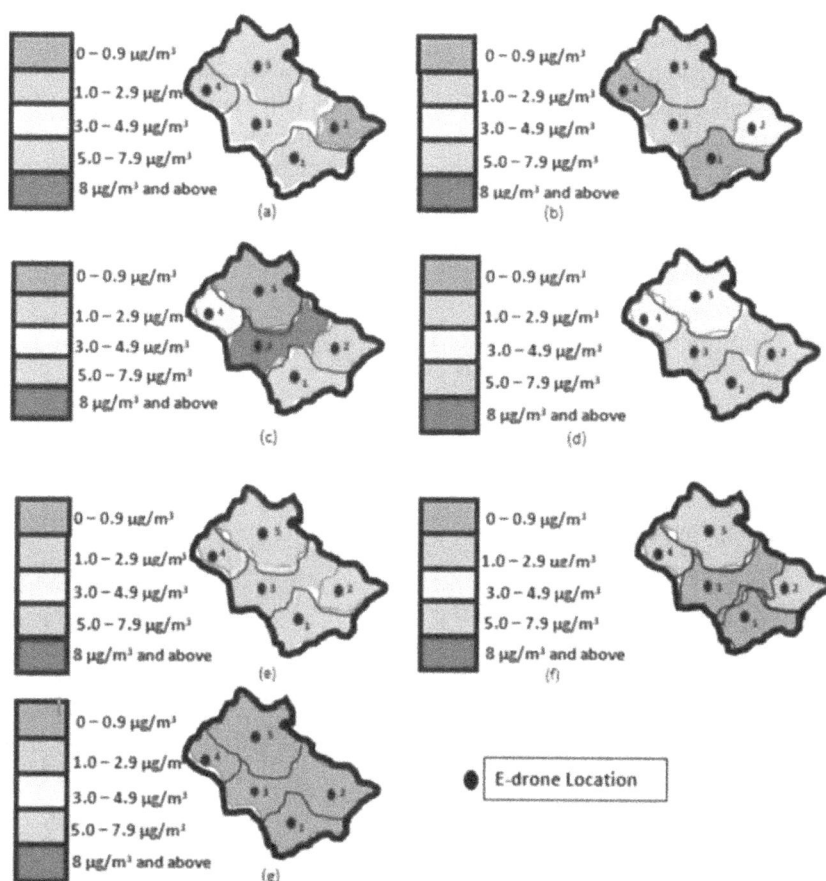

FIGURE 8.23 Heat map generated with the help of drone-sensed data.

Water pollution is one of the major problems in developing countries. Industries with chemical emissions and the discharge of harmful wastes into water are the major culprits in urban areas. They not only pollute water quality. They also destroy and affect living organisms that consume polluted water.

Monitoring water resources and ecological conditions will help us protect the aquatic environment and its organisms and ensure decisions for the conservation of resources and for preventive measures when needed.

The main causes of water pollution are (1) accidental events, like oil spills in oceans and water bodies; (2) discharge of sewage and direct incorporation of chemicals in water bodies; (3) dumping of wastes, like agricultural wastes, in water bodies.

UAVs can be used for the surface management of water resources. Underwater monitoring is done by using underwater drones. Underwater vehicles were initially used for scuba diving, analyzing corals, and conservation of aquatic ecosystems, as shown in Figure 8.24.

Thus, drones can be used as a new tool for water monitoring. An observed using drone footage include (1) marine conditions, (2) marsh and sea grass, (3) spread of oil spills, and (4) monitoring the nature of algal blooms. In these cases, UAVs capture the images, which are then processed.

This allows the concerned person to know the parameters and conditions of the event. In events like oil spills, images from drones not only show the coverage of oil in the water but also reveal the volume of the oil spill in water bodies and its pattern of pollution and measure the concentration of dissolved oxygen, pH, and so on.

On the other hand, algal blooms are monitored, and the output from UAVs will demonstrate their behavior and their structure, pattern of growth, volume, density, and quality of water in that zone. Another interesting area for UAV application is

FIGURE 8.24   Detection of wastes in water bodies

waste management. As we all can see in our day-to-day life, wastes are dumped on open spaces.

This happens not only in rural areas but also in cities. Sensors are constructed to monitor waste disposal in smart cities. The outcome of such implementations was not as planned. Drones in waste management can be used for landfill monitoring, garbage collection, and detection of progress in waste treatment plants.

8.3.3.6   Infrastructure Monitoring

Infrastructure monitoring is the aspect of inspecting existing structures for their stability, design, and maintenance. Information gathered thus can be used for correct decision-making. It also helps its efficiency and rate of performance.

Structures like roads, high buildings, bridges, pavements, tunnels, towers, dams, power plants, and so on require proper monitoring and adequate care. It is hard for someone to personally inspect a dam and other superstructures, and human errors are possible.

Data acquisition is also not clear and lacks information, which can lead to the failure of the structure. In this regard, a UAV could be a valuable addition to this field. It can be used for data acquisition and inspecting other conditions of the building. It will be an important factor as it has a close impact on the city's overall economic and physical development. The application of drones is not limited to monitoring and maintenance alone, as shown in Figure 8.25.

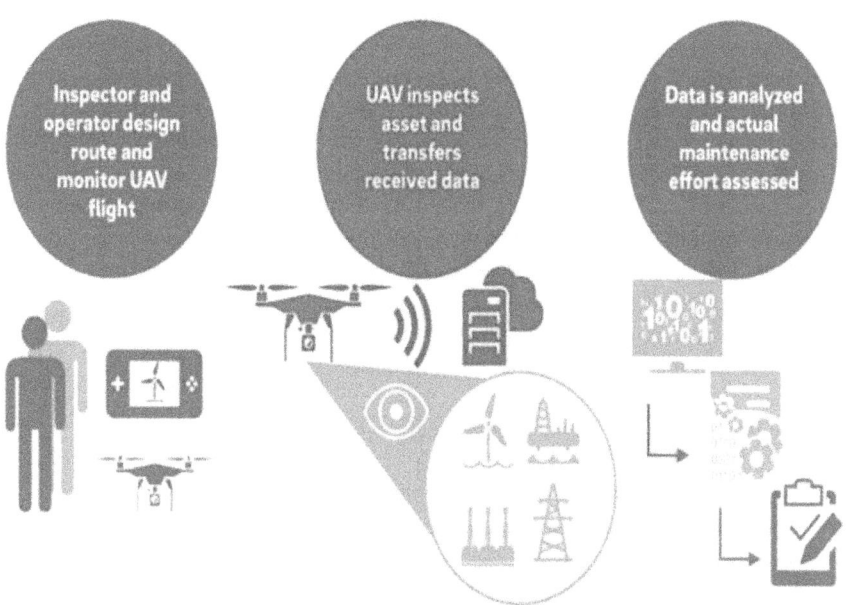

FIGURE 8.25   Drone-based asset inspection.

In addition, they can be used for designing, planning, and surveying a field, improving the building process, detecting faults in structures, and so on. Conventional methods used are time-consuming and are not cost-effective, while manual inspection cannot be undertaken in terrains and regions that can't be accessed.

The productivity and outcome of any structure can be increased by adequate sharing of its conditional factors.

To compare it with other inspecting modes, the usage of aerial drones increases the payload, collects and transmits data more accurately, and consumes less inspection time and cost while being flexible for any altitude.

Some applications of UAVs include (1) detection of cracks and deflection in bridges, (2) detection of cracks due to fatigue in steel towers, (3) inspecting the health of concrete structures in dams, and (4) construction management, which includes surveying fields autonomously and data for landscaping.

8.3.3.7 Mechanism behind Inspection

The inspection of structures can be done by three modules: module 1 is plotting the router, module 2 is the inspection phase, and module 3 involves image processing or analysis of collected data.

MODULE 1

The primary part of a drone's inspection is plotting the route and flight control. The analyzer or inspector programs the drone to fly in a specific path and motion. In this case, drones cannot be powered using a cable. When the mission is turned on, drones follow the designed path. Before taking off, the inspector attaches necessary equipment and detection sensors for efficient data acquisition. Usually, drones can fly for an hour.

In emergency situations, ground controllers can operate the drone on their own. Once it is airborne, the second module (the inspection and data collection phase) is carried out. Multiple drones can be used to inspect large structures. Data from each aerial unit is cumulatively processed to detect flaws and understand the condition of structures.

In general, any drone flight plan will ideally be uniform in every possible sense in most practical cases, which may include landmass to airspeed and the angle of the camera mounted over the drone. Improper flight planning may end up with less accurate figures in addition to encountering distortion of images captured over the drone camera.

The following components may be considered in a precise scale to obtain accurate and undistorted images in most possible cases.

- Height of flying from the target object
- Speed (ground)
- Type and specification of the sensor mounted over drones
- Image resolution (generally expressed in pixels)
- Flying time calculation
- Texture and nature of the land area to be targeted
- Data related to number and type of laps

It can be considered that flight planners should anticipate around 70% image overlap to prevent gaps and thus more amount of distortion for the images acquired from the flying procedure. But few practical cases may allow the flight planners to adjust the lap value of around ±5–10% by considering possible practical cases and related implications.

All these information should match flight notes collected in the metadata of images for the later processing stage. Document showing metadata should also include necessary data, like weather condition, altitude, extent of climatic parameters, wind speed, and so on.

MODULE 2

The most important part of monitoring is receiving proper information about the structure, as it can potentially increase the outcome of the analysis. The craft in the specified location captures all aspects of the structure. This helps for accurate monitoring, and faults, if any, can be identified easily.

Images are captured on all axis. When compared to conventional methods, drones have the upper hand for their sensing and high-quality image generation features. Real-time videos are an advantage for drones, which is not possible with conventional methods. The captured images are transmitted via network connectivity and stored using a storage unit in drones. Later, it can be transferred to the working system through card readers and USB.

Drone inspections can dramatically reduce the high costs, safety risks, and time involved with conventional inspection methods involving bridges, railways, water tanks, and industrial applications.

Another great advantage lies with the usage of drones without shutting any operations running in the building inspected. In the energy sector, drones may offer great potential, where shut-down operations may not at all necessary during inspections, thus making the entire inspection process easy, simple, and efficient.

Unlike the collection of a small amount of data, drones enable industries to collect huge volumes of data without any discontinuity of data. When pipelines are considered, any hazardous leakage issues can easily be identified with the help of drones embedded with gas-detecting sensors for long distances, thus eliminating the intervention of any human labor.

However, the following advantages can be exercised in lieu of potential advantages over conventional methods for a long period in the construction industry.

- Defined scheduled with timely reporting
- No hindrance in case of hilly areas and any hazardous places
- Manual site inspection can be avoided to a great extent
- Survey for large-scale areas without deploying any conventional and sophisticated machines
- Data-acquisition options in emergency/accident situations in most cases.

MODULE 3

At present, many innovative methods are available for data extraction. The extracted data from the craft is processed for further information about the structure. The captured images and real-time videos can be put together to give the ultimate output.

Images taken from each point of the structure undergo inspection using a tool. Thus, from the processed final data, an inspection report is readied, and detection of a failure in structures is undertaken.

Based on this data, preventive measures to ensure the sustainability of structures are carried out, as shown in Figure 8.26.

After the flight path and control operations are over, an image bundle adjustment resulting in estimated image positions and sparse point cloud reconstruction is performed using the software for post-processing.

Before the necessary cloud densification process, Visual SFM integrated manual tools were deployed to align the images. Pairs of images with weak linkages were detected, and additional feature detection and matching could be performed for the selected pairs using the forecast tools available in the software package.

The final point densification was performed using the CMVS/PMVS2 package or any other related packages available with the software module. In Visual SFM, both packages are used concurrently, although they can be divided into two parts. In the first category, data points were divided using cluster operations. Again the divided data points were then passed to the PMVS2 for point densification and finally stored in a *.PLY format, including other desirable software options.

Cloud Compare and Quantum GIS (QGIS) software packages, which can calculate DSM from dense point cloud datasets (*.PLY format), were used to generate 3D models in most cases. Georeferencing was performed with the point picking tool compared to the selection of GCP points available to the user, thus assigning their corresponding coordinates to the dense point cloud.

FIGURE 8.26 Approach of drones in a bridge inspection.

The resulting georeferenced dense point cloud was finally imported and interpolated in QGIS with an inverse distance weighted (IDW) algorithm to create the final georeferenced DSM. The linear regression model could be derived for the final step wherein the correlation between DSM and field survey were selected for random points. Due to the lack of an orthophoto export feature in Visual SFM, a combination of Cloud Compare and Microsoft Image Composite Editor (ICE) could be utilized.

Based on the parameters defined by the user, Cloud Compare provides an option for imagery corrections based on the corrections tools. In case Microsoft ICE auto-selects the rotating motion mode, this may be an indication of remaining oblique or blurred images in the data processing workflow.

In the set processed by Cloud Compare, final georeferencing of the orthomosaics was done with QGIS using the integrated georeferencer GDAL plugin and the corresponding coordinates of visible white GCPs.

8.4 CONCLUSION

In order to regulate the policies and security issues in a safer mode, a smart city encourages and facilitates the automation concepts that house the technology called ML. IoT handles a large volume of data, which may be extracted for useful insights about the phenomena and processes prevailing in the city environment (Khang, Bhambri et al., Big Data, Cloud Computing & IoT, 2022).

Authorities are looking for security policies that may protect the data of residents in the city and should be deployed for reasonable and worthy analysis. Traditional surveillance systems may lengthen the time process by means of low-level configurations and operations for tracking, recording, and doing analysis (Boccadoro et al., 2021).

Advanced video streaming by enabling scattered sensors may trigger a greater number of applications in this field. The integration of embedded systems along with the computer vision process enables the surveillance process to be easy and effective while maintaining accuracy (Chen, 2021).

Some of the review comments made in the present chapter encompass ML algorithms, smart city components, and drone surveillance systems. Such advancement in technological tools able to handle large volumes of data and their application to smart city projects ensure a comfortable and secure life for residents, which seems to be an interesting and promising technology for engineers, architects, and programmers working in the field and may open up more tracks to perform their research tasks in a fruitful sense (Indu and Singh, 2020).

REFERENCES

Boccadoro, P., Striccoli, D., & Grieco, L.A. (2021). An extensive survey on the internet of drones. *Ad Hoc Networks*, 122, 102600. doi:10.1016/j.adhoc.2021.102600

Chen, C.W. (2021). Drones as internet of video things front-end sensors: Challenges and opportunities. *Discover Internet of Things*, 1(1). doi:10.1007/s43926-021-00014-7

Indu, & Singh, R. (2020). Trajectory planning and optimization for UAV communication: A review. *Journal of Discrete Mathematical Sciences and Cryptography*, 23(2), 475–483. doi:10.1080/09720529.2020.1728901

Khang, A., Bhambri, P., Rani, S., & Kataria, A. (2022). *Big Data, Cloud Computing and Internet of Things.* CRC Press. ISBN: 978-1-032-284200, doi:10.1201/9781032284200

Khang, A., Chowdhury, S., & Sharma, S. (2022). *The Data-Driven Blockchain Ecosystem: Fundamentals, Applications and Emerging Technologies.* CRC Press. HB.ISBN: 978-1-032-21624-9 EB.ISBN: 098-1-003-26928-1 PB.ISBN: 098-1-032-21625-6, doi:10.1201/9781003269281

Khang, A., Rana, G., Khanh, H.H. (2021). The role of artificial intelligence in blockchain applications. In *Reinventing Manufacturing and Business Processes Through Artificial Intelligence* (p. 20). CRC Press. doi:10.1201/9781003145011

Rani, S., Kataria, A., & Chauhan, M. (2022). Fog computing in industry 4.0: Applications and challenges—A research roadmap. In *Energy Conservation Solutions for Fog-Edge Computing Paradigms* (pp. 173–190). Springer.

Rani, S., Kataria, A., Chauhan, M., Rattan, P., Kumar, R., & Sivaraman, A.K. (2022). Security and privacy challenges in the deployment of cyber-physical systems in smart city applications: State-of-art work. In *Materials Today:* doi:10.1016/j.matpr.2022.03.123

Rani, S., Kataria, A., Sharma, V., Ghosh, S., Karar, V., Lee, K., & Choi, C. (2021). Threats and corrective measures for IoT security with observance of cybercrime: A survey. In *Wireless Communications and Mobile Computing*, 2021. Hindawi.

Rani, S., Khang, A., Chauhan, M., & Kataria, A. (2021). IoT equipped intelligent distributed framework for smart healthcare systems. *Networking and Internet Architecture*, https://arxiv.org/abs/2110.04997v2, doi:10.48550/arXiv.2110.04997.

Tailor R K, Ranu Pareek, Khang, A. (Eds.). (2022). Robot process automation in blockchain. In *The Data-Driven Blockchain Ecosystem: Fundamentals, Applications, and Emerging Technologies* (1st ed., pp. 149–164). CRC Press. https://doi.org/10.1201/9781003269281

9 Smart Vehicle Parking Management with IoT Technology

S.K. Chaya Devi, Sai Krishna Reddy Koppula

CONTENTS

9.1 INTRODUCTION

The vehicle parking system with IoT (Internet of Things) technology is now a smart parking system. The increase in the growth of smart cities and usage of the IoT is becoming important for developing cities. IoT is a revolution in our daily world.

IoT is the interconnection of physical devices that can provide information about the surroundings with the help of sensors and computing devices.

In IoT, each thing can communicate with others through the Internet. With the advancement in electronics, developing automobiles is becoming cheaper. Though it is a good sign, it leads to several consequences like pollution and congestion in traffic (Bhambri, P. et al., 2022).

In today's growing metropolitan cities, there are many places like shopping malls, business meeting places, highly crowded tourist areas, and the like, where it is often difficult to find a free parking slot. So parking is becoming a critical issue for vehicle drivers for finding a free parking slot can be time-consuming (Rani, S. et al., 2022; Rani, S., Kataria, A., Chauhan, M. et al., 2022).

These difficulties are faced by every vehicle driver, which leads to traffic congestion. It has been observed that the traffic congestion on a worldwide scale is increasing at a very high rate every day, and 28–45% of this congestion is due to the unavailability or limited number of parking slots. This hassle is one of the high contributors to pollution and also leads to excessive use of oil.

DOI: 10.1201/9781003252542-9

Finding a slot for parking is one of the routines for people, but if it takes much time, then it becomes a frustrating task, and also a lot of effort is required from the administrative authorities to manage the parking facilities of the users (Rani, S., Kataria, A. et al., 2021).

The International Parking Institute has conducted a survey that concludes that the implementation of smart parking technologies can reduce these problems by at least 50%.

To resolve these issues, the solution for smart cities is a smart parking system that can guide vehicle drivers on free parking slots available at the complex where they are planning to visit and regulates the traffic (Rani, S., Khang, A. et al., 2021).

Through this chapter, we are proposing a smart and efficient parking system that is developed with the help of the IoT (Khang, A. and Khanh, H.H., AI & Blockchain, 2021). This proposed model will efficiently manage and improve the usability of the parking slots with as many fewer resources. This will enable the user to understand the situation at the parking location with high accuracy by providing all the information required (Rani, S., Mishra, R.K. et al., 2021).

The main advantages of the proposed model are as follows: It is a low-cost and highly efficient model in giving information about the available parking slots. It reduces the time that the user must spend to find the parking slot. It also has extra features, like calculating the fare for parking and collecting the same from the user online, and details about the number of users in the queue and other such data.

9.2   LITERATURE SURVEY

Geng, Y. and Cassandras, C.G. (2011) proposed a parking system that works on optimal resource allocation and reservations that uses a street line parking detection sensor and an LED device.

In this model, a street line gateway is used to receive data from various sensors present in the network. The gateway forwards this data to a database. This data is processed and updated to a website that is accessible publicly to users.

Cameras are used for capturing images, and some standard image-processing algorithms are used to increase the reliability of parking state estimates from those pictures. There exist multiple street line gateways in the entire system. Each street line gateway controls multiple parking slots in an area.

Benson, J. et al. (2006) proposed a car-parking management system that uses a network of wireless sensors: The application updates the information to the database when it receives the new reports from the sensors present in the network. This application can be developed using any Java-enabled web browser, or it can be a mobile application. This system gives us the feasibility to deal with several aspects of the system, like a new sensor can be added as a node, moving an existing sensor node to a different parking slot, etc.

Martens, K. et al. (2008) proposed a parking system that is both efficient and effective for parking your vehicles, which reduces traffic congestion. This is later developed into a smart parking system that can collect the bills automatically without the need for human interaction by using a multilayer parking method.

This system consists of four major components: (1) end node, which has the sensors used to collect data from the parking slots; (2) processing node, which is the most important block, which provides the processing capability to the circuit; (3) connectivity, which is dual in nature, allowing for both sending and receiving data simultaneously; (4) front end, or the user end. The user end is a mobile application. The disadvantage of this model is that the mobile application is limited to only one operating system, which makes it tough to access.

Basavaraju, S.R. (2015) designed a system that is applicable for covered or closed parking areas, open parking areas, and street-side parking all combined. This cloud-based architecture provides storage to store information on the status of parking slots. This system is a very simple and economically effective and efficient solution to reduce pollution in the atmosphere (Khang, A., Bhambri, P. et al., Big Data, Cloud Computing & IoT, 2022).

This model lacks some features like displaying the number of available parking slots. Figure 9.1 shows a view of filled parking slots in the mobile application or the web browser.

Also, it can be enhanced by send notifications to the user's smartphones when the vehicle enters a specific parking lot. Dharma Reddy, P. et al. (2013) proposed a system that features image/video processing through multiple classification algorithms. The webcam finds free parking slots.

With the help of an image- or video-processing unit, the user can know whether the parking slot is filled or not. The advantage of this device is that it is capable of showing parking slot is full only when a car is present in the allotted parking area, but not when there is any object other than a car in the parking slot area, which is not so easily possible with other sensors. It is developed by integrating all the features of both hardware and software. However, the problem with this system is that it needs a very large bandwidth.

FIGURE 9.1 A view of a filled parking slot.

Shinde, S. et al. (2016) proposed an application that uses four basic modules: LCD, GSM (Global System for Mobile), interfacing with RF module, and Android application.

Depending on the availability of parking slots, it manages the number. of vehicles that can be parked, and if the parking space is filled, vehicles are not allowed until any parking slot is available. All the commands are given using an Android application without human interference.

There are some limitations as a GSM is being used. It does not work when there are no signals and cannot be used in the models where we need immediate results.

Bi, Y. et al. (2006) suggested a parking management system based on a network of wireless sensors, web servers, and mobile applications. The current state of the parking slots is identified by using a sensor and is provided at regular intervals to the server.

The vehicle owner or the driver can detect the parking slots by accessing this information from the server through any device that has access to the Internet. The use of wireless sensor networks is not as powerful; there is a chance of failure in the devices.

From the literature survey, it can be understood that there are various ways available to implement smart parking systems, and each way has its advantages and disadvantages. It can be observed that most of the implementations lack to provide faster responses while some lack the ease of access for the user. The implementations that include mobile applications are limited to mobiles that use a particular operating system.

Our proposed model overcomes these problems by using a website instead of any mobile application, which makes it universally accessible from any device with any operating system. Also, since there is no use of any GSM, this implementation gives a very fast and accurate response to the real-time data.

9.3 DESIGN AND IMPLEMENTATION

9.3.1 COMPONENTS

The following are the components required for the implementation of the proposed system:

1. Microcontroller
2. IR sensors
3. DC motor
4. LCD
5. RFID reader
6. RFID tags
7. Wi-Fi module
8. Server
9. Machine learning

9.3.1.1 Microcontroller

The W78E052D microcontroller is used for implementing smart parking. The W78E052D is a microcontroller based on 8051 with dual clock speeds. The

W78E052D has 256 bytes of RAM, 8 kilobytes of ISP Flash, 1 power-management unit, up to 36 I/O pins, and 2 kilobytes of LDROM (Bootloader). The most important advantage of the W78E052D microcontroller is that it comes with a UART bootloader, which allows the user to directly program from the serial port.

9.3.1.2 Infrared Sensor

IR (infrared) sensors are used to detect the current state of the parking slots. An IR sensor is a device that detects any obstacle in its way with the help of infrared radiation.

There are two types of IR sensors available: passive IR sensors and active IR sensors. Passive IR sensors are used when there is no need for emitting but only detecting infrared radiation. When there is a need for both emitting and detecting infrared radiation, they are used as proximity sensors. Active IR Sensors are used in this model.

9.3.1.3 A Direct Current Motor

A direct current (DC) motor is used to convert direct electrical energy into mechanical energy. DC motors are created with an internal mechanism, which produces either an electromechanical or electronic signal to periodically change the direction of the current.

It is a 360-degree rotatable motor and rotates both in the clockwise and anticlockwise direction. It serves as a gate at the entrance or exit. When the parking slots are available or free for parking, only then is the gate opened, and the vehicle is let in.

9.3.1.4 Liquid Crystal Display

Liquid crystal display (LCD 16 × 2) is used for displaying the needed information for the user near the entry of the parking lot. This will be used when something has to be informed to the user, like whether the slots are full when they are trying to enter the parking lot. This helps the user to understand the reason for not being able to enter the parking lot.

9.3.1.5 Radio-Frequency Identification

Radio-frequency identification (RFID) technology is used to identify different users differently. RFID uses an electromagnetic field for identifying and tracking RF tags automatically.

An RFID system consists of a radio transmitter, radio receiver, and radio transponder. When any electromagnetic interrogation pulse triggers the tags from a nearby RFID reader, the cards start to transmit digital data, which is mostly an identifying number for the tag.

9.3.1.6 Radio-Frequency ID Tags

There are two different types of RFID tags: Passive tags are powered by the electromagnetic interrogation pulse from the RFID reader. Active tags contain an internal battery to power them. This internal battery helps the tag to stay active for a longer time, thus making them readable from a greater range.

9.3.1.7 Wi-Fi Module

The ESP8266 Wi-Fi module is used in the implementation of this model, which is an efficient and low-cost Wi-Fi microchip. The ESP8266 Wi-Fi Module comes with an SoC (system on a chip) that integrates with a stack of TCP/IP protocol to give Wi-Fi network accessing capability to any microcontroller.

The ESP8266 module comes with a pre-programmed set of commands called an "AT command set" firmware, with the help of which you can simply connect with any microcontroller and access the Internet.

9.3.1.8 Server System

The Ubidots server is used to support the prototype. Ubidots is an IoT platform that supports the development of prototypes very easily and improves them to the level of production.

The Ubidots server can be used as a web server that receives data from any device that can connect to the Internet. It also lets the user set up actions to be performed based on your real-time data.

9.3.2 Architecture and Implementation

The block diagram of the proposed model is shown in Figure 9.2. The central device of the model is the microcontroller. The IR sensors, Wi-Fi module, LCD, and DC motor are interfaced to the microcontroller. The IR sensors feed the data about the status of the parking slots to the microcontroller by detecting the proximity.

The microcontroller transmits the same to the server over Wi-Fi, which in turn is obtained by the client from the server.

The user who wants to park the vehicle logs in to the website, where he can see the status of the parking slots.

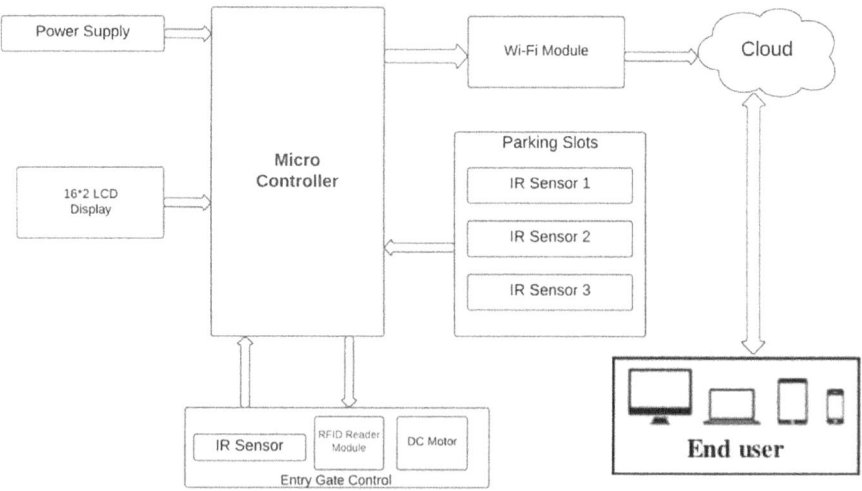

FIGURE 9.2 Block diagram of the proposed smart parking system.

The flowchart explains the different cases or situations that arise while executing the smart parking system and what actions are to be performed in such situations. The same implementation can be used to integrate multiple parking areas of a town or a city by connecting all of them to the same server.

Though they are present at different locations, they can be integrated to be part of the same server. To facilitate the user to easily understand the location of the available parking slots, they can be divided into clusters.

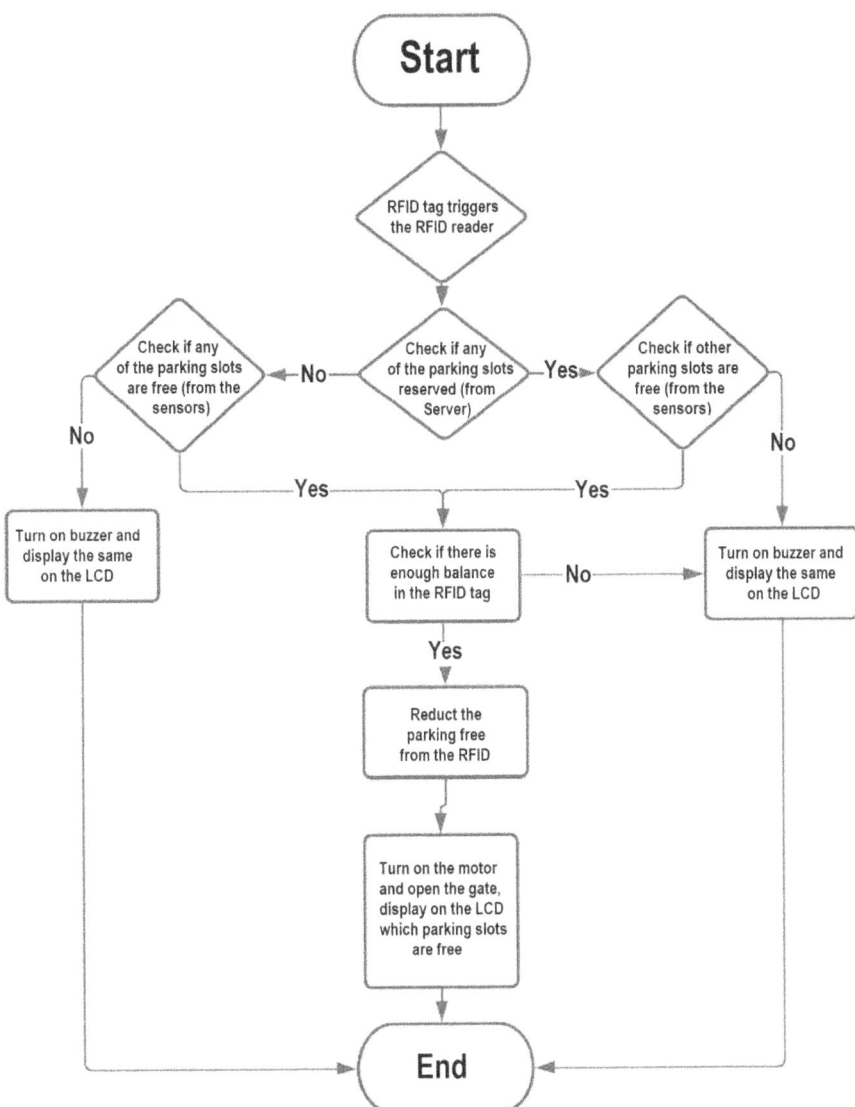

FIGURE 9.3   Workflow of the proposed method.

9.3.3 Software

The proposed software is very user-friendly, and a user with no technical knowledge can also use it on any of their devices with an Internet connection. Through this utility, the vehicle owner or the driver can search for available parking slots from anywhere. This smart parking system saves time and fuel and also reduces pollution and congestion of traffic in towns and cities.

The working model can be seen in the Figures 9.4, 9.5, and 9.6; Figure 9.4 is a picture of the login page of the Ubidots website, which is where our IoT devices can

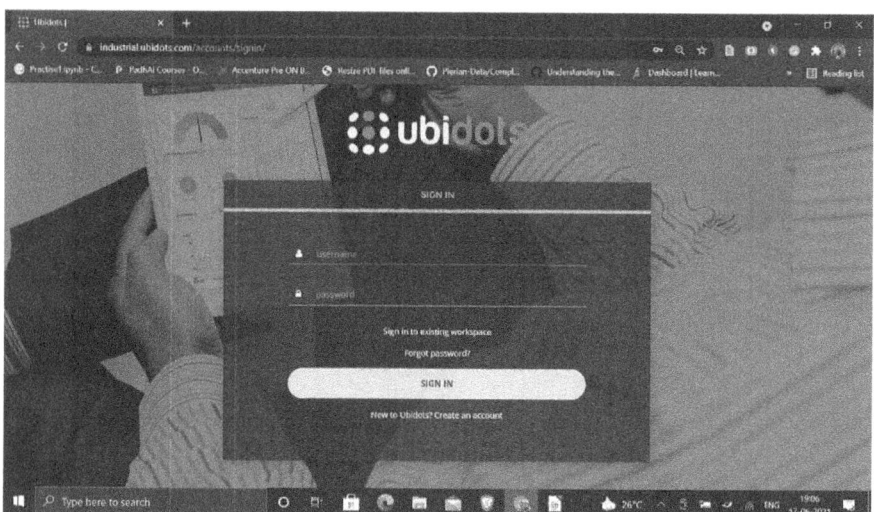

FIGURE 9.4   Ubidots login page.

FIGURE 9.5   Ubidots output showing all the slots are empty.

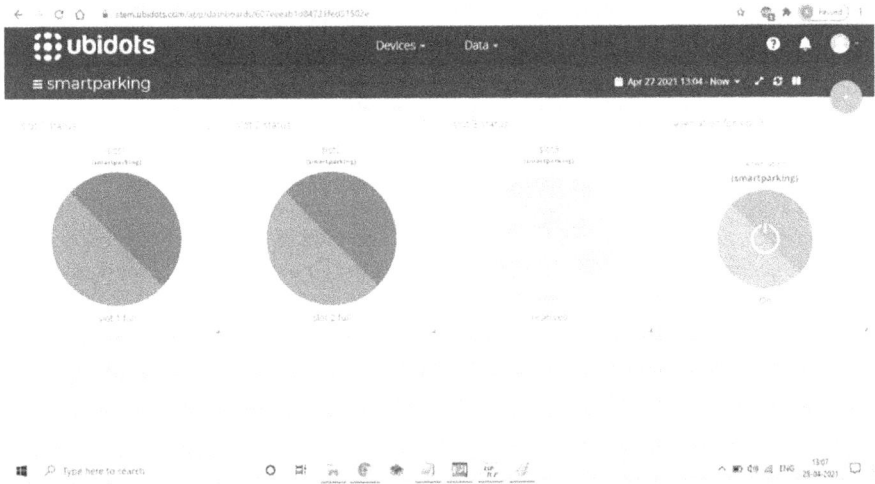

FIGURE 9.6 Demo output of Ubidots.

be monitored. You can then configure actions and alerts based on your real-time data and unlock the value of your data through visual tools.

Ubidots offers a REST API that allows you to read and write data to the resources available: data sources, variables, values, events, and insights. The API supports both HTTP and HTTPS, and an API Key is required. This provides a very user-friendly website—that is, even a non-technical character can use it via mobile device. Through this utility, the consumer can search for an unfastened parking slot from everywhere in the world.

The information collected by the microcontroller is posted to the Ubidots server, and the data is processed to be produced as the following output.

9.4 CONCLUSION AND FUTURE SCOPE

This project is a low-cost IoT-based smart parking system that deals with the issue of finding an empty parking slot, which may not be very easy for the driver (Rana, G. et al., AI & Manufacturing, 2021). The user can have knowledge of the availability of the parking slots not only when he is physically present at the parking lot but also when he is not around. The user can access the website available to check the status of the parking slots and find empty ones.

The main advantage of the system is that not only will the user know whether there are empty slots, but also the user can reserve a slot available if he is registered for it. This will lessen the work of the users of vehicles and the money can be collected online, reducing human effort. This will reduce the burden on both the user and the organizer of the parking lot.

This can be further improved by developing it for large-scale parking lots, and all those slots can be made reservable, which will increase the chance of reservation. Also, the system can be further developed to give options, like recharging online and making reservations for a limited time; for example, if the vehicle does not show up in the given time, the reservation is canceled. This way, all the slots will be used effectively, and there will be no wasting of resources.

The facility of recharging the parking card online will make it easy for the users to deal with the transactions in less time.

The same implementation can be used to integrate multiple parking areas of a town or a city by connecting all of them to the same server. Though they are present at different locations, they can be integrated to be part of the same server since the user can access the server just the same way as previously where the parking slots are mentioned to be belonging to different locations.

To make it easier for the user to understand the location of the parking slots, they can be divided into clusters. Each cluster belongs to one place, or multiple places can also be lumped into a single cluster. Each of these clusters can be shown through the folder system present in our systems.

REFERENCES

Basavaraju, S.R. 2015. Automatic smart parking system using the Internet of Things (IoT). *Int. J. Sci. Res.* Vol. 5, no. 12, pp. 2250–3153. doi:10.21786/bbrc/13.13/15

Benson, J., Donovan, T., Sullivan, P., Roedig, U. and Sreenan, C. 2006. Car-park management using wireless sensor networks. In *Proc. 31st IEEE Conf. Local Comput. Netw.*, pp. 588–595. doi:10.1109/LCN.2006.322020

Bhambri, P., Rani, S., Gupta, G. and Khang, A. 2022. *Cloud and Fog Computing Platforms for Internet of Things.* CRC Press. ISBN: 978-1-032-101507. doi:10.1201/9781032101507

Bi, Y., Sun, L., Zhu, H., Yan, T. and Luo, Z. 2006. A parking management system based on wireless sensor network. *Acta Automat. Sin.*, Vol. 32, no. 6, pp. 38–45. https://www.semantic scholar.org/paper/A-Parking-Management-System-Based-on-Wireless-Bi-Yan-zhong/d6492e38778208076ee3437af61d98634198a38f

Dharma Reddy, P., Rajeshwar Rao, A. and Ahmed, S.M. 2013. an intelligent parking guidance and information system through the use of image processing method. *IJARCCE*, Vol. 2, no. 10. https://www.ijarcce.com/upload/2013/october/59-O-dharmareddy_-AN_INTELLIGENT.pdf

Geng, Y. and Cassandras, C.G. 2011. A new "smart parking" system based on optimal resource allocation and reservations. In *Proc. IEEE Conf. Intell. Transp. Syst.*, pp. 979–984. IEEE. doi:10.1109/TITS.2013.2252428.

Khang, A. and Khanh, H.H. 2021. The role of artificial intelligence in blockchain applications. In *Reinventing Manufacturing and Business Processes Through Artificial Intelligence*, pp. 19–38. CRC Press. https://doi.org/10.1201/9781003145011

Khang, A., Bhambri, P., Rani, S. and Kataria, A. 2022. *Big Data, Cloud Computing and Internet of Things.* CRC Press. ISBN: 978-1-032-284200, doi:10.1201/9781032284200

Khang, A., Chowdhury, S. and Sharma, S. 2022. *The Data-Driven Blockchain Ecosystem: Fundamentals, Applications and Emerging Technologies.* CRC Press. ISBN: 978-1-032-21624, doi:10.1201/9781003269281

Martens, K., Birr, S. and Benenson. 2008. Parkagent: An agent-based model of parking in the city. *Comput. Environ. Urban Syst.* Vol. 32, no. 6, pp. 431–439, November.

Rana, G., Khang, A., Sharma, R., Goel, A.K. and Dubey, A.K. 2021. The role of artificial intelligence in blockchain applications. In *Reinventing Manufacturing and Business Processes Through Artificial Intelligence*. CRC Press. doi:10.1201/9781003145011.

Rani, S., Kataria, A. and Chauhan, M. 2022. Fog computing in industry 4.0: Applications and challenges—A research roadmap. In *Energy Conservation Solutions for Fog-Edge Computing Paradigms*. Springer, pp. 173–190.

Rani, S., Kataria, A., Chauhan, M., Rattan, P., Kumar, R. and Sivaraman, A. K. 2022. Security and privacy challenges in the deployment of cyber-physical systems in smart city applications: State-of-art work. In *Materials Today: Proceedings*. doi:10.1016/j.matpr.2022.03.123

Rani, S., Kataria, A., Sharma, V., Ghosh, S., Karar, V., Lee, K. et al. 2021. Threats and corrective measures for IoT security with observance of cybercrime: A survey. *Wireless Commun. Mobile Comput*. Vol. 2021., pp. 1–30. https://www.researchgate.net/publication/344757378_Threats_and_Corrective_Measures_for_IoT_Security_with_Observance_to_Cybercrime

Rani, S., Khang, A., Chauhan, M. and Kataria, A. 2021. IoT equipped intelligent distributed framework for smart healthcare systems. *Networking and Internet Architecture*, https://arxiv.org/abs/2110.04997v2, doi:10.48550/arXiv.2110.04997.

Rani, S., Mishra, R.K., Usman, M., Kataria, A., Kumar, P., Bhambri, P. et al. 2021. Amalgamation of advanced technologies for sustainable development of smart city environment: A review. *IEEE Access*, Vol. 9, pp. 150060–150087.

Shinde, S., Bhagwat, S., Pharate, S. and Paymode, V. 2016. *Prediction of Parking Availability in Car Parks Using Sensors and IoT: SPS*. ISSN 2321 3361 © 2016 IJESC. doi:10.4010/2016.1083

10 The IoT-Based Estimation of a Human Heart Condition for the Smart Healthcare System

Vikas Verma, Abhishek Vishnoi

CONTENTS

10.1 INTRODUCTION

There are a lot of developments being observed, creating a new era of smart environments with smart databases globally. To achieve this goal, the Internet of Things (IoT) contributes significantly, as it enables the exchange of information and assets through mobile or remotely things through the wireless communication system. It includes a number of specialized sensors and various computing devices. It provides a system that has a two-way suggestive integrated platform for sharing the desired data and feedback. It is a combination of hardware and software that offers storing, processing, and communicating. All communication is made with an electronic system between two users (Vishnoi et al., 2021).

DOI: 10.1201/9781003252542-10

IoT has a significant role in making the healthcare sector a smart healthcare system. With the involvement of sensors, transducers, many computers, and peripheral devices, healthcare monitoring is remotely feasible. One of the crucial sections of healthcare is the care of the heart (Verma et al., 2019).

For a healthy person, the heart is one of the most important organs, and it has to work effectively. So for its effective functioning, regular monitoring is essential, but due to busy schedule, it gets affected and neglected. This can be overcome by an IoT-based heart-monitoring system (Khang, Bhambri et al., Big Data, Cloud Computing & IoT, 2022).

This chapter aims to propose an IoT-based heart-monitoring system to check the state of the heart, adding some value to a smart healthcare system.

10.2 SMART HEALTHCARE SYSTEM

10.2.1 Smart Healthcare System Concept

The idea and concept of smart healthcare was initiated by IBM's "smart planet" in 2009. This includes an infrastructure that employs the sensors for acquiring the information, transmitting the acquired information with the help of IoT, and the processing of this information is done with cloud computing and supercomputers (Bhambri et al., Cloud & IoT, 2022).

The thought process is to integrate the social system in an organized way and coordinate their whole dynamics. This was a paradigm shift from the traditional human society management system. The smart healthcare system is the referential base of modern healthcare service management, involving sensing and acquiring devices, mobile Internet, wireless devices, and connectivity between users, materials, and healthcare institutions.

The key challenge is to coordinate and manage it in an excellent manner. This system is developed to facilitate health services more effectively and to make everyone reachable in society.

10.2.2 Why We Need Smart Healthcare System

The purpose of developing a smart healthcare system is to promote a time-driven, effective treatment to the needy person, which will save their life. This system has a wide coverage as it includes multiple users like patients, doctors, government and private hospitals, medical research institutes, drug institutes, and pharmaceutical companies (Rani et al., 2022).

The main agenda of this system is multidimensional, consisting of the patient's health monitoring, treatment, and diagnosis, decision-making from a medical specialist, and so on. This can all be done remotely with IT technologies, such as cloud computing, blockchain, artificial intelligence, IoT, telemetry, telemedicine, and so on (Rani et al., 2021).

With the advent of this, a patient can consult with any specialist instantly and remotely. Even the doctors can manage all the data of their respective patients, like medical reports, pathology reports, and daily status reports, in the system database. The communication in this system is two-way, electronic, and effective. This integrated management system is essential to diagnosis and decision-making.

10.3   IOT PLATFORM IN A HEALTHCARE SYSTEM

This is an open platform that enables the patient and doctors to meet at a common platform and communicate with each other. The user has various options to access in an effective manner.

The smart healthcare platform is a smart platform because of its reach and accessibility. This system becomes smart because of the involvement of IoT. It is an application platform where the patient, after logging in, can seek any information from various departments. They can even upload their various health concerns, conditions, parameters, and reports for consultation and book their appointment accordingly. It makes the services more flexible and also provides them with choice-based services (Khang, Rani et al., IoT & Healthcare, 2021).

The patient can schedule and book their consultant doctor according to their requirements. Even now, with more technological advancements, they can meet with specialists virtually through video meetings. Then the doctors can receive their information and can provide their feedback and suggestions.

Many healthcare companies working in the healthcare sector have developed their application software and provided a smart health system infrastructural platform, which includes all multiple options and facilities for both patients and doctors.

Besides this, it can record of a large amount of data. It also has a library of all kinds of data related to the symptoms of diseases. It also has an uploading and downloading facility on both sides to communicate findings and reports. It is attached to various hospitals, helping people in case of emergency by connecting them to the nearest hospital and ambulance facility. This application platform is accessible on both laptops and smartphones.

In this sector, much research is also conducted by various healthcare companies to make it more popular and accessible to all. With the involvement of new sensors, VLSI circuits and different ICs improve its hardware structure also by providing compact healthcare monitoring devices that are easy to operate.

It is a new revolution in the field of smart systems. It connects assets, devices, mobile phones, or remote objects with communication technologies, such as computing devices, wireless sensors, low-cost sensors, and many storages and peripheral devices. It can be considered as an advanced version of computers and the web network.

The objective of the IoT is to connect all devices and systems (Khang, Rani et al., IoT & Healthcare, 2021) existing in an environment to make a coordinated and controlled interaction with each other, as shown in Figure 10.1.

The increasing availability of sensors, smartphones, cloud computing, nanoelectronic devices, nano film chips, Wi-Fi routers, and embedded systems has enabled the development of a smart system in every sector.

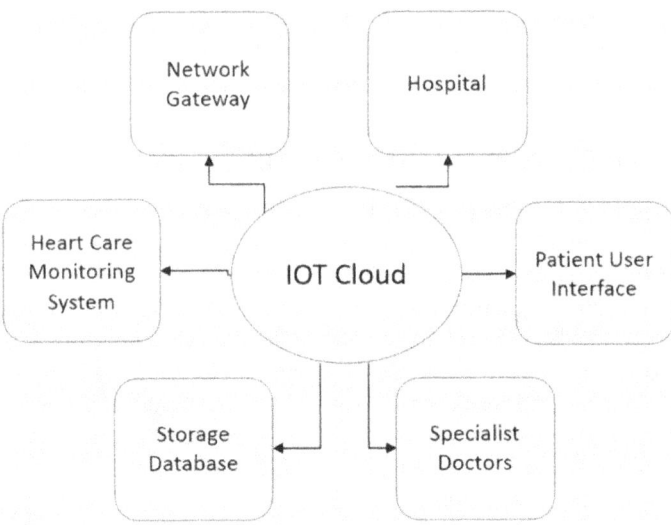

FIGURE 10.1   IoT-based Smart healthcare system.

The main areas are the following:

- **Green and clean environment:** It includes monitoring and promoting the activities related to the development of natural resources. It also helps to monitor and create protection from pollution and natural calamities, such as floods and landslides.
- **Industrial sectors:** It enables proper managerial control of the processes and commercial transactions between two other companies, like PV installation, water flow, power flow, and smart grid.
- **Society services:** For the development of an efficient human society, there must be certain initiatives that must be taken for their betterment. The major concern is health. The developments in the smart healthcare system, traffic system, and waste management system are certain examples of this.

10.4   IOT IN A SMART HEALTHCARE SYSTEM

Nowadays, every country is facing social and economic challenges. Healthcare alone represents the social aspect of society. Even the financial burden of the country is also affected by its health management system.

The advancements in information technology, wireless networks, and telecommunication systems play a significant role in the medical management system. There is a lot of pressure on healthcare services as the population is increasing day by day, and the number of unhealthy people is also growing.

The cost of the treatment of chronic diseases is already high, and getting treatment from a specialist is also difficult, so by having an integrated health management system, these shortcomings will be easily eradicated.

An IoT-based healthcare system requires an integrated, efficient, and secure health management system by using information and communication technology.

Real-time data are acquired with the help of wireless sensors and transferred to PCs and smartphones for quick access. This is beneficial to a patient who prefers to be monitored at home rather than in the hospital. This also enables us to use a variety of care devices, such as oximeters, weighing scales, and pulse meters.

The main problem in home-based monitoring is making a system through which patients can enter their accurate data to be delivered to specific healthcare workers. There is a large number of software companies developing software databases that will integrate with the source end and the receiver end.

This new care system will provide extensive support to senior citizens suffering from serious problems and not in the condition to move for their regular health checkups. IoT is a boon to the medical sector in developing an efficient society.

10.5 HUMAN HEART MONITORING SYSTEM

Heart problems are among the major health issues. On average, six out of ten people are suffering from heart problems. Even in recent years, the mortality rate increased due to unhealthy hearts. This is all due to the delay in the treatment and lack of awareness.

People are unaware of emerging heart problems in their bodies until severe symptoms appear. There are also some cases in which due to certain circumstances, they cannot follow up on their scheduled check-up with doctors. This is especially the case of older adults who are dependent on someone, as shown in Figure 10.2.

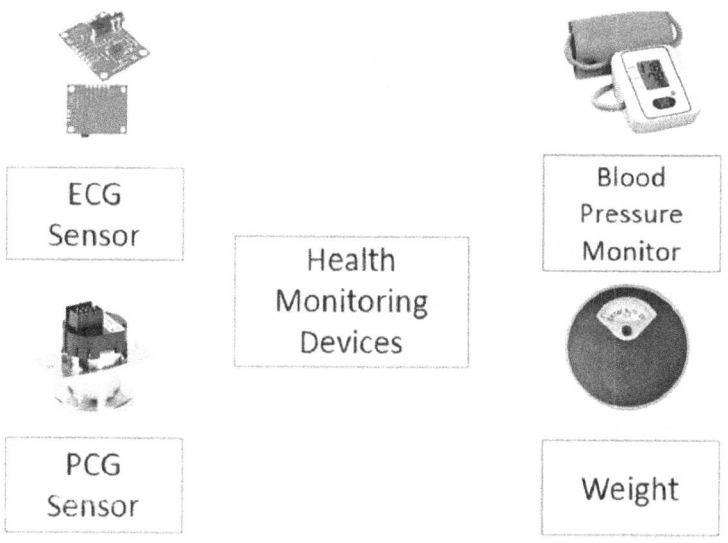

FIGURE 10.2   Heart-monitoring devices.

FIGURE 10.3   Different stages of a heart condition assessment system.

So this key issue can be resolved atin the primary level by using IoT technologies, which are the most common prevailing around us. Major health problems nowadays begin in the earlier age group (30–40), which is getting service in further stages.

The main reason behind it is improper daily habits, which include irregular food intake, lack of physical activity, high-pressure lifestyle, stress, and unhappiness.

These are the factors that affect heart activity in a very gradual manner, such as a lack of any portable monitoring device to check the state of the heart. With the involvement of IoT, heart care can be done separately in a smart healthcare system, as shown in Figure 10.3.

The whole system of IoT-based healthcare can be divided into different sections for a better view and integrated as a complete system.

10.5.1   ACQUIRING INFORMATION

The use of low-cost wireless sensors that are specifically used for the heart can be considered for gathering ofinformation. ECG and PCG sensors are used to monitor the electrical signals and sounds of the heart.

At the initial level, both signals are sufficient for determining the heart condition. These sensors are user-friendly and accurate it can easily be used by any person to check their ECG and PCG signals. Second, other parameters, like blood pressure and weight, are also important factors that can be measured by the user.

10.5.2   DATA FEEDING/STORAGE

These wireless IoT sensors are available with the software database. So the software related to it is installed on their respective computing devices.

In their software, the parametric values sensed by the sensors are recorded and saved in the respective software in various file formats. These can be converted to any file type.

10.5.3   DATA TRANSMISSION

Once the data is recorded, these data files can be shared to a heart specialist through email.

The person on the other side can download it, access it, and analyze it with the database, which includes a large amount of data on health and illness, so it can compare the obtained values. It can closely monitor into deep levels.

10.5.4 STATUS VALIDATION

After obtaining the data successfully, it is compared with the database. The doctor will validate the report and give the proper diagnosis and medicines to the patient. Even the doctor can also video-conference with the concerned patient. So at the primary level, the doctor can give consultation in a definite time frame.

10.6 HUMAN HEART ANALYSIS WITH IOT-BASED ANN SYSTEM

Many people are suffering from heart problems. Before, the numbers were low, and if a person had a heart problem, they were less likely to recover because of a lack of technology and advanced services in the healthcare sector. Today, the rate is same but can get worse because of a stressful lifestyle and bad habits (Tailor et al., 2022).

Resources were limited, but now, innovation in the health sector makes it possible to eradicate conventional problems. With advanced and smart devices, users can monitor and control their heart conditions on their own. These devices are the smart ECG sensor, smart PCG sensor, smart stethoscope, BMI device, smart blood pressure device, and so on.

A healthy or unhealthy human heart can be monitored and analyzed by its basic values (ECG, PCG, blood pressure, weight, etc.). These four are the key factors for the working of a healthy heart. These are vital signs at the initial level of diagnosis. If these levels are normal, then complications can be prevented.

In our work, these parameters are measured with the help of available smart heart-monitoring devices, such as smart stethoscopes and smart sensors, which are easily available on the market.

The measured readings are taken, and the normalization is done with the average ones. After the normalization process by using artificial intelligence techniques, these readings are optimized and provide an optimized system that will declare the status of the heart (Hasan et al., 2022).

In the whole process of training the ANN network (Rana et al., AI & Blockchain, 2021) with the proposed IoT system, as shown in Figure 10.4, data has been measured

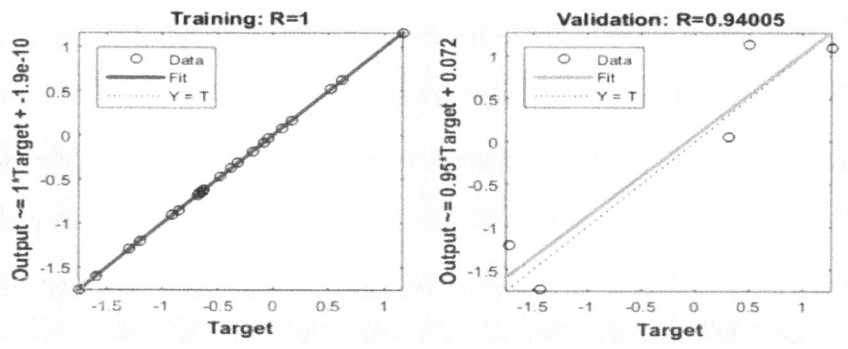

FIGURE 10.4 Training and the validation curve of the proposed IoT system.

TABLE 10.1

Dataset of Persons

S. No.	1	2	3	4	5	6	7	8
Patient ID	64	65	66	67	70	68	69	73
Age (Year)	0.97	1.09	1.46	0.58	1.04	0.78	0.87	0.53
Height (m)	1.03	0.99	0.98	0.92	0.98	0.91	1.01	1.05
Weight (kg)	0.66	0.95	0.88	0.79	0.98	0.88	1.19	1.17
BMI (kg/m2)	0.62	0.96	0.92	0.92	1.01	1.06	1.16	1.06
Heart Rate (BPM)	1.06	1.04	0.95	0.98	1.01	0.87	0.98	1.08
BP (diastolic)	1.07	0.9	0.97	0.83	1	1.19	0.92	1.19
BP (systolic)	0.96	0.98	0.86	0.9	0.96	1.06	1.04	1.06
QT Interval (ms)	0.94	0.95	1.12	0.87	0.97	1.07	1.01	0.92

TABLE 10.2

Demo of Obtained Validated Output

S. No.	1	2	3	4	5	6	7	8
Patient ID	64	65	66	67	70	68	69	73
Validated Value	0.90	0.90	1.01	0.94	0.91	1.27	0.86	1.20
Measured Value	0.90	0.88	1.01	0.93	0.91	1.27	0.85	1.20

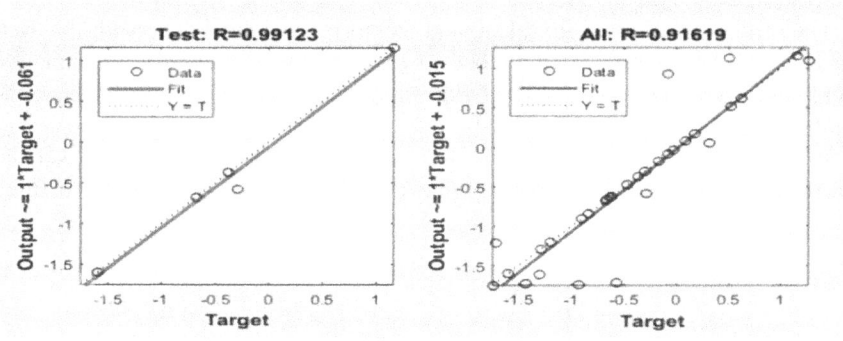

FIGURE 10.5   Status of healthy and unhealthy data curve of the proposed IoT system.

with the help of smart monitoring devices. Table 10.1 shows the datasets at different parameters of some persons to train the neural network.

Table 10.1 presents the dataset of persons used in the training and validation process for proposed IoT system.

The validation results are shown in Table 10.2. It reveals that the proposed IoT system in Figure 10.5 is perfectly trained to give the correct values of the output information.

Table 10.2 presents the obtained validated output from the proposed IoT system.

10.7 CONCLUSION

The IoT-based smart healthcare system has provided a platform for better and effective living. Through this medium, various services are being implemented in healthcare services.

Heart conditions are well monitored by specific monitoring devices that not only state the condition but also generate proper solutions. This two-way communication between the patient and the doctor is made possible by IoT.

This approach has many benefits; for example, it saves time, provides correct treatment in a particular time frame, reduces the need to rush to hospitals, and reduces extra treatment costs. It advances the healthcare system, and it will save many lives.

REFERENCES

Bhambri, P., Sita Rani, Gaurav Gupta, Khang, A., 2022. *Cloud and Fog Computing Platforms for Internet of Things,* CRC Press. ISBN: 978-1-032-101507, doi:10.1201/9781032101507

Hasan, S., T. B. Sivakumar, Khang, A., 2022. Cryptocurrency Methodologies and Techniques. In *The Data-Driven Blockchain Ecosystem: Fundamentals, Applications, and Emerging Technologies* (1st ed.), CRC Press, pp. 27–37. https://doi.org/10.1201/9781003269281

Khang, A., Pankaj Bhambri, Sita Rani, Aman Kataria. 2022. *Big Data, Cloud Computing and Internet of Things,* CRC Press. ISBN: 978-1-032-284200, doi:10.1201/9781032284200

Khang, A., Subrata Chowdhury, Seema Sharma. 2022. *The Data-Driven Blockchain Ecosystem: Fundamentals, Applications and Emerging Technologies,* CRC Press. ISBN: 978-1-032-21624, doi:10.1201/9781003269281

Khang, A., Sita Rani, Meetali Chauhan, Aman Kataria. 2021. IoT equipped intelligent distributed framework for smart healthcare systems. *Networking and Internet Architecture,* https://arxiv.org/abs/2110.04997v2, doi:10.48550/arXiv.2110.04997.

Rana, G., Khang, A., Khanh, H.H. 2021. The role of artificial intelligence in blockchain applications. In *Reinventing Manufacturing and Business Processes Through Artificial Intelligence,* CRC Press, p. 20. doi:10.1201/9781003145011

Rani, S., A. Kataria, M. Chauhan, P. Rattan, R. Kumar and A.K. Sivaraman. 2022. Security and privacy challenges in the deployment of cyber-physical systems in smart city applications: State-of-art work. In *Materials Today: Proceedings.* doi:10.1016/j.matpr.2022.03.123

Rani, S., A. Kataria, V. Sharma, S. Ghosh, V. Karar, K. Lee and C. Choi. 2021. Threats and corrective measures for IoT security with observance of cybercrime: A survey. *Wireless Communications and Mobile Computing,* Vol. 2021, pp. 1–30. doi:arXiv:2010.08793

Tailor R.K, Ranu Pareek, Khang, A., (Eds.). 2022. Robot process automation in blockchain. In *The Data-Driven Blockchain Ecosystem: Fundamentals, Applications, and Emerging Technologies* (1st ed.), CRC Press, pp. 149–164. https://doi.org/10.1201/9781003269281

Verma, Vikas, Bharti Dwivedi, Satyendra Singh. 2019. A novel approach to study electrical, mechanical and hydraulic activities of heart and their coordination based on ECG and PCG. *International Journal of Applied Engineering and Research* (IJAER 2019 edition), Vol. 14, No. 2 (Special Issue), pp. 227–231. http://www.ripublication.com

Vishnoi, Abhishek, Vikas Verma. 2021. ANN based technique for assessment of wellness of human heart. *International Journal of Future Generation and Networking.* Vol. 14, No. 1 https://sersc.org/journals/index.php/IJFGCN/article/view/37497

11 Application of Robotic Process Automation in the Loan-Sanctioning Process for a Smart Bank Office

R.K. Tailor, Nidhi Sharma

CONTENTS

DOI: 10.1201/9781003252542-11

11.1 INTRODUCTION

The term *RPA* means "robotic process automation." RPA is evolving as one of the most innovative tools in the present era. RPA is basically a software technology, or we can say robots, which is useful to automate highly repetitive and routine tasks normally performed by a knowledgeable person. Sometimes it is referred to as software robotics. RPA is based on AI (Artificial Intelligence) or machine learning (Rana et al., AI & Blockchain, 2021).

RPA is showing tremendous growth since 2016. It plays a vital role for businesses to automate their process, which in turn increases their profits.

In this competitive era, RPA is beneficial for businesses to run their business very efficiently. It helps to integrate human action on digital platforms to automate different processes. RPA uses the user interface to capture data and manipulate applications just like humans do (Bhambri et al., Cloud & IoT, 2022).

11.2 OVERVIEW OF ROBOTIC PROCESS AUTOMATION

RPA enables organizations to achieve their capabilities effectively and accurately as it enables employees to concentrate on creative and skillful tasks rather than repetitive tasks. The RPA technique is capable of doing most tasks, like manual tasks, data extraction, invoice processing, portal queries and prices, maintaining customer data, validating files, and many other tasks, but not all that a human can do.

With the help of RPA technology, employees can use their time on other highly important and discerning tasks. The implementation of RPA helps to make the business processes more effective, error-free, efficient, and with reduced operating costs. It is the most relevant technology for industries, banking and financial services, insurance sector, healthcare sector, manufacturing sector, technology and telecom sector, and so on (Rani et al., 2022; Rani, Katraia et al., 2021).

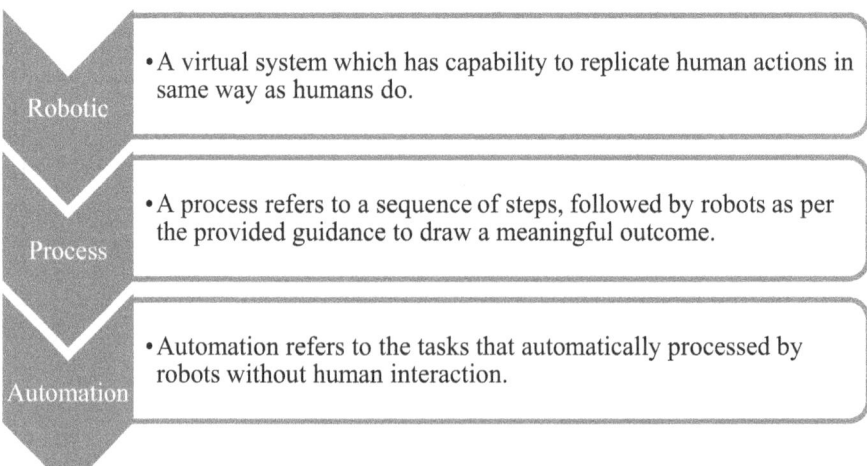

Robotic
• A virtual system which has capability to replicate human actions in same way as humans do.

Process
• A process refers to a sequence of steps, followed by robots as per the provided guidance to draw a meaningful outcome.

Automation
• Automation refers to the tasks that automatically processed by robots without human interaction.

FIGURE 11.1 RPA's definition.

In industries, RPA is useful to maintain their accounts, procurement, human resource management, customer service, and many other facilities. In the banking sector, it is useful for banks for their operating activities as well as their financial and customer-based services.

The adoption of the RPA technique is useful for those companies that are highly labor-intensive and performs high-volume transactions.

11.2.1 BACKGROUND

RPA is a group of several technologies that comes under one tool for different auto-mation purposes. Although the term RPA emerged in the early 2000s, the initial steps were taken after the 1990s. RPA was first coined in 1959 by Arthur Samuel, a pioneer in the field of artificial intelligence, who at that time was working for a famous computer company. However, it remained relatively unknown for some time till 2015. Afterward, RPA was widely adopted by businesses.

In 2018, the impact of RPA became significant, and in 2019, the technology was used by many companies for a complete wide range of processes. Machine learning is a tool for innovation, which eventually led to the creation of RPA. NLP (natural language processing) is a tool or software that enables computers to understand human language accurately.

In the 1960s, NLP combined with AI to establish interaction between computers and human language; this combined technology moved further toward the innovation of RPA.

There were a few more developments that took place in the 1990s. After the continuous development, it was felt that there was a need for a new technology, which has the functions of different technologies that can automate the task without any interference. There were three key predecessors of RPA:

1. **Screen scraping:** Screen-scraping technology has a significant role in the creation of RPA. This technology is used to find out data from the web, programs, and documents, which is further displayed by another application.
2. **Workflow automation and management tool:** Workflow automation is a process that comprises the sequence of automated tasks, which helps eliminate the need for human interaction. The management tool is helpful for automation so that tasks can be automated. Workflow automation decides which steps have been completed and which ones will be the next.
3. **AI:** AI is basically a technology or software program that has the ability to perform tasks, with the help of computers and robots, that typically require human intelligence. The programming of AI is based on three basic techniques: learning, reasoning, and self-correction (Rani, Mishra et al., 2021). The uses of AI are endless and can be applied to most sectors and businesses. Image recognition, speech recognition, sentiment analysis, and natural language generation are some of AI's technologies. They all together made RPA an impactful technological platform.

11.2.2 Concept

RPA is a tool to create software or robot that can automatically perform clerical tasks. It is a quick and effective automation tool. The purpose of RPA is to transfer the process execution from humans to robots. RPA is a software program that runs on PCs, laptops, or mobile devices. It is a set of rules and commands that are performed by robots according to the set of business rules.

The main goal of RPA is to replace repetitive and boring clerical tasks performed by humans, in a virtual workplace. RPA enables organizations to complete their work with minimal or no employee intervention. A front-end automation robot can read and write data with the help of a user interface and apply analytics and provide solutions within a minute.

11.2.3 Technology

RPA technology has become a necessity for businesses and organizations due to increasing competition, productivity, and returns on investment. It eliminates the need for employees for routine tasks and rule-based tasks (Tailor et al., 2022). Organizations can adopt RPA for simple and complex tasks. It does not require too much programming language, so anyone can use these automation techniques with the specified tools for it. Some of the popular RPA tools are follows:

1. UiPath
2. PEGA
3. Redwood
4. WorkFusion
5. Blue Prism
6. Automation Anywhere

These are some most popular RPA tools available in the market that a business and organization can choose anyone among them according to their business process and tasks. RPA works by accessing information from existing IT systems (Hasan et al., 2022). There are numerous ways that the RPA tool can integrate with applications. One is through connections to the database and enterprise web services in the back end. Another is through the front end, or desktop connection, which takes multiple forms. Which way is best will depend upon the needs, targets, tasks, and problems of the organization. With the backend connectivity, the automation server controls the system access automatically (Vugar et al., Autonomous Robots, 2022). Once a business firm implements an RPA tool for different activities, RPA starts work automatically according to the given information, reduces human error, and increases productivity.

RPA is a very easy technology to operate that anyone can work on it after getting the training. Before implementing RPA in business process, various steps should be analyzed:

1. First, the organization needs to analyze the whole business process and design automation around it.

2. Then the organization should develop an understanding of which task there is a need to implement or apply RPA.
3. Then there is a need to get knowledge about various tools available in the market and choose an appropriate tool according to the business purpose.
4. Then the automation design on the selected process can be implemented.
5. Last, the automation process is tested if it is performing properly or not.

It is assumed that RPA software does not require human supervision or command, but this is not true. To operate robots, there is a need for humans to command and guide them to complete the tasks automatically and to manage and update these robots from time to time. A small, medium, or large organization can implement RPA in their business process.

However, initial costing will be high but can be recovered in four to five years because it increases productivity and reduces cost. There are various concepts available for starting RPA, but here we are discussing some key concepts of RPA to understand it:

1. RPA is a software that has been developed to automate everyday processes and tasks that once required human actions. Robot replicates human tasks automatically with the help of applications or systems. In RPA, the process robot performs their tasks using a combination of automation, computer vision, and machine learning.
2. Robots are not intelligent like human beings; they cannot deal with any mental tasks on their own. Robots perform only transactional and standardized processes.
3. The major role of RPA is to automate labor-intensive tasks without changing the existing process. It is also useful for security tasks.
4. To apply RPA, there are a few needs required to learn development skills. Anyone can learn automation tasks with a few hours of training and experience with day-to-day tasks.
5. RPA-enabled automation tasks can eliminate process waste and human efforts. Robots increase the process quality, reduce losses, promote better customer experience, reduce their operating costs, and result in a profitable business.

11.3   IMPLEMENTATION OF RPA METHODOLOGY

There are four states in the implementation of RPA technology, as shown in Figure 11.2.

FIGURE 11.2   Stages of implementation of RPA technology.

1. **Planning:** Planning refers to gathering all the processes that need to be automated. In this phase the organization should identify the processes or areas, where is a need to implement automation tool. First, the need is to check that, is the process manual and repetitive? Is the input data being in electronic format and is readable? In planning phase before implementing RPA, a clear roadmap should be designed.
2. **Development:** In this phase the organization start preparing the automation workflow as per the designed roadmap and implement the automation process as discussed in planning phase.
3. **Testing:** In this phase the organization start RPA testing to identify the problem if occur and make correction if any error detects.
4. **Support and maintenance:** In this phase the need is to provide continuous support and maintenance whenever there is requirement to run the process properly with the help of IT & support team.

11.3.1 Best Practices while Implementing RPA Technology

1. The organization should consider the impact of the application of RPA before opting for any automation technology.
2. The organization should focus on the desired return on investment after the implementation of RPA.
3. The organization should focus on the large groups and areas and impact on automating the process before implementing RPA.
4. The organization should study the impact of RPA application on management and employees.
5. Operation of the project is the most important thing in the RPA process.
6. Policy, corporate, and government compliance should be ensured by the organization.

11.3.2 General Uses of RPA Technology

1. Through various applications and systems, RPA reduces the human execution of repetitive tasks.
2. RPA can easily transfer or collect data from one source to another source. It performs tasks like data entry and copying and pasting automatically.
3. RPA performs multiple and complex tasks across different applications or systems. It also helps in analyzing records and data, manipulating data, and preparing reports.
4. RPA virtually integrates data by connecting them to the user interface.
5. RPA is helpful in preparing an accurate, effective, and efficient report on time.
6. RPA provides an authentic and valid report, which can be verified by different sources.
7. RPA helps to bridge the gap between the IT system and product management platform by automatically updating both systems.

8. RPA is helpful in providing information about customer preferences or choices and customer scenarios about services, which is beneficial for their quality assurance process.
9. RPA allows automated data migration through systems, which is not possible to use through traditional mediums like documentation, spreadsheet, and other source data files.
10. RPA can automatically predict the revenue for casting according to the updated financial statements.

11.4 APPLICATION OF RPA TECHNOLOGY

The applications of RPA technology in different sectors and industries are presented here:

1. **Healthcare industry:** In the healthcare industry, RPA can be used in registering or preparing bills, maintaining records, and accepting payments.
2. **Human resource (HR) industry:** In the HR industry, RPA is helpful in maintaining records, joining new employees, processing payroll, paying salary, and providing training also.
3. **Insurance industry:** In the insurance sector, RPA provides all information about all policies, premium information, and claims and clearing process.
4. **Manufacturing and retail industry:** RPA enables companies to maintain their bills and calculate of sales, profit, and costs. At the same time, RPA is helpful in providing information about customer preferences and satisfaction and in seeking out their queries.
5. **Telecom industry:** RPA enables the telecom sector to provide quality service, manage services, solve customer complaints, and get all required information about their customers.
6. **Travel and logistic industry:** With the help of RPA, passengers can get all information about the time and charges on their travel. They can book their tickets and make payments online through RPA from their home. It is also helpful for travel companies to maintain their accounts, provide their passengers details, and receive payment.
7. **Banking and financial services:** In the banking sector, RPA is applicable in their operating and administrative activities, such as opening accounts, card activation, account details, 24/7 customer queries, cash transactions, and so on.
8. **Education sector:** Nowadays, RPA is a key platform for schools and colleges to provide classes during the COVID-19 pandemic. The education sector is becoming a major user of RPA, which is able to provide meetings, conferences, and digital learning platforms,
9. **Corporate sector:** RPA is becoming the most usable technology for the corporate sector nowadays. In this pandemic, where it is important to maintain social distancing, RPA is playing an essential role in running businesses. Through RPA, employees can work from home without visiting the office.

10. **Income tax department:** RPA is the most applicable technology for the income tax department as taxpayers can fill their return online through the digital website. RPA enables their system to provide education and information about IT rules and processes.

Here the major areas of robotics application have been discussed, but still, there are more options available where RPA is applicable and plays a vital role in automating processes efficiently.

The most important RPA skills that an organization keeps in mind before implementing RPA technology are as follows:

1. Analyze the requirement of the business to automate the processes or tasks. At the same time, the organization must be good at understanding the flow of control.
2. The organization should have experience or knowledge about different RPA tools available in the market.
3. The organization should have knowledge about how to integrate various components which are applicable.
4. There are a few coding skills required to implement RPA in the business process because it does not require much programming language.
5. While implementing RPA, one should consider the SWOT (strengths, weaknesses, opportunities, and threats) of the technology and must be careful about security issues (Vladimir Hahanov et al., 2022).
6. To implement RPA in an organization or automate the business process, there is a need to appoint new employees or train the existing employees.

11.4.1 MERITS OF RPA TECHNOLOGY

In case of a change in technology or business process with RPA, it is always faster and cheaper than appointing hundreds of employees. Here are some major vibrant benefits of RPA technology:

1. **Employee productivity:** RPA executes tasks as just humans do and free employees from their routine tasks so that they can focus on more complex and discerning tasks, which increases their productivity.
2. **Customer satisfaction:** RPA enables businesses to provide customer services 24/7 with better quality and higher accuracy compared to human actions, resulting in customer satisfaction. Better service quality and customer satisfaction increase the goodwill of the company in the market.
3. **Speed:** Robots can perform lots of work faster and more accurately than humans, which increases productivity and reduces losses.
4. **Consistency:** Robots work according to the specified guidelines and do not make any errors. So every time, they provide consistent results for the same task and provide an authentic report.
5. **Versatility:** RPA has the ability to perform different functions or activities across industries or organizations.

6. **Security:** RPA is a secure technology that collects data, makes analyses, prepares reports, and keeps data secure from outside attacks and hacking with the help of coding or password protection (Khang, Bhambri et al., Big Data, Cloud Computing & IoT, 2022).

7. **Multitasking:** Robots are able to complete multiple tasks at a time more effectively compared to human employees.

8. **Less effort:** A human can work for eight to ten hours a day only, get tired, make mistakes, and charge additional wages for extra work, but a robot can work 24/7 without any tiredness.

9. **Efficient work:** Robots are more efficient and productive in completing tasks without making any errors. Humans can be biased in their jobs, but robots always work with full authenticity and objectivity without any bias.

11.4.2 DEMERITS OF RPA TECHNOLOGY

Some of the major drawbacks of RPA software are discussed here:

1. **Potential job losses:** Robots can work at a fast pace without any error with more consistency. It creates a major threat for employees to lose their jobs. However, this thinking is not acute in this context. The current example of this is amazon. Amazon has implemented RPA technology in its business, and its profit is increasing day by day. At the same time, the demand for more employees is increasing with the increasing number of services.

2. **Investment cost:** Businesses are facing many challenges in this competitive environment. Implementing RPA in their business is one of the major challenges because the cost of initial investment is high. So before implementing RPA, they have to consider its outcome. If there are a few differences between and after return on investment, they should not apply RPA in their business because it may lead to high cost rather than profit.

3. **Skilled employees:** To implement RPA in business process, there is a requirement to appoint skilled employees who have programming skills or must be trained in their existing staff to expand their skills so that they can operate new technology.

4. **Employee resistance:** To apply technology in an organization, there is a requirement to change the working style. Human always resists implementing changes in the organization because it generates new responsibilities, new skills, or new styles of working. New technology creates fear in the mind of existing employees.

5. **Selective tasks:** RPA is applicable only to a repetitive, rule-based, and standardized process. The non-standardized and judgment basis tasks are difficult to automate. So there are limited tasks that can be automated through RPA.

6. **Time and money constraint:** The process of automating the business tasks is a time- and money-consuming process because implementing and making it convenient for employees takes time, and the initial cost for implementation is very high, which takes almost three-four years if your business is running in profit.

7. **Security threat:** RPA technology provides convenience to perform routine tasks, but at the same time, it may encourage hacking and stealing of confidential information, which can be dangerous for businesses and for customers also.

8. **Network issue:** RPA can only be applicable in places that have a full network. Places where networks do not work properly will be difficult for businesses to implement properly.

9. **Awareness:** To incorporate new technology in businesses, they have to be aware of every aspect, like the pros and cons, importance, applicability, and requirement of this technology.

11.5   ROLE OF RPA IN THE BANKING SECTOR TO REDUCE THE NPA LEVEL

A bank is a financial institution that provides different finance-related services, such as accepting deposits, lending money, and processing funds and loans. The Indian economy is the world's sixth-largest economy by nominal GDP (gross domestic production) and third-largest by purchasing power parity.

The banking industry is an important sector of our economy. It plays a vital role in the development of the economy and the increase in GDP. The major role of banks is to provide funds to their customers. Banks accept deposits from their customers and use these deposits to provide loans and finance their customers. Banks pay interest on the deposit accepted from the customers and charge interest from borrowers on loan facilities.

The difference between interest paid and interest received is the actual income of banks to run their business. These loans and investments are the key factors for economic growth. The Indian banking sector is regulated by the Reserve Bank of India (RBI). Banks have to work under the guidelines of the RBI.

The Indian banking system consists of 12 public sector banks, 21 private sector banks, 46 foreign banks, 43 regional rural banks, 1,485 urban cooperative banks, and 96,000 rural cooperative banks.

11.5.1   Non-Performing Assets (NPA)

Per the RBI, an asset, including a leased asset, becomes non-performing when it ceases to generate income for the banks. NPA generally refers to the loan amounts due or non-payable for a specified time period. This time period is for 90 days or more. If a person takes loan from a bank and does not pay the principal and interest amount for more than 90 days, this account is considered an NPA account.

NPAs are basically assets or deposits of banks that are generating any income or does not perform any financial activity. When a person takes a loan from a bank and is not able to make payments on time, this account is considered a default account or can be classified as an NPA account.

According to the bank norms or RBI guidelines, when a person takes out a loan, he becomes a borrower of the bank. If he is not able to pay the advances on time, he can be termed a defaulter. NPAs place financial burden on banks, which affects banks' financial stability and profit. NPA is significantly affecting the economic growth of India.

The level of NPA amount is directly related to the development of a nation. NPA has an adverse effect on the profitability, cash flow, and financial situation of the banks. NPA has a negative impact on the banking industry.

11.5.2 TYPES OF NPA ACCOUNTS

Loans can be classified into four different categories. Out of them, three are considered NPA accounts:

1. **Substandard assets:** An asset is categorized as a substandard asset when the amount is due for a period of 12 months or less than 12 months.
2. **Doubtful assets:** When a substandard asset remains outstanding for more than 12 months or one year, it turns into a doubtful asset.
3. **Loss assets:** An asset is considered a loss asset when recovery of an outstanding amount seems to be not feasible in the future, termed as a bad debt, and the bank does not write off this amount fully or partially.

NPAs can be classified into two types:

1. **GNPAs (gross non-performing assets):** the total amount of all NPA accounts.
2. **NNPAs (net non-performing assets):** the amount after subtraction of the provision from Gross Non-Performing Assets (GNPAs) made by banks.

11.6 APPLICATION OF RPA IN THE BANKING SECTOR

The banking sector today is under tremendous pressure to optimize costs and boost productivity. The other challenges that banks are facing include scarcity of knowledge or skilled person, increased efficiency in process, and increased in cost altogether, giving way to the adoption of robotic process assets (RPAs).

RPAs are gaining traction, with adoption rates increasing since 2016. The banking sector is working as a traditional branch and a virtual or digital banking system. After the implementation of RPAs, the banking sector will provide faster, more secure, and more reliable services.

There is a need felt to implement RPAs in the banking sector to run their businesses efficiently in this competitive market. RPAs have been significantly adopted in the banking sector to make their operating activities more organized and automated. In India, RPAs are evolving as innovative tools in the banking sector. It is an important technique to automate routine tasks. Most of the banking activity is based on routine work that can be automated through the technique.

RPAs have implemented operating activities in many banks, but still, there are lots of areas remaining that are untouched by this technique. It will be helpful for banks to manage their routine task, have better customer satisfaction, reduce their costs, and increase profitability without human interaction.

RPAs are helpful for banks to automate their routine works, which maximizes efficiency, reduces operating costs, and maintains security, resulting in higher profits. Banks are incorporating RPAs for faster process execution and operational efficiency. It is a technical tool for banks to reduce overworking bankers by automating their back office and routine tasks so that bankers can utilize their time in more important and analytical work to provide more effective services to their customers.

RPAs are also helpful in resolving the issues or complaints of customers on time without error 24/7, which results in higher customer satisfaction. It enables bankers to get information about their customers so that they can know what kind of services they are looking for and offer them a customized product or service according to it.

RPAs have been implemented into various operating activities of banks:

1. Better customer service
2. Account updating facility
3. Opening a new account from anywhere
4. Online mortgage facility
5. Cash transaction facility
6. KYC report
7. Account statement and information
8. Deposits and more
9. Credit card processing
10. Fraud detection
11. Account closer processing
12. Issuance of a checkbook
13. Making an international payment
14. Getting information on a bank's policy

According to the RBI's new announcement, RBI is going to implement RPA technology to automate the process of printing new currency and collecting, counting, and maintaining records of currency circulation. Adopting RPA technology is new in the financial service sector.

There are many processes, operations, transactions, and tasks, which can get benefit from this advanced technology. Other areas where the banks are planning to expand RPA integration include wholesale lending, mortgage, treasury implementation, and imaging services.

11.6.1 DIGITAL BANKING

With the increasing adoption of AI, customers' banking-related habits are changing rapidly. Earlier banks used to work only from brick-and-mortar places, which is a traditional way of working for banks, but nowadays, banks have stepped into digital technology.

Now banks provide services traditionally and virtually. The traditional way of working is now slowly being replaced by digital technology and provides services like debit and credit card, ATM, Internet banking, point of sale machines, and so on, 24/7, with the help of any electronic device. Among the different delivery channels of banking services, mobile banking is becoming the most suitable delivery channel (Khanh and Khang, 2021).

RPAs play an important role for banks in this advanced technological era, with the adoption of digital services. Bankers are able to devote their quality time toward other important tasks, like credit monitoring, maintaining assets quality, loan recovery, and customer satisfaction.

Digital banking is also beneficial in the field of financial inclusion. With this technology, customers who live in remote areas can get the benefits of financial services from anywhere and from their home location. Digital technology eliminates the need to go to a bank branch for their routine work, which is time- and cost-saving for customers. They can operate their account from their homes with their mobile phone or any other electronic device.

Although there is a need to educate them to operate a digital account and better network connectivity. Implementation of RPAs through digital technology can be useful in customer education and product awareness.

Customers should not only be educated with the use of different products but also have knowledge about safety or security issues while using the digital platform to make banking services more secure and safe. Banks look toward making their banking and other financial and operating activities more comfortable and convenient for their customers, which led to the adoption of RPA technology.

RPAs are now used as a digital banking platform in the banking sector. Digital banking is an evolving concept; electronic banking services use new innovative technologies and banking tools. In this modern era of technology, it is very difficult for the banking sector to run its business without digitalization.

If banks do not provide digital services to their customers, no bank will be able to survive in this competitive market. Electronic banking services such as money transfer, statement inquiries, deposit creation, online trading settlement, and fees and tax payments are possible with digital banking without visiting a bank branch.

Digital banking is a paperless concept, which is a step toward a green environment. Implementation of RPA technology to provide a digital platform is beneficial for customers as well as for bankers. It enables the bank to reduce their operational and service cost, which in turn enhance their profitability and directly increases their goodwill in the market, which is helpful to gaining their customer's trust and having an impact on their businesses.

11.6.2 BENEFITS OF DIGITAL BANKING

Most of the benefits of digital banking are listed here:

1. Digital banking is a platform that provides 24/7/365 a year service.
2. Digital banking provides quick and efficient service to its customers.
3. Digital banking allows privacy to deal with their accounts.
4. Digital banking works through robots that provide services without any error.
5. Digital banking enables customers to access cash at any location regardless of where and in which bank.

6. Digital banking offers flexibility to operate their account.
7. Digital banking facilitates their customers to use banking activities from anywhere and anytime.
8. Digital banking enables fund transfer across bank branches through the card linked with their account.

The banking industry has innovated many technologies in its system. Nowadays, banks are providing mobile banking services through different mobile banking apps. Mobile banking gives immense convenience to its customers, who avail banking services at their fingertips (Khang et al., IoT & Healthcare, 2021).

11.7   IMPLEMENTATION OF RPA IN LSP

RPAs are important tools for banks to manage and control their NPA to a significant level. Although RPAs have been implemented in the banking process, they are limited to their operational activities and still do not touch the loan-sanctioning process. This can play a significant role in the recovery of present NPA and controlling future NPA. When RPAs are applicable in the loan-sanctioning process, it may be easier for bankers to get all the information (like A/C details, accounts with other banks, loan information with other banks, credit history, earning sources, paying capacity, mortgage property, KYC details, etc.) about the borrowers.

With the help of RPAs, bankers can get all the information about their customers with a single click automatically. Here is a need to develop software that will be capable of concentrating all information of a particular person in a single place. To operate this kind of software, there is a need to train the staff for the successful implementation of this technique.

When a person applies for a loan in any branch of any bank, bankers can get all information regarding their accounts, loans, credit history with other banks, and the earning source and paying capacity of their customers.

This information will be helpful for bankers to make the right decision on whether they should sanction their loan applications or not. If the bankers are not sure about the timely recovery of the loan amount on the basis of the information, they can reject the loan application. These precautions will be helpful for banks to control their future NPA, and they can concentrate on to recovery of their current NPA.

Although we cannot recover the whole amount, it can be reduced to some extent. Although the cost of implementation of RPAs in the loan-sanctioning process (LSP) may be high once, it will result in high profitability, financial soundness, and high trust from their investors, and finally, it will be helpful for the growth of the GDP of the nation.

Figure 11.3 shows the proposed stages of implementation of RPA in LSP in the banking sector.

RPAs can be implemented in banks at their LSP for controlling the loan accounts to be downgraded as NPA accounts. Through the different phases, the LSP can be automated with the help of customized software for it.

Out of these three phases, RPAs can be implemented, from phase 1 to 3, in the process of automating the banking sector.

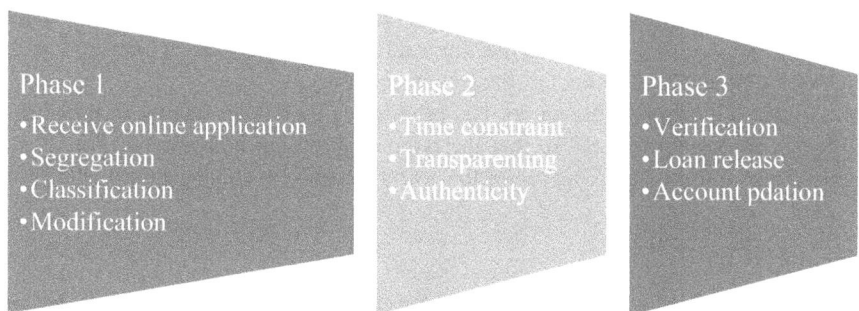

Phase 1
• Receive online application
• Segregation
• Classification
• Modification

Phase 2
• Time constraint
• Transparenting
• Authenticity

Phase 3
• Verification
• Loan release
• Account pdation

FIGURE 11.3   Stages of implementation of RPA.

11.7.1 PHASE 1

Phase 1 is the initial step in applying for a loan online or through digital banking. Nowadays, banks provide online loan application services, which is convenient for their customers.

In this phase, there are four steps: receive online applications, segregate these applications according to their category, classify these applications according to their prescribed limits, and check the required documents.

11.7.2 PHASE 2

Phase 2 involves checking the authenticity and transparency of the loan documents on time. Humans sometimes may be biased and can make mistakes while checking these documents, which can be time-consuming, but through the RPAs, the loan application process can be completed on time with full accuracy.

11.7.3 PHASE 3

Phase 3, the last step, involves the verification of loan documents. RPAs can check the possession of the assets that are mentioned as a security of the loan and the actual value of that assets. With all the requirements satisfied, RPAs will release the loan application for sanctioning, and after sanctioning the loan, RPAs will keep their eyes on the regularity of loan installment. If the three installments become overdue, then RPAs will automatically cease the account.

11.8 CONCLUSION

The financial stability of the banking sector is a basic requirement for the financial growth of the Indian economy because banks are the main wheel of the Indian economy. The study found that RPA is becoming a necessity for the Indian banking sector as this technology is helpful for banks to automate the process of their routine work with full accuracy and efficiency.

In this chapter, the author has discussed all aspects of this advanced technology. The role of RPA technology at the time of LSP in the banking sector has been discussed. The current number of NPAs is becoming a major threat to the Indian economy because it is affecting the financial position of banks, which affects the GDP growth and development of the Indian economy.

Also, in this chapter, the author has proposed a model (in stages) to implement RPA technology in the LSP so that the issue of NPA can be resolved. This model can be helpful for the banking sector to control the present and the future NPA level.

REFERENCES

Bhambri, Pankaj, Sita Rani, Gaurav Gupta, and Khang, A., *Cloud and Fog Computing Platforms for Internet of Things*. ISBN:978-1-032-101507, 2022, doi:10.1201/9781032101507.

Hasan, S., Khang, A., and T. B. Sivakumar, "Cryptocurrency Methodologies and Techniques", in *The Data-Driven Blockchain Ecosystem: Fundamentals, Applications, and Emerging Technologies* (1st ed., pp. 27–37). CRC Press, 2022. https://doi.org/10.1201/9781003269281.

Khanh, H.H., and Khang, A., "The Role of Artificial Intelligence in Blockchain Applications", in *Reinventing Manufacturing and Business Processes Through Artificial Intelligence*, pp. 20–40. CRC Press, 2021. https://doi.org/10.1201/9781003145011.

Khang, A., Pankaj Bhambri, Sita Rani, and Aman Kataria, *Big Data, Cloud Computing and Internet of Things*. CRC Press, 2022. ISBN:978-1-032-284200, doi:10.1201/9781032284200.

Khang, A., Subrata Chowdhury, and Seema Sharma, *The Data-Driven Blockchain Ecosystem: Fundamentals, Applications and Emerging Technologies*. CRC Press, 2022. ISBN:978-1-032-21624, doi:10.1201/9781003269281.

Khang, A., Sita Rani, Meetali Chauhan, and Aman Kataria, "IoT Equipped Intelligent Distributed Framework for Smart Healthcare Systems," *Networking and Internet Architecture*, 2021, https://arxiv.org/abs/2110.04997v2, doi: 10.48550/arXiv.2110.04997.

Rana, G., Khang, A., and Khanh, H.H., "The Role of Artificial Intelligence in Blockchain Applications", in *Reinventing Manufacturing and Business Processes Through Artificial Intelligence* (p. 20). CRC Press, 2021, doi:10.1201/9781003145011.

Rani, S., A. Kataria, and M. Chauhan, "Fog Computing in Industry 4.0: Applications and Challenges—A Research Roadmap," in *Energy Conservation Solutions for Fog-Edge Computing Paradigms* (pp. 173–190). Springer, Singapore, 2022.

Rani, S., A. Kataria, V. Sharma, S. Ghosh, V. Karar, K. Lee, and C. Choi, "Threats and Corrective Measures for IoT Security with Observance of Cybercrime: A Survey," *Wireless Communications and Mobile Computing*, 2021.

Rani, S., R. K. Mishra, M. Usman, A. Kataria, P. Kumar, P. Bhambri, and A. K. Mishra, "Amalgamation of Advanced Technologies for Sustainable Development of Smart City Environment: A Review," *IEEE Access*, 9, 150060–150087, 2021.

Vugar, A., Khang A. Vladimir Hahanov, Eugenia Litvinova, Svetlana Chumachenko, and Abuzarova Vusala Alyar, "Autonomous Robots for Smart City: Closer to Augmented Humanity", in *AI-Centric Smart City Ecosystems: Technologies, Design and Implementation* (1st ed.). CRC Press, 2022. https://doi.org/10.1201/9781003252542.

Tailor, R.K, Khang, A., and Ranu Pareek (Eds.), "RPA in Blockchain", in *The Data-Driven Blockchain Ecosystem: Fundamentals, Applications, and Emerging Technologies* (1st ed.), pp. 149–164. CRC Press, 2022. https://doi.org/10.1201/9781003269281.

Vladimir Hahanov, Khang A., Gardashova Latafat Abbas, and Vugar Abdullayev Hajimahmud, "Cyber-Physical-Social System and İncident Management", in *AI-Centric Smart City Ecosystems: Technologies, Design and Implementation* (1st ed.). CRC Press, 2022. https://doi.org/10.1201/9781003252542.

12 Epidemic Awareness Spreading in Smart Cities Using the Artificial Neural Network

R. Dhanalakshmi, N. Vijayaraghavan,
Arun Kumar Sivaraman, Sita Rani

CONTENTS

12.1 INTRODUCTION

The objective of this idea is to spread awareness among the people in smart cities about a lethal epidemic. This method involves the collection of data from various sources, mining the data for useful information that can be given to the application, creating a notification, and connecting the application to firebase for targeted messaging and data storage (Khang, Chowdhury et al., Blockchain, 2022).

Data collection is done using reliable sources of information only, such as news websites and government sites. The collected data is then formatted, and the details are sent to the users' devices as notifications. Prediction algorithms are used to formulate the risk index.

By using this application, the end users will be more educated about the epidemic and can assess the risk factor of the COVID-19 epidemics at their location (Wong et al., 2020).

With the start of COVID-19, the government of India executed a few arrangements and limitations to reduce the blowout. An ideal choice occupied with public authority aided in slowing the extent of COVID-19 generally.

DOI: 10.1201/9781003252542-12

In spite of these choices, the pandemic keeps on spreading. Future forecasts on this disease will be useful for upcoming policy creation (i.e., to suggest ways to reduce the spread of COVID-19 disease).

Additionally, it is noticed in the universe, the irregular crown cases play a significant part in the blowout of the sickness (Acuna Zegarra et al., 2020). This propelled to incorporate similar situations for precise pattern forecast. India was picked for examination because of the dense population in the country, which rapidly spread the illness. Thus, an improved SEIRD (susceptible, exposed, infected, recovered, deceased) model was formed to anticipate the pattern and pinnacle of COVID-19 cases in India and its four most exceedingly severe states (Santhosh, 2020).

The improved SEIRD model depends on an asymptomatic uncovered populace and asymptomatic however irresistible for the expectations.

Moreover, a profound learning-based long short-term memory (LSTM) unit is likewise utilized for pattern forecast. Forecasts of the LSTM model are contrasted, and the expectations are acquired from the proposed modified SEIRD unit for the following 30 days (Gupta and Rani, 2010; Bhambri et al., 2022).

This epidemiological information up to September 6, 2020, have been utilized for doing forecasts. Various lockdowns enforced by the government of India have additionally been utilized in demonstrating and breaking down the suggested improved SEIRD unit (Shakil et al., 2020).

The world saw the COVID-19 situation negatively affected the lives of numerous individuals worldwide (Rani and Gupta, 2017). The extent of infection happened quickly. Accordingly, on March 11, 2020, the World Health Organization (WHO) reported that the viral infection was now a worldwide pandemic (Li et al., 2020).

Because of this disease, unavoidable choices were made so that the specialists of different nations needed to restrict the movement of people and inessential activities (Rani et al., 2022). Some of these limitations included lockdowns, social distancing, online classes, and business continuity plans.

Hence, the spread of COVID-19 affects the environment just as on the way of life of people (Wang et al., 2020). Practically, all schools were closed, sports classes were dropped, and individuals were encouraged to do virtual financial exchanges by utilizing different computerized stages. COVID-19 was first announced in China at Wuhan City on November 17, 2019, and from that moment onward, it spread rapidly (Dhanalakshmi et al., 2021).

This disease by human contact, due to which an enormous number of cases have been accounted overall. A healthy individual can be infected by interacting with a sick individual or by touching the surfaces touched by an infected individual.

The virus is detectable after a certain period of time. During this time, the infected individual is a carrier of the virus and ca infect healthy people (Rani and Kaur, 2012).

On September 6, 2020, more than 27 million individuals were infected, and more than 8 million individuals passed on from this earth. Consequently, this virus has developed a major danger to individuals and the environment (Sahoo and Sapra, 2020).

To handle this problematic circumstance, the initial phase was to take prudent steps to forestall the spread of the disease; infected individuals should isolate themselves and get clinical assistance.

Safety precautions are also practiced, such as using sanitizers, wearing face masks, and physical distancing (Gayathri et al., 2020).

Items such as sanitizers and face masks are required now. Notwithstanding, appropriately discard the pre-owned covers to protect the surroundings. Governments have likewise found numerous ways to organize the essential items to offer better clinical assistance to infected individuals. An inventory of these items is made the needs of patients can be met on time. Scientists are utilizing different strategies to assess assets (Wang et al., 2020).

A nation like India, with huge population of nearly 1.38 billion, is a challenge to handle professionally amid a pandemic. In India, the first COVID-19 case was discovered in Thrissur region of Kerala on January 30, 2020, from a person got back from Wuhan University, China.

Initially, the spread of the disease was not significant in India, and by March 15, the number of infected individuals barely increased to 100 (Singh et al., 2017). On September 6, 2020, the virus spread to 215 nations, with over seven million active cases worldwide.

In India, the COVID-19 cases reached 4.2 million, and there are 0.8 million active cases, 3.3 million recoveries, and thus 72,816 as the complete number of passings, which makes India the most noticeably severely affected country in Asia. There is a variance in the absolute number of active cases in India, with the highest number of cases in Maharashtra and the lowest in Mizoram (Peninah et al., 2016).

As India is a thick nation as far as populace and culture are concerned, patteren forecast may play a very significant role to predict pandemic cases and help to control it.

The number of events around the world, just as in India, is expanding at an extremely rapid rate. It is clear from the information that Maharashtra is the most severely affected as far as confirmed cases go, which represent about 21% of the cases in India. The following three the next most severely affected states/regions— Andhra Pradesh, Tamil Nadu, and Karnataka, with 32% of the confirmed disease— and the remainder of Indian states/regions have 47%.

The Northeastern territories of India are greatly better, such as Mizoram and Sikkim, each having under 2,000 diseases as of now. To predict the spread of the pandemic and to develop methodologies, different models have been planned that may give real-time data of contamination, the number of infected cases, and the required clinical resources to deal with the pandemic (Naveen and Gustafsson, 2020).

Also, it is seen that asymptomatic cases are the major contributor to a spread of the disease. Because of their lack of symptoms, these individuals do not follow the pandemic guidelines. This causes a remarkable spreading of the disease locally.

This chapter proposes a modified SEIRD model to foresee the pattern of disease in India and its four most affected states. It utilizes a boundary, epsilon for disease forecasts by considering the extent of the uncovered populace that is asymptomatic and is accidentally scattering the contamination.

The proposed model uses the data on infections, recoveries, and deaths brought about by disease.

The data was gathered from the start of the pandemic until September 6, 2020, from Maharashtra, Andhra Pradesh, Tamil Nadu, and Karnataka, which are the four states with the most significant cases. This chapter forecasts the number of susceptible, exposed, infected, recovered, and deceased cases in smart cities using modified SEIRD model.

SEIRD model can be utilized to any nation with smart cities and its province utilizing the detailed information of that nation.

12.2   LITERATURE SURVEY

From the start of COVID-19, different scientists have predicted its its dispersal pattern in different nations and their provinces (Bhatia and Abraham, 2020). Ahmed (Ahmed, 2020) studied the effects of the disease in relation to the person's age and sexual orientation and the area of the pandemic. In his work, the number of inhabitants in India that has shown up from different districts is considered. These individuals were isolated into six groups to study the local effects of COVID-19.

In each group, various age ranges were examined, and the recovery rates were determined. Ceylan (2020) used auto-backward ARIMA to foresee the upcoming pattern of the COVID-19 illness in the three most exceedingly awful pretentious nations of Europe, specifically Italy, Spain, and France.

The creator formed a few ARIMA models utilizing different boundaries. The best models were utilized for assessing the spread of the illness in every one of the three nations. Patrikar et al. (2020) have utilized the altered SEIR strategy in their work to anticipate the spread of virus in India.

The creators played out a momentary gauge of corona virus for different provinces in India. SEIR is additionally used to contemplate the effect of temperature on the coronavirus outbreak in China (Shi et al., 2020).

Writers consolidated climatic components in the first SEIR unit to break down the effect on the spread of COVID-19. Because of the unique idea of the virus expansion, numerous scientists have done momentary anticipating of things to come out of this dangerous infection.

Roosa et al. (2020) had done 5-and 10-day gauges of the absolute established patients for Guangdong and Zhejiang regions of China. The creators utilized the calculated development unit, Richards's development unit, and a sub-pestilence wave model to produce expectations.

A bootstrap method is used to register the vulnerability limits for foreseeing aggregate cases in a matter of seconds. Arora et al. (2020) utilized different varieties of LTSM models for momentary forecast of diseases in India.

Profound LSTM, bi-directional LSTM (Bi-LSTM), and convolutional LSTM (Conv-LSTM) were utilized to ascertain week forecasts for different states and association domains of India. Similar algorithms are used in the biomedical applications with the hotspot algorithm (Johnson and Dhanalakshmi, 2019), genetic algorithm model (Dhanalakshmi et al., 2020), particle swarm optimization models (Nath et al., 2020), Q-learning approach (Kothandaraman et al., 2021), and trident optimization technique (Tomar et al., 2020).

Tomar et al. (2020) additionally utilized LSTM alongside power-law bend fitting to anticipate the pattern of coronavirus in India. The developer's gauges all out count of complete, recuperated, and perished diseases for a brief time of next one month.

The effect of different upsides is examined with broadcast rate on the quantity of anticipated tainted cases.

Santhosh et al. (2020) examined the utilization of dynamic education with the assistance of multimodal information to battle the virus flare-up. A contextual investigation utilizing CMC was amended by profound erudition organization and fluffy standard enlistment for acquiring restored stochastic experiences on the scourge advancement experiments.

The spread of this disease was seen late in India, unlike in other nations such as China, United States, Italy, Russia, Spain, and France. Hence, the focal point of the earlier investigation was restricted to these nations, and the studies in India are negligible.

The country is the second most populated country in the world. Furthermore, most individuals in India are living in ghettos. Accordingly to anticipate the pattern of corona for India with momentary expectations, yet the asymptomatic irresistible diseases are not counted in the previously stated writing.

Hence, to fill the exploration hole, this work proposed to study the disease transmission technique which is most noticeably awful.

This work helps to deal with the increase in cases because of any celebration or social event during that period (Sebastian et al., 2009). The purpose of this study is to analyze the contributing risk factors of the epidemic and to narrow down the possible significant parameters.

From this study, requirements and modifications were made to the initially proposed methodology. The trend that is observed for any epidemic awareness software is that great importance is given to the factors that are used in calculating the risk.

From the papers referred to, the factors and their values can be approximately determined. The study was conducted based on the stages of software development, and each paper corresponds to a particular stage in the software development life cycle.

Naveen et al. (2020) stated that events related to COVID-19 are related to the past and, possibly, future occurrences. Although, it is possible to prevent the disease. But it is a must to reduce the effect to the public. This has reached to huge economic hazard not to a specific country or continent but to the world (Nicholas et al., 2016).

Researchers stated that when WHO announced that COVID-19 is a global pandemic, restrictions were made mandatory, including wearing masks and social distancing, making things less convenient.

But many death trolls have been raised enormously even before the experts announced this as the form of the pandemic. This made the world to announce lockdowns as a mandatory precaution. Many regions started to get some active cases, and they are monitored by the coloring the hotspots red, orange, and green. Orange zones denote there is some infection, and green zones indicate an area with no infection.

In this chapter, authors proposed an Android application. With this application, a person can be traced by Google Maps if they enter a quarantined zone, and the user's IMEI number was uploaded into cloud database (Neha et al., 2017).

Neha et al. (2017) discussed Firebase Cloud Messaging (FCM), a real-time database that produces applications at a faster rate and helps transmit messages more reliably.

FCM is cross-platform application that helps users develop their applications on different platforms, like Android, iOS, and the web (Bhambri et al., Cloud & IoT, 2022). FCM can be used to deliver push notifications and other data in real-time at no cost. This is considered as backend as a service (BaaS) as it deals with more back-end resources. There are two types of messages used in FCM notification messages: display messages and data messages (Huijuan et al., 2017).

HTTP and XMPP are two reliable protocols used in FCM for communication between the sender and the receiver. Notification messages include title, message icon, and so on. This is sent from the Firebase console, and the user does not have much command over such messages. It appears as a notification on the user's device while the app is running in the background.

Data messages can carry customized key-value pairs to transfer a data payload. A maximum of 4 KB payload can be sent. Thus, Firebase is a suitable and simpler platform for building applications faster with enhanced reliability (Hadil et al., 2020).

Hadil et al. (2020), in their paper "An analytical study on the awareness, attitude and practice during the COVID-19 pandemic in Riyadh, Saudi Arabia," stated that on March 12, 2020, the WHO announced that COVID-19 is a global pandemic, and one can be easily get infected from a sneeze or cough or from touching COVID-19-contaminated surfaces.

Saudi Arabia was among the first countries to take precautions. Quarantines and restrictions were made, and the Ministry of Health created awareness regarding virus transmission.

Apart from the monitoring, the key-value was to provide awareness to people in the direction of a wide range of virus infections. A cross-section study of 1767 contributors was directed to reconnoiter the consciousness, attitude, and awareness of COVID-19 regarding socioeconomic statistics amid inhabitants in the city of Riyadh.

Following which the spread of viral diseases was analyzed by Rina et al. (Rina et al., 2020). Regarding the pandemic situation, good measures are taken in the residence of southwest Saudi Arabia (age ≥ 12 years).

The measures were based on the community and cross-sectional inspection using a self—developed structured questionnaire randomly for the residents and health-care workers. Under such circumstances, people's observance of defensive procedures is significantly exaggerated by their alertness of the disease (Muhannad and Yaser, 2016).

To calculate risks associated with, the COVID-19 disease varying scenery of strategies and novel measures are used.

From standard and various means, this chapter offers the issues and difficulties that may affect the industries and the nation due to COVID-19. From the perspective of 12 major subjects, an analysis was taken by the experts that focused on the key domains to study the practical and theoretical aspects of this pandemic.

Sharnil et al. (2020) proposes an IoT-related work to sanitize the non-human-motion detection using a smart epidemic tunnel to protect people from COVID-19 infections.

A fusion-based sensor detects a human automatically by an ultrasonic sensor kept at a height of 1.5 feet to sanitize. The smart tunnel works by using solar cells in the daytime, and during at night, it functions by using energy from the power bank charged by the solar cells.

Wong et al. (2020) proposes the idea that social media act as a communication terminal in various platforms like education and healthcare, and many advantages and disadvantages were frequently discussed with its outcomes.

Since the COVID-19 outbreak, social media has spread awareness in various forms as if the medical system has become a part of our living. Any tool in medical diagnosis has pros and cons that have to be considered. With past experiences from the organization of the medical system, this chapter provides communal drawbacks (Rina et al., 2020).

Implementations with embedded devices to support biomedical applications in the field of cardiac monitoring (Sudevan et al., 2021), spinal cord correct position maintaining (K. N. et al., 2020), and device to observe the physiological condition of asthma patients (Booma et al., 2014) were developed. The illnesses like COVID-19, SARS, the plague, and so forth are obtained infections. It implies that illnesses blowout over pathogenic specialists. A conventional method for irresistible infection is defined, named as Epidemiologic Triad.

The four significant components associated with the epidemiologic group are ecological variables, transporter specialist, contaminated hosts and the microbes. The specialist is normally the transporter of the disease. The disease is communicated to the host when a specialist interacts with the host under a specific climate.

A microbe can be carried by a vector. A vector is a creature that spreads infectious microscopic organisms from one host to the next (Huijuan et al., 2017).

In recent years, it has been seen that because of changing lifestyles, more accessible world travel, and urbanization, infectious illnesses easily turn into pandemics.

To forestall these diseases, solid precautions should be made, or else the situation can quickly worsen. Since the beginning, humankind has dealt with plagues and pandemics.

The first plague that humankind suffered was in mid-1300s called Black Death. This was one of the worst epidemics ever. This scourge caused a multitude of deaths (Kumaresan and Vijayaraghavan, 2015).

The least pandemic was in mid-1500s called smallpox, where only half of the death cases were noticed (Yogesh et al., 2020). Perhaps, humankind's deadliest pandemic was the fifth cholera pandemic, which took 1.5 million lives (Lixia and Maoxing, 2014).

In 1918, Spanish influenza took 20–110 million lives. In 1957, Asian influenza took 0.7–1.5 million lives.

In 1981, the world saw another pandemic: HIV/AIDS. Over 70 million patients were infected. As indicated by the WHO information, 36.7 million died because of HIV/AIDS (Naveen and Gustafsson, 2020). After this, the world saw another rush of diverse pandemics, beginning with SARS in 2003, which affected four continents and 37 nations (Neha et al., 2017).

In 2009, swine flu occurred, with 151,700–575,500 deaths (Rina et al., 2020), and was followed by MERS in 2012, affecting 22 nations (Gayathri et al., 2020). Two pandemics followed the MERS, Ebola and Zika, in 2013 and 2015, respectively.

Both the pandemics brought thousands of deaths Sharnil et al. (2020). Now entire world is affected with COVID-19 pandemic. More than 100 or more nations to date are significantly affected by COVID-19, growing day by day. Later, these plagues swelled into pandemics or commonly alluded to as the episode of the infection/ illness.

The vulnerability that wins with respect to the sickness has led to a ton of tales in regards to its whereabouts. Individuals are hazy about the preclinical indications and the methods to deal with it. This can help save a many lives. In the event that the first episode in any country is detected, the circumstance is kept from growing into a pandemic.

When the pandemics happen, ecosphere economics are badly affected. Billions of dollars are put resources into resources for the improvement of an antibody for the novel illness (Yogesh et al., 2020).

Fabon (2016) have taken augmentations of the traditional Ross-Mc Kendrick Mac Donald methods which are joined with segment and spatial conditions of the infection on the host just to blowout of sickness.

This exploration talks about the retro forecast model to study the spread of the virus. To anticipate the spread of HIV/AIDS, Kaplan's method was utilized. The forecast focused on drug addicts utilizing needles. Henceforth, the examination was focused on the spread example relating to the specific gathering of individuals. MERS was another pandemic looked by the world.

To break down the transmission course of the MERS, choice trees and priority calculations were utilized in (Kumar and Roy, 2020). A greatest probability strategy was utilized to evaluate the spread of SARS, utilizing a phylogenetic tree. COVID-19 is an illness that developed into a pandemic. It was identified by the WHO on December 31, 2019, in Wuhan, China.

After the outbreak in China, many nations were affected by the virus. According to the WHO, all throughout the planet, 634,835 active cases and 29,891 deaths were recorded (Zhou et al., 2020).

In China, the coronavirus spread very quickly, rapidly turning into a pandemic. In this dynamic cycle, past information are examined to get viewpoint.

However, the information gathered in a brief time frame was not sufficient to be used in AI models. Effectually, teachable AI methods for time-series information are needed. The time series information gives progress of competence of estimating Prediction and Forecast Modeling, assuming an essential part in dynamic during the rise circumstances of infection spread.

To know the worldwide effect of COVID-19, we need exact gauging, inescapable populace data, affirmed cases and investigation on recuperations and passing's. Because of vulnerability in the registered assessments and theoretical mediation of specialists, the activities on arising microbe consistently get delay.

At the end of 2019, from the SARS group, another strain of COVID-19 was identified in Wuhan, the area in China where the pandemic originated. The WHO, on March 11, 2020, proclaimed that this incredibly infectious sickness as a pandemic illness Fabon (2016).

Considering the degree of its spread around the world, governments of various nations have mandated restrictions on movement and flight, enforced thorough cleanliness, and social distancing. However, the infection rate remained exceptionally quick and high.

Though the greater part of the people contaminated with the COVID-19, prepared fragile to direct respiratory sickness, there are also some fostered with a lethal respiratory disease called pneumonia.

Older people with illnesses like diabetes, cardiovascular disease, or chronic respiratory, renal, or hepatic diseases are all the more likely to manifest severe symptoms. Up to this point, no vaccine COVID-19 had been developed (Michael et al., 2015).

There are many advancing clinical trials assessing possible treatments. At the time of writing, almost 33.4 million active cases were found in over 215 nations as of September 2020, among which around 1 million deaths, 23.2 million recoveries, 6.8 million mild cases, and 60,000 new cases were accounted for.

Lixia and Maoxing (2014) gave thoughts on the most proficient method to look for a balance between the models of "analyzable," "straightforward," and "unsolvable."

However, there are different assets, in particular World Health Organization and archive of Johns Hopkins University (Muhannad and Yaser, 2016), which give refreshed exploration information as dominate pages (Michael et al., 2015) that are utilized in forecast methods, and this has led to very difficulties in visualizing situation for the pandemic.

In SEIR method, a sort of repetitive neural organization, called LSTM is utilized to anticipate the quantity of new diseases over the long haul (Yang et al., 2020). A momentary high goal estimating dependent on Deep Learning Epidemic Forecasting with Synthetic Information (DEFSI), is suggested (Wang et al., 2019) to sum up and feasible scourge system.

A blended technique for displaying epidemiological conditions and neural organizations is proposed for proliferation number assessment.

12.3 PREDICTION MODEL

The prediction of risk index or severity is done based on an artificial neural network model. The ANN (artificial neural network) model is an iterative process of learning. In common ANN implementation (Rana et al., AI & Blockchain, 2021), the signal, in this case the parameters, is the real numbers, and the output of every artificial neuron is considered by means of sum of the inputs.

Artificial neurons and networks naturally have a weight that corrects as learning continues. Artificial neurons refer to the various constituent factors that affect the output. The goal of this network is to predict the risk index. The process first starts based on human learning. It starts with an approximation of the result. The weight for each parameter is then calculated with the result in mind, and the process is iterated.

In the second iteration, a report from the training set is then taken and the weights are substituted, then the new data for the parameters are then used. On substitution, if the result is close to the expected result, further testing is done with the rest of the reports in the training set.

However, when the result is not accurate, the weights are reformulated such that it closely adheres to all the test cases.

The weighted average is calculated for the risk index as follows:

$$\Sigma a_i b_i / 6 = rating \qquad (1)$$

Hence,

$$\Sigma a_i b_i = 6 * rating \qquad (2)$$

$$\text{In general, } \Sigma a_i b_i = n * rating \text{ (if there are "n" parameters)} \qquad (3)$$

Where ai = severity coefficient, bi = parameter used. The rating is given based on the highest priority parameter.

$$\text{Number of training set reports} = 5$$
$$a_1 b_1 + a_2 b_2 + a3b3 + a4b4 + a5b5 + a6b6 = 6 * 4 = 24$$

The rating for the first training set is 4, based on that result and over six iterations; the formula after the learning process has been completed is as follows:

$$\text{Risk Index} = 17.64\ b0 + 72.22\ b1 + 0.012\ b2 + 0.18\ b3 + 1.26\ b4 + \alpha \qquad (4)$$

Where,
b0 = Weather
b1 = Fraction of reports
b2 = Population density
b3 = Number of deaths
b4 = Wind
α = Constant for error reduction

The formula is first computed from the first set of reports. The risk index that is to be presented for that report is assigned to the right-hand side of the equation. The value can be split in different ways.

In this case, the first product, a0b0, contributes the maximum to the total value. Product a1b1 will contribute the second greatest to the total value. This is how the priority of the parameters is obtained.

The weather and fraction of reports parameters are given the highest priority (a small change in these values will contribute to a large change in the total sum), and the speed of the wind is the least important factor in the equation.

To split the right-hand side value, which is equal to 6 * (assigned rating), the product of a0b0 is equal to 0.35 * (value of RHS). Then, a1b1 will be 0.225 * (value of RHS). a2b2 is 0.175 * (value of RHS) will has lesser priority. a3b3 is 0.15 * (value of rhs) and a4b4 is 0.1 * (value of rhs).

The input layer nodes are passive, meaning they do not alter the data. The outputs from the hidden layer are represented in the flow diagram by the input variables.

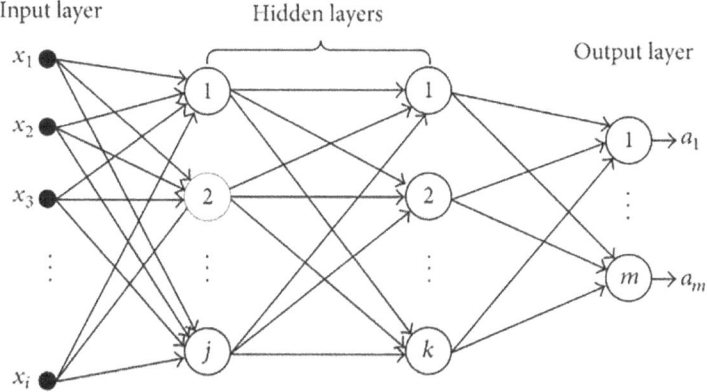

FIGURE 12.1 Architecture of an artificial neural network.

TABLE 12.1
Dataset Used

Iteration no.	Weather	Fraction of reports	Population density	People affected	Wind speed
1	–	–	–	–	–
2	2	0.04	602	50	5
3	1	0.089	823	47	9
4	1	0.0112	705	34	1
5	1	0.157	732	35	6
6	1	0.08	823	46	4
7	1	0.157	731	31	2
8	1	0.056	460	50	9
9	1	0.0112	556	18	2
10	2	0.157	731	65	1
11	2	0.112	648	101	4
12	1	0.112	648	44	6
13	2	0.089	823	85	6
14	1	0.157	731	30	1
15	1	0.056	460	38	9

Every value is replicated and given to the next layer. Active nodes in the output layer are combined and alter the data to yield the network output values.

The dataset used is illustrated in Table 12.1.

12.3.1 INPUT LAYER

Input for the first iteration is a rough estimate of the rating of the severity. The output of the first iteration becomes the input for the next layer, which helps to reformulate AI parameters.

12.3.2 Hidden Layer

The hidden layer includes pivotal data that produce the output. In this case, the hidden layer data is available and is used along with the input layer to find the output. The hidden layer values change according to the training set used.

12.3.3 Output Layer

The output of the ANN is the weightage for each parameter involved. The output of one layer is given as the input to the next layer. Once the weight has been changed, the new weights are used as the input layers.

This process is done until the weightage hits a plateau or stable point where there is not much change in the weights as the process continues. Output is represented by the set bi. Then bi is set to input (i + 1).

12.4 MATHEMATICAL CALCULATIONS

By using the algorithm and also with the help of iterative learning, the coefficients for each of the parameters have been determined. The coefficients are as follows:

a0 = 17.64
a1 = 72.22
a2 = 0.012
a3 = 0.18
a4 = 1.26
α = 2.53

The equation, therefore, is

$\sum$ an bn = a0 b0 + a1 b1 + a2 b2 + a3 b3 + a4 b4 + α

1ST ITERATION

($\sum$ an bn) / 6 = 42/5
$\sum$ an bn = 50.4

a0 b0 = 0.35 * 50.4
a0 (1) = 17.64
a0 = 17.64
a1 b1 = 0.225 * 50.4
a1 (0.157) = 11.34
a1 = 72.22
a2 b2 = 0.175 * 50.4
a2 (731) = 8.82
a2 = 0.012
a3 b3 = 0.15 * 50.4
a3 (42) = 7.56
a3 = 0.18

a4 b4 = 0.1 * 50.4
a4 (4) = 5.04
a4 = 1.26

Using such iterations, all of the initial coefficients are determined.

2ND ITERATION

The value of b0 is 2, b1 is 0.04, b2 is 602, b3 is 50, and b4 is 5. We also determine the values of Σ an bn = 60 based on the predefined rating, while the values and coefficient are substituted in the following equation:

17.64 * 2 + 72.22 * 0.04 + 0.012 * 602 + 0.18 * 50 + 1.26 * 5 + 2.53 = 63.28

The expected value = 60
The actual value = 63.28

3RD ITERATION

The value of b0 is 1, b1 is 0.089, b2 is 823, b3 is 47, and b4 is 9. We also determine the values of Σ an bn = 56.4 based on the predefined rating, while the values and coefficient are substituted in the following equation:

17.64 * 1 + 72.22 * 0.089 + 0.012 * 823 + 0.18 * 47 + 1.26 * 9 + 2.53 = 56.28

The expected value = 56.4 and the Actual Value = 56.28

4TH ITERATION

The value of b0 is 1, b1 is 0.0112, b2 is 705, b3 is 34, and b4 is 1. We also determine the values of Σ an bn = 40.8 based on the predefined rating, while the values and coefficient are substituted in the following equation:

17.64 * 1 + 72.22 * 0.0112 + 0.012 * 705 + 0.18 * 34 + 1.26 * 1 + 2.53 = 36.81

The expected value = 40.8
The actual value = 36.81

5TH ITERATION

The value of b0 is 1, b1 is 0.157, b2 is 732, b3 is 35, and b4 is 6. We also determine the values of Σ an bn = 42 based on the predefined rating, while the values and coefficient are substituted in the following equation:

17.64 * 1 + 72.22 * 0.157 + 0.012 * 732 + 0.18 * 35 + 1.26 * 6 + 2.53 = 54.15

The expected value = 42
The actual value = 54.15

6TH ITERATION

The value of b0 is 1, b1 is 0.08, b2 is 823, b3 is 46, and b4 is 4. We also determine the values of Σ an bn = 55.2 based on the predefined rating, while the values and coefficient are substituted in the following equation:

$17.64 * 1 + 72.22 * 0.08 + 0.012 * 823 + 0.18 * 46 + 1.26 * 4 + 2.53 = 49.14$

The expected value = 55.2
The actual value = 49.14

7TH ITERATION

The value of b0 is 2, b1 is 0.157, b2 is 731, b3 is 31, and b4 is 2. We also determine the values of Σ an bn = 37.2 based on the predefined rating, while the values and coefficient are substituted in the following equation:

$17.64 * 1 + 72.22 * 0.157 + 0.012 * 731 + 0.18 * 31 + 1.26 * 2 + 2.53 = 49.08$

The expected value = 37.2
The actual value = 49.08

8TH ITERATION

The value of b0 is 1, b1 is 0.056, b2 is 460, b3 is 50, and b4 is 9. We also determine the values of Σ an bn = 60 based on the predefined rating, while the values and coefficient are substituted in the following equation:

$17.64 * 2 + 72.22 * 0.056 + 0.012 * 460 + 0.18 * 50 + 1.26 * 9 + 2.53 = 67.17$

The expected value = 60
The actual value = 67.17

9TH ITERATION

The value of b0 is 1, b1 is 0.112, b2 is 556, b3 is 18, and b4 is 2. We also determine the values of Σ an bn = 34.8 based on the predefined rating, while the values and coefficient are substituted in the following equation:

$17.64 * 1 + 72.22 * 0.112 + 0.012 * 556 + 0.18 * 18 + 1.26 * 2 + 2.53 = 40.69$

The expected value = 34.8
The actual value = 40.69

10TH ITERATION

The value of b0 is 2, b1 is 0.157, b2 is 731, b3 is 65, and b4 is 1. We also determine the values of Σ an bn = 60 based on the predefined rating, while the values and coefficient are substituted in the following equation:

17.64 * 2 + 72.22 * 0.157 + 0.012 * 731 + 0.18 * 50 + 1.26 * 1 +2.53 = 68.18

The expected value = 60
The actual value = 68.18

11TH ITERATION

The value of b0 is 2, b1 is 0.112, b2 is 648, b3 is 101, and b4 is 4. We also determine the values of Σ an bn = 60 based on the predefined rating, while the values and coefficient are substituted in the following equation:

17.64 * 2 + 72.22 * 0.112 + 0.012 * 648 + 0.18 * 50 + 1.26 * 4 + 2.53 = 67.71

The expected value = 60
The actual value = 67.71

12TH ITERATION

The value of b0 is 1, b1 is 0.112, b2 is 648, b3 is 44, and b4 is 6. We also determine the values of Σ an bn = 52.8 based on the predefined rating, while the values and coefficient are substituted in the following equation:

17.64 * 1 + 72.22 * 0.112 + 0.012 * 648 + 0.18 * 44 + 1.26 * 6 + 2.53 = 51.51

The expected value = 52.8
The actual value = 51.51

13TH ITERATION

The value of b0 is 2, b1 is 0.089, b2 is 823, b3 is 85, and b4 is 6. We also determine the values of Σ an bn = 60 based on the predefined rating, while the values and coefficient are substituted in the following equation:

17.64 * 2 + 72.22 * 0.089 + 0.012 * 823 + 0.18 * 50 + 1.26 * 6 + 2.53 = 70.67

The expected value = 60
The actual value = 70.67

14TH ITERATION

The value of b0 is 1, b1 is 0.157, b2 is 731, b3 is 30, and b4 is 1. We also determine the values of Σ an bn = 36 based on the predefined rating, while the values and coefficient are substituted in the following equation:

17.64 * 1 + 72.22 * 0.157 + 0.012 * 731 + 0.18 * 30 + 1.26 * 1 + 2.53 = 46.94

The expected value = 36
The actual value = 46.94

15TH ITERATION

The value of b0 is 1, b1 is 0.056, b2 is 460, b3 is 38, and b4 is 9. We also determine the values of $\sum$ an bn = 45.6 based on the predefined rating, while the values and coefficient are substituted in the following equation:

$$17.64 * 1 + 72.22 * 0.056 + 0.012 * 460 + 0.18 * 38 + 1.26 * 9 + 2.53 = 47.91$$

The expected value = 45.6
The actual value = 47.91

Figure 12.2 shows the graph representing the difference between the actual values and expected values. Figure 12.3 shows the graphical representation of variations in error percentages. Table 12.2 lists the evaluation results with all 15 iterations.

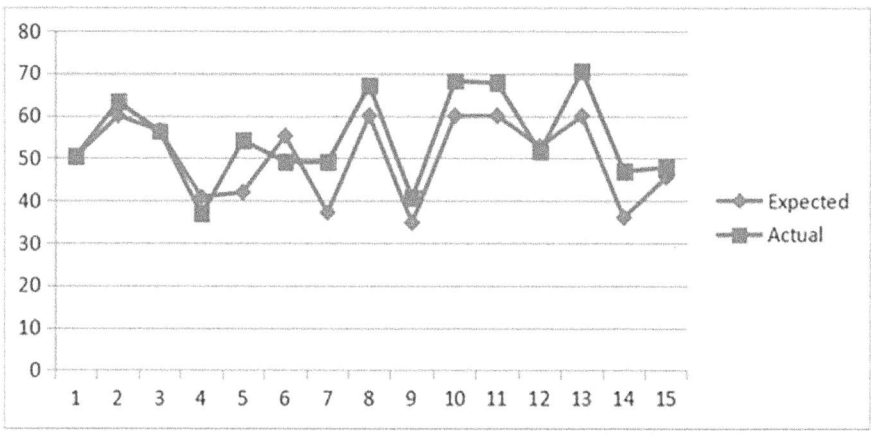

FIGURE 12.2   Actual values and expected values.

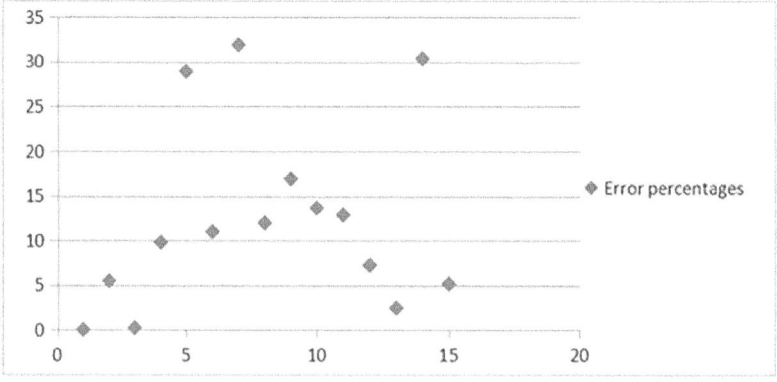

FIGURE 12.3   Error percentages

TABLE 12.2
Calculation Results

Iteration Number	Expected Value	Actual Value	Error	% of Error
1	50.4	50.4	0	0
2	60	63.28	3.28	5.46
3	56.4	56.28	0.12	0.21
4	40.8	36.81	3.99	9.77
5	42	54.15	12.15	28.92
6	55.2	49.14	6.06	10.97
7	37.2	49.08	11.88	31.93
8	60	67.17	7.17	11.95
9	34.8	40.69	5.89	16.92
10	60	68.18	8.18	13.63
11	60	67.71	7.71	12.85
12	52.8	51.51	1.29	7.23
13	60	70.67	10.67	2.44
14	36	46.94	10.94	30.38
15	45.6	47.91	2.31	5.13

12.5 CONCLUSION

The proposed system has been implemented in such a way that it has completely fulfilled all the mentioned objectives. The main aim is to spread awareness among the public in smart cities during the outbreak of the COVID-19 pandemic, and this is done with the help of the developed Android application (Rani, Khang et al., IoT & Healthcare, 2021).

The application also provides periodic notifications regarding the spread of this pandemic in the location specified by the user. The application works based on artificial neural networks to predict the severity of an epidemic and also its direction of spread.

As stated, the objective of the system is to improve the recall rate of users with special features like periodic alerts and prediction of spread. The degree of awareness about any epidemic in smart cities has a direct influence on the severity of the epidemic.

The system provides all required information regarding the epidemic, like symptoms, precautions, and treatment information, which help people to take the necessary measures to protect themselves from the epidemic. Therefore, the system has achieved its social cause and all the stated objectives.

The current system predicts the spread of an epidemic in the state of Tamil Nadu. It can further be extended to a national level, which includes predicting the spread of an epidemic in any state in India, as the mechanism used in the current system is scalable to a larger level as well. Thus, at a larger scale, the fully functional system will hinder the spread of epidemics in an efficient manner.

REFERENCES

Acuna Zegarra, M.A., M. Santana Cibrian, and J.X. Velasco Hernandez, "Modeling Behavioral Change and COVID-19 Containment in Mexico: A Trade OFF Between Lockdown and Compliance", *Math Bioscience*, 325, 108370, 2020. doi:10.1016/j.mbs.2020.108370.

Ahmed, S., "Potential of Age Distribution Profiles for the Prediction of COVID-19 Infection Origin in a Patient Group", *Inform Med Unlocked*, 20, 100364, 2020.

Arora, P., H. Kumar, and B.K. Panigrahi, "Prediction and Analysis of COVID-19 Positive Cases using Deep Learning Models: A Descriptive Case Study of India", *Chaos Solitons Fractals*, 139, 1–9, 2020.

Bhambri, P., Sita Rani, Gaurav Gupta, and Khang, A., "Cloud and Fog Computing Platforms for Internet of Things", ISBN:978-1-032-101507, 2022, doi:10.1201/9781032101507.

Bhatia, R., and P. Abraham, "Lessons Learnt During the First 100 Days of COVID-19 Pandemic in India", *Indian Journal of Medical Research*, 151(5), 387–391, 2020.

Booma, P.M., S. Prabhakaran, and R. Dhanalakshmi, "An Improved Pearson's Correlation Proximity-Based Hierarchical Clustering for Mining Biological Association between Genes", *The Scientific World Journal*, (6), 357873, 2014. doi:10.1155/2014/357873.

Ceylan, Z., "Estimation of COVID-19 Prevalence in Italy, Spain and France", *Science of the Total Environment*, 729, 138817, 2020. doi:10.1016/j.scitotenv.2020.138817.

Dhanalakshmi, R., A.M. Aloy, D. Shrijith, and N. Vijayaraghavan, "A Study on COVID-19, Impacting Indian Education", *Materials Today Proceedings*, March 2021. doi:10.1016/j.matpr.2021.02.786.

Dhanalakshmi, R., and T. Sri Devi. "Adaptive Cognitive Intelligence in Analyzing Employee Feedback Using LSTM", *Journal of Intelligent and Fuzzy Systems*, 39(6), 8069–8078, 2020. doi:10.3233/JIFS-189129.

Fabon, D., Thomas Lansdall-Welfare, FindMyPast Newspaper Team, NelloCristianini, "Discovering Periodic Patterns in Historical News", *PLoS one*, 11, 2016.

Gayathri, R., Rajiv Vincent, M. Rajesh, Arun Kumar Sivaraman, and A. Muralidhar,"Web-Acl Based Dos Mitigation Solution for Cloud", *Advances in Mathematics: Scientific Journal*, 9(7), 5105–5113, 2020.

Gupta, O.P., and S. Rani, "Bioinformatics Applications and Tools: An Overview", *CiiT-International Journal of Biometrics and Bioinformatics*, 3(3), 107–110, 2010.

Hadil, A., B. Fatemah, and A. Reem, "An Analytical Study on the Awareness, Attitude and Practice During the COVID-19 Pandemic in Riyadh, Saudi Arabia", *Journal of Infection and Public Health*, 13(10), 1446–1452, October 2020.

Huijuan, W., C. Chuyi, Q. Bo, L. Daqing, and H. Shlomo, "Epidemic Mitigation via Awareness Propagation in Communication Networks: The Role of Time Scales", *New Journal Physics*, 19, 2017. doi:10.1088/1367-2630/aa79b7.

Johnson, M., and R. Dhanalakshmi, "Predictive Analysis Based Efficient Routing of Smart Garbage Bins for Effective Waste Management", *International Journal of Recent Technology and Engineering*, 8(3), 5733–5739, 2019.

Khang, A., Pankaj Bhambri, Sita Rani, and Aman Kataria, *Big Data, Cloud Computing and Internet of Things*. CRC Press, 2022. ISBN:978-1-032-284200, 2022, doi:10.1201/9781032284200.

Khang, A., Subrata Chowdhury, and Seema Sharma, *The Data-Driven Blockchain Ecosystem: Fundamentals, Applications and Emerging Technologies*. CRC Press, 2022. ISBN:978-1-032-21624, doi:10.1201/9781003269281.

K.N., R.V., S.S.S., and D.R., "Intelligent Personal Assistant—Implementing Voice Commands enabling Speech Recognition", 2020 International Conference on System, Computation, Automation and Networking (ICSCAN), 2020, 1–5, doi:10.1109/ICSCAN49426.2020.9262279.

Kothandaraman, D., A. Balasundaram, R. Dhanalakshmi, Arun Kumar Sivaraman, S. Ashok Kumar, Rajiv Vincent, and M. Rajesh, "Energy and Bandwidth Based Link Stability Routing Algorithm for IoT", *Computers, Materials & Continua (CMC)*, 70(2), 3875–3890, 2021.

Kumar, A., and R. Roy, "Application of Mathematical Modeling in Public Health Decision Making Pertaining to Control of COVID-19 Pandemic in India", *Special Issue SARS-CoV-2 (COVID-19) Epidemiology International*, 5(2), 23–36, 2020.

Kumaresan, B., and N. Vijayaraghavan, "A Study on the Sectors that Increase the Percentage of Carbon Dioxide Which Causes Global Warming Using Fuzzy Cognitive Maps", *International Journal of Applied Engineering Research*, 10(5), 11581–11586, 2015.

Li, Y., B. Wang, R. Peng, C. Zhou, Y. Zhan, Z. Liu, et al., "Mathematical Modeling and Epidemic Prediction of COVID-19, and Its Significance to Epidemic Prevention and Control Measures", *Annals of Infectious Disease and Epidemiology*, 5(1), 1–9, 2020.

Li, Y., Z. Yang, Z. Dang, C. Meng, J. Huang, H. Meng, et al., "Propagation Analysis and Prediction of the COVID-19", *Infectious Disease Modelling*, 5(7), 282–292, 2020.

Lixia, Z., and L. Maoxing, "Effect of Awareness Programs on the Epidemic Outbreaks with Time Delay", *Abstract and Applied Analysis*, 2014.

Michael, C.S., A.B. David, J.P. Michael, and D. Mark, "Towards Real-Time Measurement of Public Epidemic Awareness: Monitoring Influenza Awareness through Twitter", *Association for the Advancement of Artificial Intelligence*, 2015. Corpus ID: 14806767. https://www.cs.jhu.edu/~mdredze/publications/2016_ossm.pdf

Muhannad, Q., and J. Yaser, "Multi-Tier Cloud Infrastructure Support for Reliable Global Health Awareness System", *Simulation Modelling Practice and Theory*, 67, 44-58, 2016. doi:10.1016/j.simpat.2016.06.005.

Nath, K., R. Dhanalakshmi, V. Vijayakumar, B. Aremu, K.H. Kumar Reddy, and X.Z. Gao. "Uncovering Hidden Community Structures in Evolving Networks Based on Neighborhood Similarity", *Journal of Intelligent & Fuzzy Systems*, 39(6), 8315–8324, 2020.

Naveen, D., and Anders Gustafsson, "Effects of COVID-19 on Business and Research", *Journal of Business Research*, 117, 284–289, September 2020. doi:10.1016/j.jbusres.2020.06.008.

Neha, S., S. Uma, R.C. Nupa, and K.T. Dinesh, "Firebase Cloud Messaging (ANDROID)", *International Journal of Innovative Research in Science, Engineering and Technology*, 6(Special Issue 9), 2017. http://www.ijirset.com/upload/2017/cotii/3_CS_COTII_2017_Firebase_cloud.pdf

Nicholas, T., S. Donal, H. Chris, and G. Joseph, "DEFENDER: Detecting and Forecasting Epidemics Using Novel Data-Analytics for Enhanced Response", *Journal PLOS One*, 2016. doi:10.1371/journal.pone.0155417.

Patrikar, S., D. Poojary, D.R. Basannar, and R. Kunte, "Projections for Novel Coronavirus (COVID-19) and Evaluation of Epidemic Response Strategies for India", *Medical Journal Armed Forces India*, 76(3), 268–275, 2020.

Peninah, M.M., R. MbabuMurithi, I. Peter, H. Allen, M.T. Samuel, A.A. Samuel, Jusper Kiplimo, B. Bernard, V. Anton, F.B. Robert, and M. Kariuki Njenga, "Predictive Factors and Risk Mapping for Rift Valley Fever Epidemics in Kenya", *Journal PLOS One*, 2016. doi:10.1371/journal.pone.0144570.

Rana, G., Khang, A., Khanh, H.H., "The Role of Artificial Intelligence in Blockchain Applications", in *Reinventing Manufacturing and Business Processes Through Artificial Intelligence* (20). CRC Press, 2021. doi:10.1201/9781003145011.

Rani, S., and O.P. Gupta. "CLUS_GPU-BLASTP: Accelerated Protein Sequence Alignment Using GPU-Enabled Cluster", *The Journal of Supercomputing*, 73(10), 4580–4595, 2017.

Rani, S., A. Kataria, and M. Chauhan, "Fog Computing in Industry 4.0: Applications and Challenges—A Research Roadmap", in *Energy Conservation Solutions for Fog-Edge Computing Paradigms* (173–190). Springer, Singapore, 2022.

Rani, S., A. Kataria, V. Sharma, S. Ghosh, V. Karar, K. Lee, and C. Choi, "Threats and Corrective Measures for IoT Security with Observance of Cybercrime: A Survey", *Wireless Communications and Mobile Computing*, 2021, 1–30, 2021.

Rani, S., and S. Kaur. "Cluster Analysis Method for Multiple Sequence Alignment", *International Journal of Computer Applications*, 43(14), 19–25, 2012.

Rani, S., Khang, A., Meetali Chauhan, and Aman Kataria, "IoT Equipped Intelligent Distributed Framework for Smart Healthcare Systems", *Networking and Internet Architecture*, 2021. https://arxiv.org/abs/2110.04997v2, doi:10.48550/arXiv.2110.04997.

Rani, S., R.K. Mishra, M. Usman, A. Kataria, P. Kumar, P. Bhambri, and A.K. Mishra, "Amalgamation of Advanced Technologies for Sustainable Development of Smart City Environment: A Review", *IEEE Access*, 9, 150060–150087, 2021.

Rina, T., S.A. Saad, A.A. Ahmed, M.M. Abdulkarim, et al., "Awareness and Preparedness of COVID-19 Outbreak Among Healthcare Workers and Other Residents of South-West Saudi Arabia: A Cross-Sectional Survey", *Frontiers in Public Health*, 18 August 2020. doi:10.3389/fpubh.2020.00482.

Roosa, K., Y. Lee, R. Luo, et al., "Short Term Forecasts of the COVID-19 Epidemic in Guangdong and Zhejiang", *Journal of Clinical Medicine*, China 13–23 February 2020. doi:10.3390/jcm9020596.

Sahoo, B.K., and B.K. Sapra, "A Data Driven Epidemic Model to Analyse the Lockdown Effect and Predict the Course of COVID-19 Progress in India", *Chaos Solitons Fractals*, 139, 110034, 2020.

Santhosh, K.C., "AI Driven Tools for Coronavirus Outbreak: Need of Active Learning and Cross Population Train/Test Models on Multitudinal/Multimodal Data", *Journal of Medical Systems*, 44–93, 2020.

Sebastian, F., G. Erez, W. Chris, A. Vincent, and A. Jansen, "Spread of Awareness and its Impact on Epidemic Outbreaks", *PNAS*, 106(16), 6872–6877, April 2009.

Shakil, M.H., Z.H. Munim, M. Tasnia, and S. Sarowar, "COVID-19 and the Environment: A Critical Review and Research Agenda", *Science Total Environment*, 745, 141022, 2020.

Sharnil, P., S. Anirban, and K. Ketan, "Smart Epidemic Tunnel: IoT-based Sensor-Fusion Assistive Technology for COVID-19 Disinfection", *International Journal of Pervasive Computing and Communications*, 31 August 2020. doi:10.1108/IJPCC-07-2020-0091.

Shi, P., Y. Dong, H. Yan, C. Zhao, X. Li, W. Liu, et al., "Impact of Temperature on the Dynamics of the COVID-19 Outbreak in China", *Science of the Total Environment*, 728, 138890, 2020.

Simon, D., R. Rebecca, and G.P. Oliver, "Explaining the Geographic Spread of Emerging Epidemics: a Framework for Comparing Viral Phylogenies and Environmental Landscape Data", *BMC Bioinformatics*, 2016. doi:10.1186/s12859-016-0924-x.

Singh, P., O.P. Gupta, and S. Saini. "A Brief Research Study of Wireless Sensor Network", *Advances in Computational Sciences and Technology*, 10(5), 733–739, 2017.

Sudevan, S., Bushra Al Barwani, Esraa Al Maani, Sita Rani, and Arun Kumar Sivaraman, "Impact of Blended Learning during Covid-19 in Sultanate of Oman", *Annals of the Romanian Society for Cell Biology*, 25(4), 14978–14987, 2021.

Tomar, A., and N. Gupta, "Prediction for the Spread of COVID-19 in India and Effectiveness of Preventive Measures", *Science of the Total Environment*, 728, 1–6, 2020.

Vijayaraghavan, N., and B. Kumaresan, "A Study on Significance of Globalization of Higher Education in India Using Fuzzy Cognitive maps", *Global Journal of Pure and Applied Mathematics*, 1(5), 2993–2999, 2015.

Wang, L., J. Chen, and M. Marathe, "DEFSI: Deep Learning Based Epidemic Forecasting with Synthetic Information", *Proceedings of the AAAI Conference on Artificial Intelligence*, 33(01), 9607–9612, 2019.

Wang, S., Y. Pan, Q. Wang, H. Miao, A.N. Brown, and L. Rong, "Modeling the Viral Dynamics of SARS, COV-2 Infection", *Math Bioscience*, 328,108438, 2020.

Wong, A., Serene Ho, Olusegun Olusanya, Marta Velia Antonini, and David Lyness, "The Use of Social Media and Online Communications in Times of Pandemic COVID-19", *Journal of the Intensive Care Society*, 22 October 2020. doi:10.1177/1751143720966280.

Yang, Z., Z. Zeng, K. Wang, S.S. Wong, W. Liang, M. Zanin, et al., "Modified SEIR and AI Prediction of the Epidemics Trend of COVID-19 in China under Public Health Interventions", *Journal of Thoracic Disease*, 12(3), 165–174, 2020.

Yogesh, K.D., D.L. Hughes, C. Crispin, C. Ioann, D.J. Yanqing, and S. Edwardseand Babita G., "Impact of COVID-19 Pandemic on Information Management Research and Practice: Transforming Education, Work and Life", *International Journal of Information Management*, 55, 102211, December 2020.

13 Smart Libraries as Promoters of Development in the Smart City Context
A Smart Library Infrastructure Model Proposal

Humberto Martínez-Camacho

CONTENTS

DOI: 10.1201/9781003252542-13

13.1 INTRODUCTION

Smart cities are an industry that promises to reinvent the way we live. The smart city concept arises from the need to rationalize and optimize through technologies the use of resources susceptible to end and suffer social dilemmas, such as water and energy, among others, which have consequences and problems, such as congestion, pollution, and overexploitation.

According to Galve-montore (2019), there are different models of economic development of cities and territories. The smart city archetype is one of them. He argues that this archetype is increasingly consolidating because it is a global understanding of current cities and its multiple approaches that ultimately aims to improve the future of citizenship (Rani & Gupta, 2017).

Any smart city project is based on an information system composed of infrastructure, data processing, and management platforms. Smart cities seek to add values such as inclusion and transparency and improve the distribution of resources and encourage the specialization of a city in some specific aspect.

In this sense, the library, as one of the most trusted spaces and sources of information, is playing an increasingly important role in smart cities, and according to Crowe (2020), as cities partner with libraries to disseminate information about municipal services, close digital divides, and drive civic engagement, this role is likely to continue to evolve (Rani, Kataria et al., 2021).

The library is a growing organism, and so is its relationship with its users. As Ranganathan pointed out in his fifth law, libraries are a growing organism, which gives importance to the trinity, which contemplates the user, services, and resources.

The 21st century has presented an explosion of technological innovation, wireless networks, and the Internet (García-Marco, 2011), and for this reason, the library is reinventing itself at an accelerated pace. Regarding the concept of a smart library, there is still no clear definition. However, most agree that the smart library aims to provide better services to users by making use of technology and communication, for which Cao et al. (2018) present a three-dimensional model: the technological dimension, service dimension, and user-oriented dimension (Rani, Chauhan, Kataria, & Khang, 2021).

In addition, in order for a library to become intelligent, Baryshev et al. (2015) mention some basic requirements for constructing a smart library: the creation of intelligent environments, which refers to the sustainable management of resources; the adoption of intelligent technologies to give way to an intelligent atmosphere; mobile access, which implies the accessibility of all library services to all corners of the world; the creation of new knowledge in a collaborative environment and expert systems; and other important aspects, such as adaptability and the use of intelligent technologies for content formation, intelligent detection of knowledge involving the use of metrics information, and impact factor to acquire new sources of knowledge and finally intelligent and innovative services.

13.2 IOT DEVICES FOR SMART LIBRARIES

The Internet is composed of interactive technologies that make sense of each other. IoT is a group of interconnected technological devices or systems that collect data from the environment and share it with each other over the Internet. IoT acts as an enabler, is easy to use for smart library users, and offers the possibility of providing a solution that improves the efficiency of service in libraries (Gul and Bano, 2019).

It is important to know the current context of smart libraries. For this purpose, the following are some of the concepts, definitions, and practices that smart libraries are implementing in order to optimize their services (Rani & Kumar, 2022).

13.2.1 CLOUD COMPUTING

One of the emerging tools for IoT is cloud computing, which is just a collection of software and computing services that can be accessed directly through the Internet rather than from the desktop or internal server, and its availability to multiple users at the same time is one of the main features. With cloud computing, various applications, such as creating repositories, searching library data and resources, automating libraries, hosting websites, searching academic content, storing files and modifying data easily from any terminal, building the power of community, and enabling innovation, can be realized (Gul & Bano, 2019).

Currently, different entities are seeking to migrate their management to other ways of working, such as cloud computing. Sun et al. (2019), in their work, show a cloud library platform model that was developed and implemented using the resource retrieval function. At first, the resource retrieval function was achieved under the cloud library platform and also achieved the sharing of electronic library resources (Rani, Khang et al., IoT & Healthcare, 2021).

However, the authors point out that this cloud library resource sharing system still faces many problems, such as responsibilities, rights, obligations, contract execution time of each library, payment methods, amounts, and so on (Rani, Mishra et al., 2021).

13.2.2 FACIAL RECOGNITION TECHNOLOGY

As described Yalagi and Mane (2021), generally this service is used to authenticate the user in order to provide access to the library. Another application that we can find in smart libraries, according to Gul and Bano (2019), is the "magic mirror," which consists of a camera that includes Wi-Fi-enabled sensors, allowing interaction between people and computers.

An example of this is the system proposed by Bagal and Saindane (2019), consisting of the user having to stand in front of a camera and show his face and the QR code of the book to make the loan. For the return, the book is placed in a mailbox so that the QR code is facing upward so that the camera located at the top can scan the code and update the database. Shi et al. (2021) suggest that character recognition technology can be useful when QR or barcodes cannot be read.

13.2.3 RFID TECHNOLOGY

The RFID (radio frequency identification) systems include tags that are electronically programmed with unique information, readers or sensors to query the tags, and the antenna and the server on which the software is loaded and interacts with the integrated library software, helping library staff to save time in various processes within the library (Shahid, 2005).

One application in smart libraries is the remote monitoring of seat occupancy status and location of patrons' seats in the library, as proposed by García-Marco et al. (2011) in their paper, which exposes an IoT-based solution by integrating under-seat pressure sensors and RFID reader sensors on the side of the seat base that is connected to a node. The readings from the seat sensors were sent as a request to a remote database server, and then to a web application in real time.

The prototype not only saves time for students searching for vacant (unoccupied) seats or fellow students in the library but also enables proper utilization of library resources, monitoring of seat availability and location of library users by staff.

RFID technology helps, among other things, to detect the incorrect locations of books in the library, identify if a book is lying down, and even, as Du and Liu (2019) suggest, analyze the interaction between readers and books, with their model reaching 92.2% accuracy, and this allows to take book recommendation and space design to another level.

13.2.4 Data Mining

Gul and Bano (2019) state that data mining is an important component of information retrieval, also known as knowledge discovery in databases; it is the process of analyzing large repositories of information and discovering embedded but potentially useful information.

On the other hand, F. Chen et al. (2015) argue that data mining involves applying algorithms to extract "hidden" information. Data mining in libraries is called bibliomining and has been used in libraries to map the tracking of user and staff behavior and also the use of information resources. According to some studies (Chang & Chang, 2009; Chen & Chen, 2007; Kovacevic et al., 2010; Nicholson, 2006; Renaud et al., 2015; Shieh, 2010), data mining in smart libraries exists because of the increasing volume of data, the limitations of human analysis, low-cost machine learning, the information explosion, the global business environment, and the availability of global systems software.

13.2.5 AI, Robots, and Chatbots

In the literature, we can find concepts of AI that are generally associated with the use of computational models that allow computers to perform cognitive processes, learn new concepts and tasks, and reason and make useful conclusions. These models can perceive and understand natural language and comprehend visual scenes in a similar way to human intelligence.

The intelligent robot is an indispensable part of AI. Even as its tangible embodiment, it is important to mention that not all robots can be called AI when only performing mechanical operations. So according to the differences mentioned, AI and robotics have a strong link with computer science and mathematics, but AI is based more on psychology, philosophy, linguistics, and cognitive sciences, and robotics requires physics and engineering in some way (Zheng, 2019).

Robots can potentially replace, in intensive fields, knowledge and labor. In the library field, knowledge and labor are combined so that robotics naturally involves in its functions. Such is the case of Nanjing University's reception robot, Tu Bao, which informs the reader of the precise location on the shelf of a book, updated in real time with a laser sensor for autonomous navigation and position of the library space located by magnetic navigation and can say simple voice greetings (Yuke, 2017).

Another example is Pepper, which was the first robotic helper introduced into the library system in the United States by the Roanoke County Public Library. Pepper can answer a list of pre-programmed questions, tell jokes, and reveal some dance steps. Both robots have limited learning capabilities, unlike Alpha Dog, which is self-evolving (Petska, 2018).

Another type of robot that is used on a large scale in libraries is the robot that repeats routine actions; in most cases, such robots do not have advanced cognitive decision-making functions and are similar to industrial robots in warehouses and factories.

There are also other types of robots better known as chatbots, such as Pixel, which was developed by the University of Nebraska-Lincoln and provides immediate answers to questions about library services and resources to users in natural language, which emulates a human conversation (Allison, 2012). Another chatbot mentioned is Xiao Tu, from Tsinghua University, to provide real-time virtual reference services online 27/7 throughout the year in an intelligent and highly interactive way (Yao et al., 2015).

13.2.6 BLOCKCHAIN

This technology can be used in libraries to store information in temperate-resistant environments, and this technology can be used in the world of scientific publishing. It has great potential in creating an enhanced metadata system (Khang et al., AI & Blockchain, 2021) for libraries to connect with different libraries and maintain digital rights.

13.2.7 AUGMENTED REALITY

As mentioned Pence (2010), augmented reality (AR) is a computer-generated component added to the real environment, while in true virtual reality, the entire experience is computer-generated. AR involves the senses. In libraries, they use it to tell stories and make users experience real sensations. Arroyo-Vázquez (2016) mentions some examples of augmented reality in libraries, starting with the Spanish geolocation project Bibliotecas de España, in which more than 8,200 Spanish libraries are located along with their address, contact details, and how to get to them. She mentions that while maps force us to translate what we see around us into a graphic representation in order to consist of them, with AR we directly visualize the direction we are heading and do not have to make the mental effort of interpreting the representation that makes up the map.

Another application of AR is to show places or historical events in their context, in the place where they happened. It points out that this type of initiative involves two elements, geolocation and mobility. An example of this is the GeoStoryteller Project, developed by the Pratt Institute School of Information and Library Science and the librarians of the Goethe Institute.

Augmented reality is a powerful ally to dynamizing exhibitions, among other cultural activities. An example of this is the Archive LAPL application of the Los Angeles Public Library. Finally, it is mentioned that digital objects can also be superimposed on printed publications and physical spaces of the library, among others (ArroyoVazquez, 2016).

13.2.8 VIRTUAL REALITY

According to Xu et al. (2019), virtual reality (VR) is a computer simulation technology that integrates computer graphics, human-machine interface technology, network technology, stereoscopic display technology, and simulation technology, among others, which can generate a highly realistic virtual environment in which through auditory and tactile senses the user can interact with the virtual world and produce immersive sensations. This technology applied in libraries can be mainly classified into three categories: library virtual scene design, library resource management, and library service innovation. The author argues that most of the relevant research focused on virtual libraries developed on the Second Life (SL) platform. Some of them are the SL project of Stanford University, the Beaver Tracks project of Oregon State University Library, the virtual digital library of the National Library of China, the 3D Roving Library project of Wuhan University Library, and the 360° Panoramic Library project of Sun Yat (Luo, 2020).

The implementation of virtual reality technology can help improve the current situation of building library information resources. Aurasma and other software can convert static graphics into dynamic videos or animations. In this sense, special devices such as Oculus or Xbox One can provide users with a three-dimensional and fully immersive reading environment.

Regarding services, in addition to reading, some libraries use this technology to display the entire library building and the location of each book in detail. The system generates the optimal navigation path at the same time, helping users to quickly find the books they need (Hu et al., 2019). Users can also consult the reference desk and librarians in SL libraries, while librarians can provide services such as answering users' questions and guiding them through the library (Ford et al., 2008).

13.2.9 VOICE ASSISTANT

According to Laurel (1997), voice assistants, or virtual assistants, are computer programs designed with human characteristics that act on behalf of users in digital environments through voice interfaces.

Examples of these digital assistants include Amazon's Alexa, Microsoft's Cortana, Apple's Siri, and Google Assistant. Sweeney and Davis (2020) argue that voice assistants can be run on various IoT devices and numerous third-party applications, allowing libraries to create specialized uses for these technologies as part of their regular information services. The authors mention that vendors, such as

Overdrive (for e-book lending) and Hoopla (multimedia lending) are configured to connect to voice.

In this way, Crowe (2020) agrees that library providers recognize the potential of these technologies and point out various applications in their writing, such as Beanstack Tracker for reading challenges. Libro, from the company ConverSight. ai., allows one to search a library catalog and identify the availability of resources, retrieve and renew library materials, and find the location of an item or inquire about library schedules and events. An example is the University of IOWA which has Parks Libro.

EBSCO's interface that allows users to access content from their discovery service through Alexa and Google Home. Communico can interact with Alexa for user account management. Demco allows users to use Alexa to check library hours and services, place holds and renew items, discover and register for library events, or reserve meeting rooms. Finally, Ex Libris features Hey Primo, which is a voice search assistant that allows users to enter a search using their device's microphone.

Some other integrations of voice assistants within libraries are the enhancement of reading, along with music and effects, and consultation services for elderly users or people with disabilities, such as title search, reservations, and information about library events, among others.

13.2.10 Electronic Resource Management

They arise with the advent of ICT and include mainly e-journals, e-books, academic databases, value-added information portals, online e-book libraries, other online publications, and others (Cao et al., 2018) and, according to Patra (2017), consist of five components: procurement management, access management, administration, management, support management, and evaluation monitor management.

Some works (Oderuth et al., 2019; Younis, 2012) propose the development of an RFID-based smart library management system (SLMS) that they claim would significantly save users and library staff time and effort by performing their tasks with a high degree of automation.

13.2.11 Smart Buildings

As mentioned, Hoy (2016) goes beyond simply turning things on and off. Smart buildings also collect data on when and how a building is used via sensors and cameras, revealing data such as the number of occupants in a building, peak hours of use, and congestion within the building to predict the future. The author comments that for the management of these buildings, management systems are used that can alert staff to problems, such as burned-out lights, plumbing leaks, flooding, doors left open, etc. Finally, he mentions that these technologies are also being applied to energy efficiency through smart buildings.

Examples of smart building applications include new energy-efficient LED lighting fixtures that can be powered via ethernet cables, which can also transmit control information and data to the lighting devices; these lighting appliances are ideal for integration with solar panels (Castle, 2013).

Building dashboards are publicly available websites that provide real-time data on how much energy a building consumes and produces and how many occupants are in the building. Currently, Salt Lake County Library Services has a dashboard for several of its buildings (Hoy, 2016).

Another application within smart buildings is electrochromic windows, or "smart glass," thin films and an electrolyte sandwiched within the glass, which reduces glare and heat transfer while allowing visible light to pass through the windows. The darkness of the tint can be adjusted by passing an electric current through the glass. These windows can reduce air conditioning costs by 50% in large buildings (Khang, Chowdhury et al., Blockchain, 2022).

According to Su and He (2014), the location of a building is important and the same is in harmony with the environment. They also point out that the design of the interiors is essential to provide a comfortable learning environment by appealing to humanistic needs for color, ornaments, furniture, and artwork, among others. They suggest that an intelligent building automation system covers the power distribution system, lighting system, elevator monitoring system, air conditioning system, water supply, and drainage system.

13.2.12 POSITIONING TECHNOLOGY

It is another alternative use within intelligent buildings. Its relationship with libraries, according to Hawkins (1994), is from the 1990s. Geographic information systems (GIS) in libraries are used more frequently in the United States than in Europe. Moore (2005) points out that this is largely due to the lack of specialized librarians. In this sense, Aguilar-Moreno and Granell-Canut (2015) argue that, in addition, their value is not appreciated, and geographic data is not considered a management priority.

On the other hand, Aguilar-Moreno and Granell-Canut (2013) point out that GIS can be useful for the optimal management and development of collections up to the design of space (Xia, 2004). Some of these systems also have the ability to locate specific people within the building, using ID cards or cell phones.

13.2.13 INTELLIGENT ENVIRONMENTS

As indicated by Johnson (2013), one of the most important factors in these smart environments is the inclination to natural conditions, the development of sustainable resource management and, the adoption of smart technologies in the library.

Monti et al. (2019) present an experiment in which they use IoT to monitor environmental conditions, such as temperature, relative humidity, and contaminants that may adversely affect a storage facility dedicated to preserving books and paper documents using a low-cost platform called Canarin II.

These and some other technologies, such as 5G Internet connection and virtual assistants, among others, combine to offer better user and librarian experiences, reducing time and effort.

Burgess (2010) argues that smart libraries focus on the use of technology and are designed to provide a highly collaborative learning environment in which participants are encouraged to contribute ideas and information. Further comments that smart libraries enhance traditional and non-traditional library services and the user experience and improve opportunities for student learning.

According to Baryshev et al. (2018), a new type of service development may require rethinking the overall mission of the library with a focus on new information technology, emphasizing the need for the library to be flexible to adapt quickly to these changing needs.

Blewitt (2014) mentions that new libraries must have architecturally imaginative and aesthetically intense designs that attract public interest and use and thus quickly become an important element of a locality's cultural economy.

According to Cao et al. (2018), the challenges are mainly in three areas: rapid technological changes in the environment, extensive and rapid data growth, and the increase and diversification of user needs. Smart libraries have three important components: smart technologies, smart people, and smart services (Gul & Bano, 2019).

In this sense, Liao (2020) defines a hierarchical framework in which the smart city comes first, then the smart library, followed by smart services, and finally the smart user, who is the target and guide of these services.

According to Yu and Huang (2020), consumer intent is affected by their peers, so it is important to retain existing users, improve their perception of the usefulness of smart libraries, and provide incentives for their continued use, so library design should seek to maximize intelligent interaction that can set subjective standards and increase user readiness, these designs can be tailored to different user segments by adding specific interfaces.

At present, the application of IoT technology in the library is still in the initial stage and has many problems, such as the cost of sensor nodes, construction and design, technical standards, and security issues (Du & Liu, 2014)

13.3   METHODOLOGY

One hundred twenty-three scientific articles were analyzed for the preparation of this chapter. Some of the analyzed papers, in turn, study other sets of publications, as in the case of Barishev, who analyzes more than 200 publications on smart library services. Some works (Cao et al., 2018; Gul & Bano, 2018; Zimmerman & Chang, 2018) also examine various papers with the aim of conceptualizing the smart library.

These documents were obtained from multiple databases, especially Scopus and WoS, by searching for keywords such as "smart library," "smart city," "technology," and "libraries." The main technological implementations in libraries were categorized, and the aim was to conceptualize these technologies and show the best practices carried out.

A model with the architecture of an intelligent library based on the technologies and services identified in the literature review is also proposed. This model adopts various structures from other models with the intention of showing an overview of infrastructure for intelligent libraries.

The model is elaborated in Lucid Chart and is available in Figshare for its consultation and edition, as well as the list of the bibliography consulted (Martínez-Camacho, 2022).

13.4   PROPOSED MODEL FOR A SMART LIBRARY INFRASTRUCTURE

From the literature review, a wide variety of products and services that can be implemented in a library to transform it into a smart library were identified.

In addition, there are multiple versions of smart library models, such as the one by Baryshev et al. (2018), which is perhaps one of the simplest, as shown in Figure 13.1.

Later, the same author proposes a more elaborate model in which he first shows an outline of how the smart library works. His model shows that a reader has several opportunities to access resources: traditionally, by directing the consultation to a librarian, through a computer in the reading room, or through any device that allows access to a personal account regardless of where the user is physically located.

This resource access system includes the electronic directory of digital and print resources, electronic directories of partner libraries, as well as subscriptions to other resources and databases, as shown in Figure 13.2 and Figure 13.3.

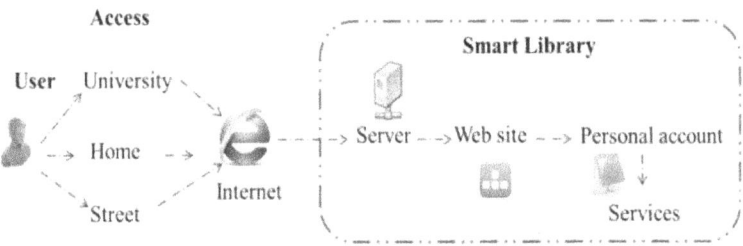

FIGURE 13.1   Philosophy of the smart library of a university.

Note: Adapted from "The smart library project Development of information and library services for educational and scientific activity" (p. 541), by Baryshev, R. A. (2018), *Electronic Library*, 36 (3).

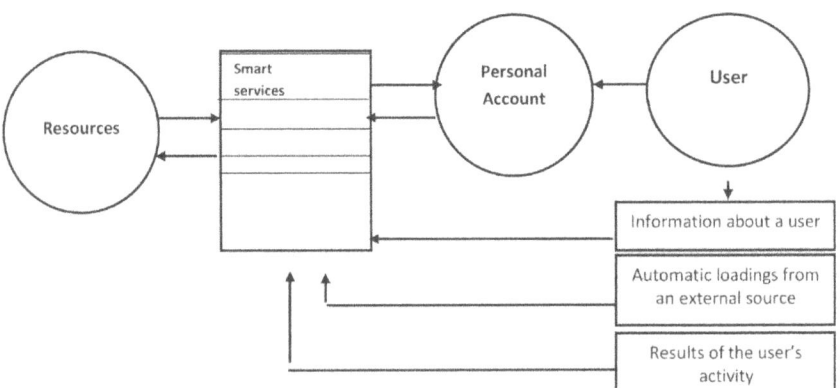

FIGURE 13.2   Smart library.

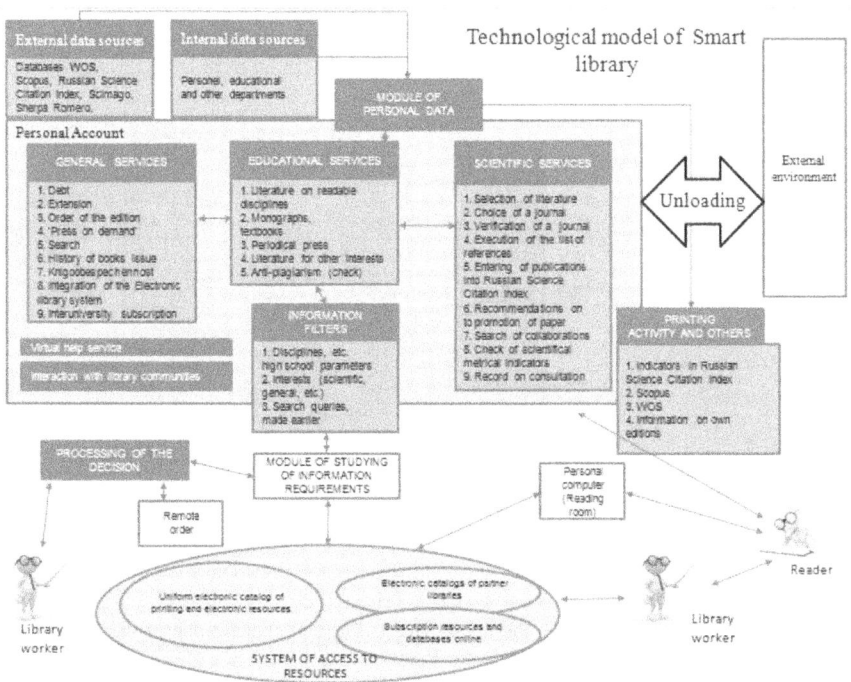

FIGURE 13.3   Smart library model.

Note: Adapted from "From electronic to smart library systems: Concept, classification of services, scheme of work and model" (pp. 435–436), by Baryshev, R. A. (2021), *Journal of Siberian Federal University—Humanities and Social Sciences*, 14 (3).

On the other hand, Cao et al. (2018) suggest that a traditional library cannot become a smart library using a single technology. Rather, integration and interaction with a variety of technologies are required. They propose that the smart library technology integration framework can be divided into three layers: perceptual, computing, and communication, as shown in Figure 13.4.

There are different types of frameworks for various items of the smart library (Luo et al., 2020), proposing an architecture based on cloud computing, as shown in Figure 13.5.

Chen et al. (2021) propose a model based on big data and artificial intelligence, as shown in Figure 13.6.

Zhao (2019) shows us a service-centric model of the smart library, which mainly includes the local service layer, the industry service layer, the service solution layer, and the upper and lower external layers as Figure 13.7 and Figure 13.8.

A year later, Liao (2020) also presented a model ofn smart library services in which the main line is a smart city, smart library, smart services, and target users, starting with a smart city and ending with a target user, as shown in Figure 13.9.

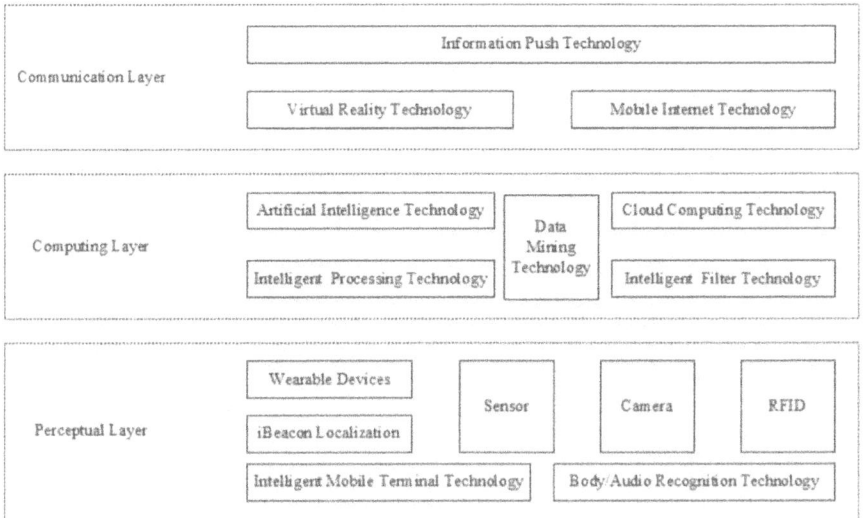

FIGURE 13.4   Smart library.

Note: Adapted from "How to make the library smart? The conceptualization of the smart library" (p. 818), by Cao et al., 2021, *Electronic Library*, 36 (5).

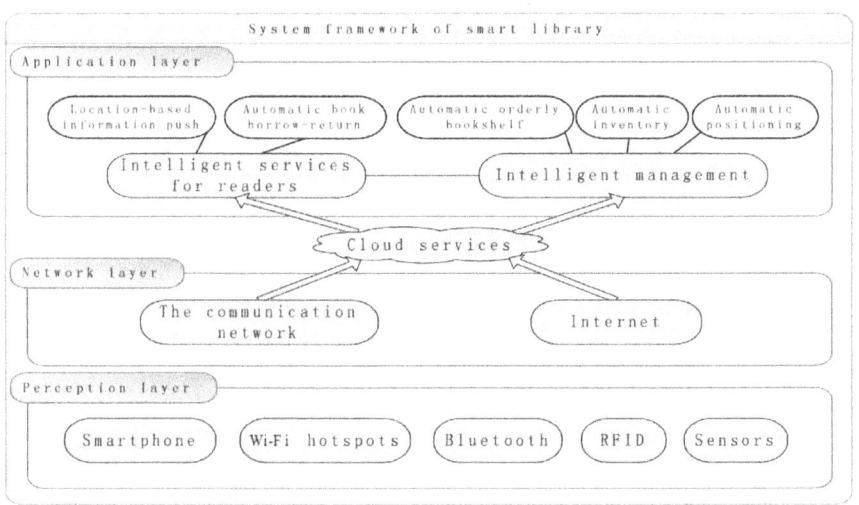

FIGURE 13.5   The cloud-based smart library architecture.

Note: Adapted from "Exploration and construction of smart library based on RFID technology" (p. 1744), by Luo et al., 2021, *Advanced Materials Research*, (765–767).

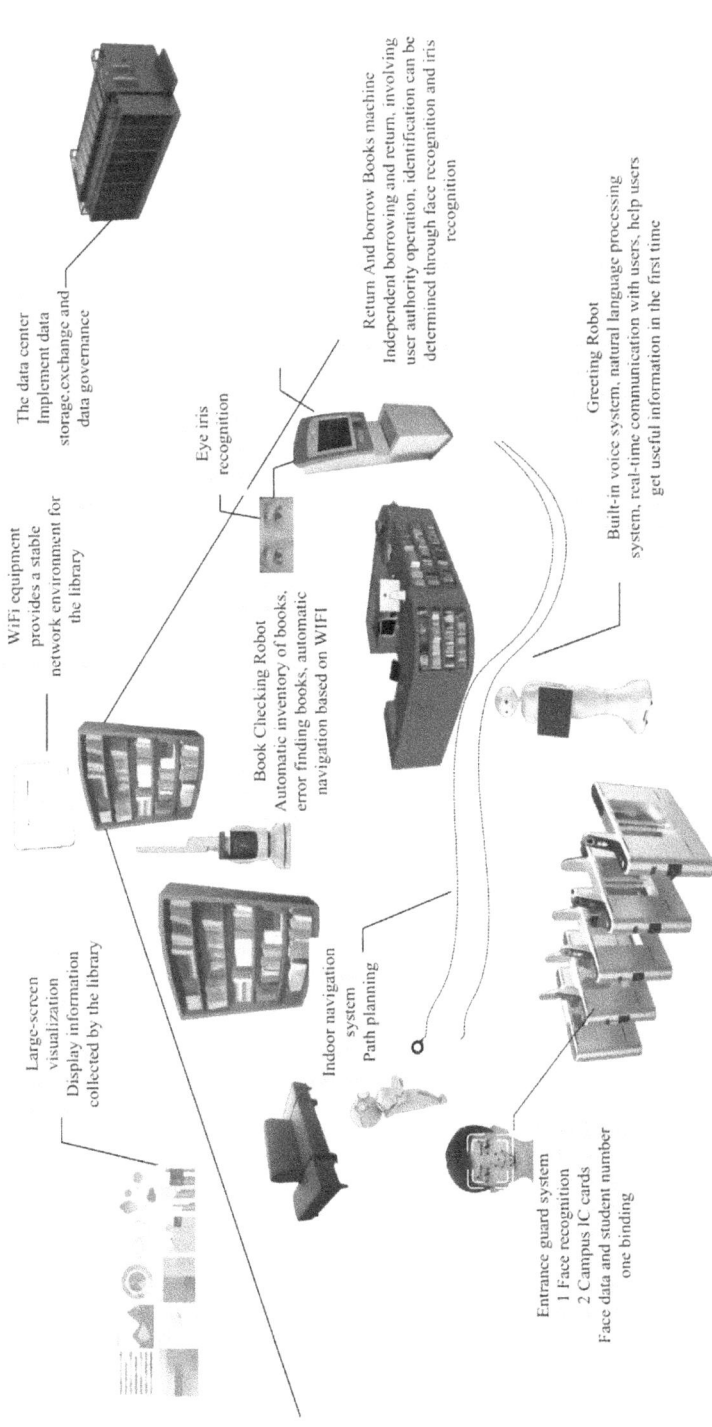

FIGURE 13.6   Smart library simulation scene.

Note: Adapted from "Analysis of the Smart Library Construction in Colleges Based Big Data and Artificial Intelligence" (p. 7), by Z. Chen et al., 2021, *Journal of Physics: Conference Series*, 1955 (I).

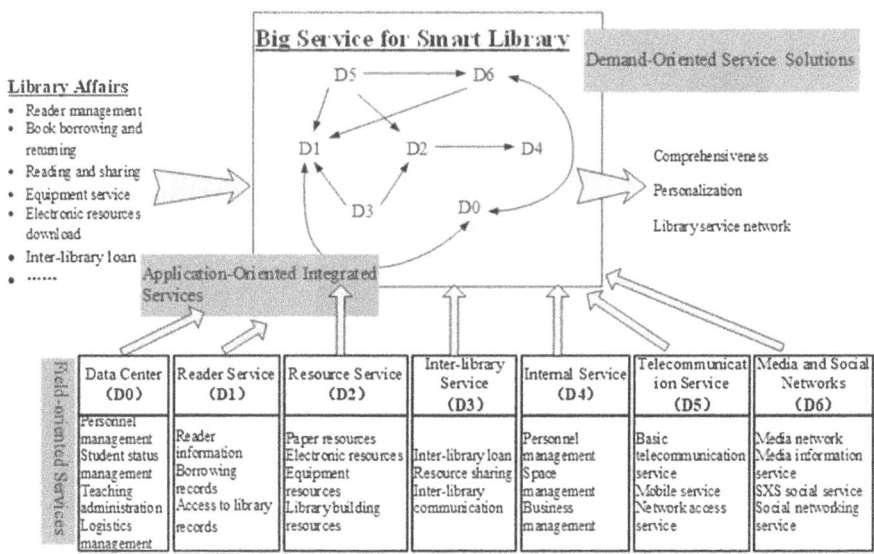

FIGURE 13.7   Big service for smart library

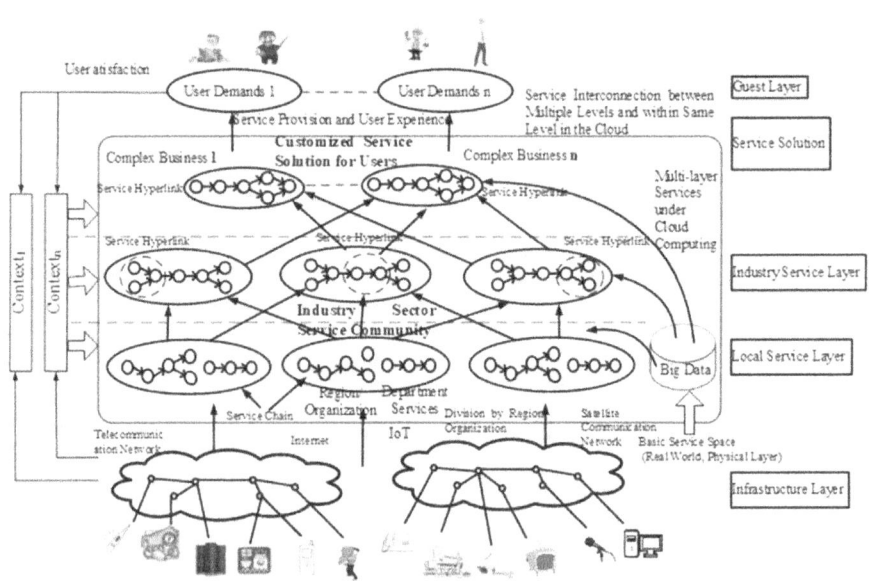

FIGURE 13.8   Architecture of big service for smart library

Note: Adapted from "Research on Smart Library Big Service Application in Big Data Environment" (p. 242–243), by Zhao et al., 2019, *Advances in Computational Science and Computing* (877).

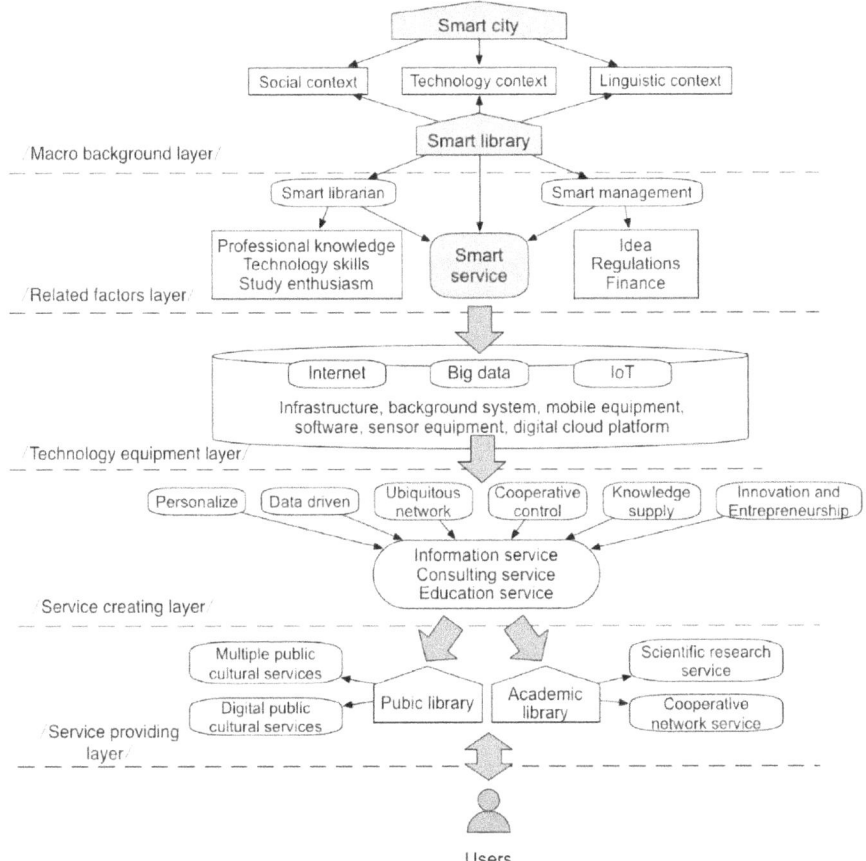

FIGURE 13.9   Theoretical framework of a smart library service.

Note: Adapted from "The Theoretical Framework of Library Smart Service" (p. 472), by Liao (2020), *Literature Review and Content Analysis.*

Leorke and Wyatt (2019) suggest an interesting postulate: they propose that, in addition to technological advances, it is plausible to assume that users continue to value the (third good) space of their community that they may go to for leisure, for exchange, for intellectual work, or for introspection.

Schöpfel (2018) presents four dimensions for the smart library, including smart services, smart users, smart building, and smart governance, as shown in Figure 13.10.

Sayogo et al. (2021) also propose four dimensions coincide in three dimensions with the Scöpfel model, except for the smart place dimension; instead, the authors suggest smart technology, as shown in Figure 13.11.

The following is a proposed model for an intelligent library infrastructure. As mentioned, the criteria for its elaboration are based on the technologies and services described in the literature review, and this model also adopts various structures from

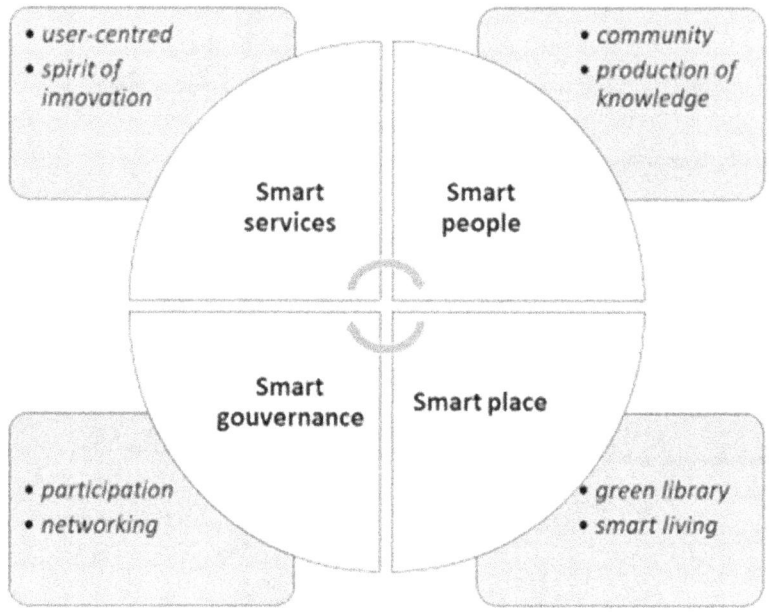

FIGURE 13.10   Four dimensions of a smart library.

Note: Adapted from "Smart Libraries" (p. 6), by Schöpfel, 2020, *Infrastructures.* 3 (43).

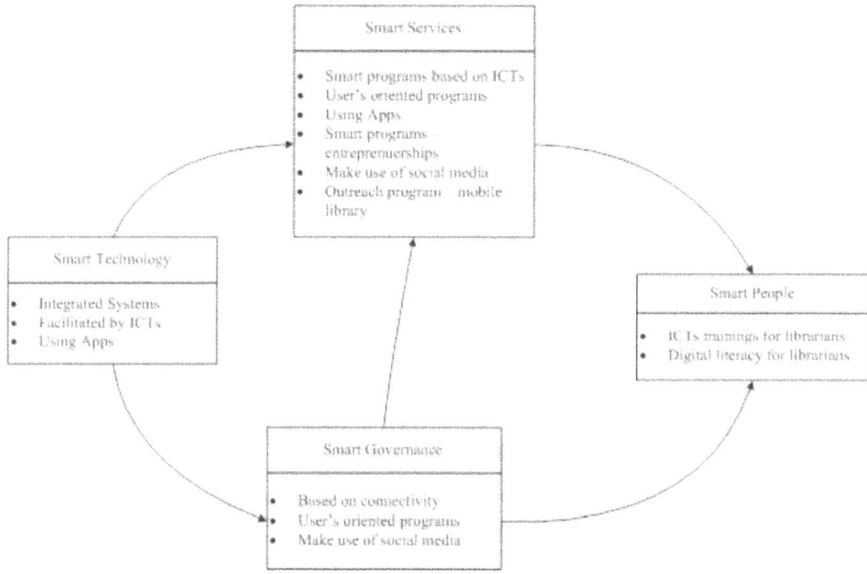

FIGURE 13.11   Components of a smart library.

Note: Adapted from "Determinants of Smart Library Readiness in Indonesia" (p. 65), by Sayogo et al., 2020, *International Information and Library Review.* 3 (43).

other models with the intention of showing an overview of an intelligent library infrastructure.

For this model, the models of Schöpfel (2018) and Sayogo et al. (2021) have been integrated into one resulting in a five-dimensional model (smart place, smart technology, smart services, smart governance, and smart people) necessary for a strategic investment model.

The general model includes four layers. The first layer is the background layer, which consists of the immediate actors and context of a smart city and a smart library. This layer is directly related to the dimensions: smart place, smart people, and smart governance.

The second layer (computing layer) contains cloud computing, such as big data, data mining, artificial intelligence, and blockchain. The third layer (communication layer) involves the Internet, servers, websites, institutional accounts, sensors, cameras, microphones, and smart devices (Rani, Khang et al., IoT & Healthcare, 2021).

Finally, the perceptual layer points to multiple products and services. These last three layers are related to the dimensions of smart place, smart technology, and smart services (Khang, Bhambri et al., Big Data, Cloud Computing & IoT, 2022).

These layers interact with layers of hierarchical character, starting with smart city, followed by smart library, which integrates the layers of cloud computing (Bhambri et al., Cloud & IoT, 2022), then smart building and smart environment, as shown in Figure 13.12.

13.5   EXPECTED OUTPUTS OF THIS PROPOSED MODEL

This work gathers the latest updates on intelligent libraries. However, technological implementations are a subject that is constantly evolving.

The proposed model on the infrastructure of a smart library aims to be a model that also evolves. In this sense, colleagues are invited to develop this model by proposing modifications to the structure of the current model or adding segments, technologies, and new services in order to strengthen this research and consolidate a model that helps libraries to make implementations that allow them to migrate to the new model of a smart library.

13.6   CONCLUSION

Libraries are undoubtedly being forced to provide better products and services in order to meet the needs of their users. Unfortunately, the smart library model is not always easy to implement, and most of the time, it is usually costly. This chapter provides an overview of the options that currently exist for a library to become a smart library.

During the review of the literature, it can be seen that some of these implementations can be carried out with little money. It can also take advantage of the large amount of free software that currently exists for technological implementations, such as virtual tours and augmented reality references, among others.

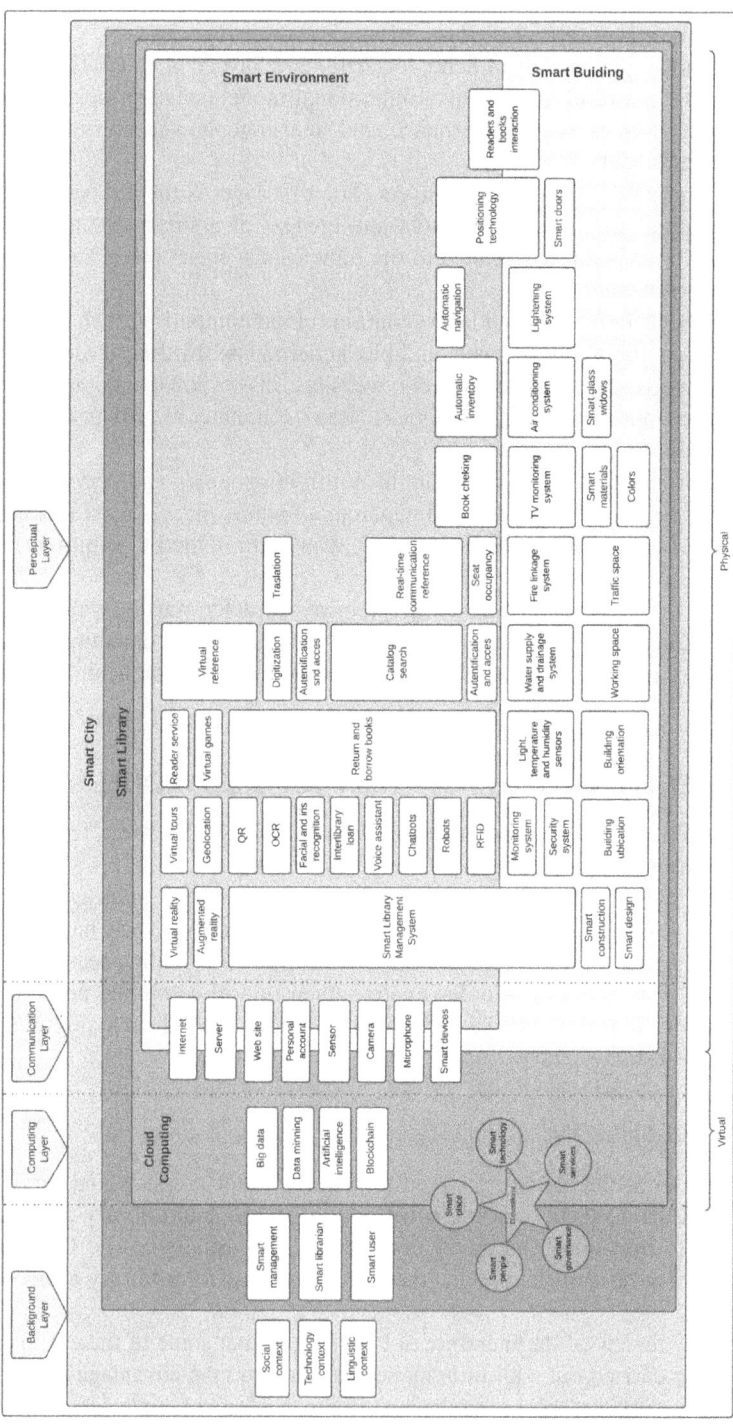

FIGURE 13.12 Smart library infrastructure.

An intelligent library also involves humanistic issues such as design, colors of the walls, decorations, furniture, and optimization of natural resources, such as light.

REFERENCES

Aguilar-Moreno, Estefanía, and Carlos Granell-Canut. 2013. "Sistemas de información geográfica para unidades de información." *El Profesional de la Informacion* 22(1):80–86. doi:10.3145/epi.2013.ene.11

Aguilar-Moreno, Estefania, and Carlos Granell-Canut. 2015. "Gestión de datos geográficos en bibliotecas universitarias españolas: Estado de la cuestión." *Revista española de Documentación Científica* 38(2):e085. doi:10.3989/redc.2015.2.1193

Allison, Deeann. 2012. "Chatbots in the Library: Is It Time?" *Library Hi Tech* 30(1):95–107. doi:10.1108/07378831211213238

Arroyo-Vázquez, N. 2016. "Experiencias de realidad aumentada en bibliotecas: estado de la cuestión." *BiD: textos universitaris de biblioteconomia i documentación* 36. doi:10.1344/BiD2016.36.4

Bagal, D., and P. Saindane. 2019. "Librany—A Face Recognition and QR Code Technology based Smart Library System." Pp. 253–258, in *Proceedings of the 4th International Conference on Communication and Electronics Systems, ICCES 2019*. IEEE. doi:10.1109/ICCES45898.2019.9002530

Baryshev, Ruslan Aleksandrovich. 2021. "From Electronic to Smart Library Systems: Concept, Classification of Services, Scheme of Work and Model." *Journal of Siberian Federal University—Humanities and Social Sciences*, 14(3): 426–443.

Baryshev, Ruslan Aleksandrovich, Olga Ivanovna Babina, P. A. Zakharov, Vera Kazantseva, and Nikita Pikov. 2015. "Electronic Library: Genesis, Trends. From Electronic Library to Smart Library." *Journal of Siberian Federal University—Humanities and Social Sciences* 6(8):1043–1051. doi:10.17516/1997-1370-2015-8-6-1043-1051

Baryshev, R. A., S. V. Verkhovets, and O. I. Babina. 2018. "The Smart Library Project: Development of Information and Library Services for Educational and Scientific Activity." *Electronic Library* 36(3):535–549. doi:10.1108/EL-012017–0017

Bhambri, Pankaj, Sita Rani, Gaurav Gupta, and Khang, A., 2022. *Cloud and Fog Computing PlatformsforInternetofThings.*CRCPress.ISBN:978-1-032-101507,doi:10.1201/978103 2101507

Blewitt, J. 2014. "Public Libraries and the Right to the [Smart] City." *International Journal of Social Ecology and Sustainable Development* 5(2):55–68. doi:10.4018/ ijsesd.2014040105

Burgess, J. T. F. 2010. "Smart-World Technologies and the Value of Librarianship." *Computers in Libraries* 30(10):12–16. https://eric.ed.gov/?id=EJ907244

Cao, Gaohui, Mengli Liang, and Xuguang Li. 2018. "How to Make the Library Smart? The Conceptualization of the Smart Library." *The Electronic Library* 36(5):811–825. doi:10.1108/EL-11–2017–0248

Castle, Steven. 2013. "Next Wave in LED Lighting?" EH Network. Recuperado el 22 de noviembre de 2021. www.electronichouse.com/home-energymanagement/next-wave-in-led-lighting/

Chang, Kuochung, and Chiao-Chen Chang. 2009. "Library Self-Service." *The Electronic Library* 27(6):938–949. doi:10.1108/02640470911004048

Chen, Chia-Chen, and An-Pin Chen. 2007. "Using Data Mining Technology to Provide a Recommendation Service in the Digital Library." *The Electronic Library* 25(6):711–724. doi:10.1108/02640470710837137

Chen, Feng, Pan Deng, Jiafu Wan, Daqiang Zhang, Athanasios V. Vasilakos, and Xiaohui Rong. 2015. "Data Mining for the Internet of Things: Literature Review and Challenges." *International Journal of Distributed Sensor Networks* 11(8):431047. doi:10.1155/2015/431047

Chen, Z., M. Zhou, and L. Feng. 2021. "Analysis of the Smart Library Construction in Colleges Based Big Data and Artificial Intelligence." *Journal of Physics: Conference Series* 1955.

Crowe, Cailin. 2020. "La biblioteca es el 'centro de la inteligencia digital' de una ciudad inteligente." *Smart Cities Dive*. Recuperado el 29 de noviembre de 2021. www.smartcitiesdive.com/news/library-smart-city-hub-digitalintelligence-inlcusion/569012/

Du, L., and T. Liu. 2014. "Study on the Development of Smart Library Under Internet of Things." *Applied Mechanics and Materials* 529:716–720. doi:10.4028/www.scientific.net/AMM.529.716

Ford, Charlotte E., Julie Gerardin, Michelle Yamamoto, and Kelly Gordon. 2008. "Fresh Perspectives on Reference Work in Second Life." *Reference and User Services Quarterly* 47(4):324–330. doi:10.5860/rusq.47n4.324

Galve-montore, Carme. 2019. *Smart Cities: An Opportunity for Libraries to Be Part of Future Urban Management*, pp. 1–14. Biblioteca Jaume Fuster, Biblioteques de Barcelona, Barcelona, Spain, Corpus ID: 204846267. http://library.ifla.org/id/eprint/2479/7/100-galve-en.pdf

García-Marco, Francisco-Javier. 2011. "Libraries in the Digital Ecology: Reflections and Trends." *The Electronic Library* 29(1):105–120. doi:10.1108/02640471111111460

Gul, Sumeer, and Shohar Bano. 2019. "Smart Libraries: An Emerging and Innovative Technological Habitat of 21st Century." *Electronic Library* 37(5):764–783. doi:10.1108/EL-02–2019–0052

Hawkins, Andrew M. 1994. "Geographical Information Systems (GIS): Their Use as Decision Support Tools in Public Libraries and the Integration of GIS with Other Computer Technology." *New Library World* 95(7):4–13. doi:10.1108/03074809410070502

Hoy, M. B. 2016. "Smart Buildings: An Introduction to the Library of the Future." *Medical Reference Services Quarterly* 35(3):326–331. doi:10.1080/02763869.2016.1189787

Hu, Jianfei, Bin Zhang, and Di Wang. 2019. "Application of Virtual Reality Technology in Library Visual Information Retrieval." *IOP Conference Series: Materials Science and Engineering* 569(3). doi:10.1088/1757899X/569/3/032062

Johnson, Ian M. 2013. "Smart Cities, Smart Libraries, and Smart Librarians." *Libraly Journal* 32(1):4–7. Corpus ID: 107376469. http://eprints.rclis.org/20429/

Khang, A., Pankaj Bhambri, Sita Rani, and Aman Kataria. 2022. *Big Data, Cloud Computing and Internet of Things*. CRC Press. ISBN:978-1-032-284200, doi:10.1201/9781032284200.

Khang, A., Subrata Chowdhury, and Seema Sharma. 2022. *The Data-Driven Blockchain Ecosystem: Fundamentals, Applications and Emerging Technologies*. CRC Press. ISBN:978-1-032-21624, doi:10.1201/9781003269281

Khang, A., Geeta Rana, Khanh, H.H., 2021. "The Role of Artificial Intelligence in Blockchain Applications." In *Reinventing Manufacturing and Business Processes Through Artificial Intelligence,* p. 20. CRC Press. doi:10.1201/9781003145011.

Kovacevic, Ana, Vladan Devedzic, and Viktor Pocajt. 2010. "Using Data Mining to Improve Digital Library Services." *The Electronic Library* 28(6):829–843. doi:10.1108/02640471011093525

Laurel, B. 1997. *Interface Agents: Metaphors with Character. En Human Values and the Design of Computer Technology,* pp. 207–219. Center for the Study of Language and Information. doi:10.5555/278213.278227

Leorke, D., and D. Wyatt. 2019. *Public Libraries in the Smart City*. Springer, Singapore. doi:10.1007/978-981-13-2805-3

Liao, J. 2020. "The Theoretical Framework of Library Smart Service Literature Review and Content Analysis." Pp. 471–72, in *Proceedings of the ACM/IEEE Joint Conference on Digital Libraries*. doi:10.1145/3383583.3398569

Luo, Wei. 2020. "Research on Library Reader Service Application and Innovation Based on VR." 416(Iccese):1529–1533. doi:10.2991/assehr.k.200316.325

Martínez-Camacho, H. 2022. "Model Proposal for Smart Library Infrastructure." *FigShare* 1(1):693–703. doi:10.6084/m9.figshare.17096840.v3

Monti, L., S. Mirri, C. Prandi, and P. Salomoni. 2019. "Preservation in Smart Libraries: An Experiment Involving IoT and Indoor Environmental Sensing." In *2019 IEEE Global Communications Conference, GLOBECOM 2019*. doi:10.1109/GLOBECOM38437.2019.9014149.

Moore, John. 2005. "Digital Map Soup: What's Cooking in British Academic Libraries and Are We Helping Our Users?" *LIBER Quarterly: The Journal of the Association of European Research Libraries* 15(1). doi:10.18352/lq.7800.

Nicholson, Scott. 2006. "The Basis for Bibliomining: Frameworks for Bringing Together Usage-Based Data Mining and Bibliometrics Through Data Warehousing in Digital Library Services." *Information Processing & Management* 42(3):785–804. doi:10.1016/j.ipm.2005.05.008

Oderuth, B. R., K. Ramkissoon, and R. K. Sungkur. 2019. "Smart Campus Library System." 2nd *International Conference on Next Generation Computing Applications 2019, NextComp 2019*. IEEE. doi:10.1109/NEXTCOMP.2019.8883636

Patra, Nihar K. 2017. "Lifecycle of Electronic Resource Management." Pp. 27–43, in *Digital Disruption and Electronic Resource Management in Libraries*. Elsevier. doi:10.1016/B978-0-08-102045-6.00003-0

Pence, Harry E. 2010. "Smartphones, Smart Objects, and Augmented Reality." *The Reference Librarian* 52(1–2):136–145. doi:10.1080/02763877.2011.528281

Petska, Alicia. 2018. *"Roanoke County Library Adds Pepper*, a 'Community Robot'." *The Roanoke Times*. https://roanoke.com/news/local/roanoke-county-library-adds-pepper-a-community-robot/article_198827e4-e357-545a-9f14-2e8137f83208.html

Rani, S., and O. P. Gupta 2017. "CLUS_GPU-BLASTP: Accelerated Protein Sequence Alignment Using GPU-Enabled Cluster." *The Journal of Supercomputing* 73(10):4580–4595. doi:10.1007/s11227-017-2036-4

Rani, S., A. Kataria, V. Sharma, S. Ghosh, V. Karar, K. Lee, and C. Choi. 2021. "Threats and Corrective Measures for IoT Security with Observance of Cybercrime: A Survey." *Journal of Wireless Communications and Mobile Computing* 2021:1–30.

Rani, S., Khang, A., Meetali Chauhan, Aman Kataria. 2021. "IoT Equipped Intelligent Distributed Framework for Smart Healthcare Systems." *Networking and Internet Architecture*. https://arxiv.org/abs/2110.04997v2, doi:10.48550/arXiv.2110.04997

Rani, S., and R. Kumar. 2022. "Bibliometric Review of Actuators: Key Automation Technology in a Smart City Framework." In *Materials Today: Proceedings*. doi:10.1016/j.matpr.2021.12.469

Rani, S., R. K. Mishra, M. Usman, A. Kataria, P. Kumar, P. Bhambri, and A. K. Mishra. 2021. "Amalgamation of Advanced Technologies for Sustainable Development of Smart City Environment: A Review." *IEEE Access* 9:150060–150087.

Renaud, John, Scott Britton, Dingding Wang, and Mitsunori Ogihara. 2015. "Mining Library and University Data to Understand Library Use Patterns." *The Electronic Library* 33(3):355–372. doi:10.1108/EL-07-2013-0136

Sayogo, D. S., S. B. C. Yuli, and W. Wiyono. 2021. "The Determinants of Smart Public Library Roles in Promoting Open Government in Indonesia." *International Information and Library Review* 54(2):155–170. doi:10.1080/10572317.2021.1936380

Schöpfel, J. 2018. "Smart Libraries." *Infrastructures* 3(4). doi:10.3390/infrastructures3040043

Shahid, Syed Md. 2005. "Use of RFID Technology in Libraries: A New Approach to Circulation, Tracking, Inventorying, and Security of Library Materials." *Library Philosophy and Practice* 8(1). Corpus ID: 61816296.

Shi, X., K. Tang, and H. Lu. 2021. "Smart Library Book Sorting Application with Intelligence Computer Vision Technology." *Library Hi Tech* 39(1):220–232. doi:10.1108/LHT-10-2019-0211

Shieh, Jiann-Cherng. 2010. "The Integration System for Librarians' Bibliomining." *The Electronic Library* 28(5):709–721. doi:10.1108/02640471011081988

Su, X. M., and G. X. He. 2014. "Study on Smart Materials of Library Buildings." *Applied Mechanics and Materials*: 484–485. doi:10.4028/www.scientific.net/AMM.484-485.691

Sun, N., Y. Li, L. Ma, W. Chen, and D. Cynthia. 2019. "Research on Cloud Computing in the Resource Sharing System of University Library Services." *Evolutionary Intelligence* 12(3):377–384. doi:10.1007/s12065-018-0195-8

Sweeney, M. E., and E. Davis. 2020. "Alexa, Are You Listening?: An Exploration of Smart Voice Assistant Use and Privacy in Libraries." *Information Technology and Libraries* 39(4). doi:10.6017/ITAL.V39I4.12363

Xia, Jingfeng. 2004. "GIS in the Management of Library Pick-Up Books." *Library Hi Tech* 22(2):209–216. doi:10.1108/07378830410543520

Xu, Y., J. Li, B. Wang, Y. Bu, and C. Ji. 2019. "Analysis of Automation Systems and Virtual Reality Applications in Smart Libraries." Pp. 161–66, in *Proceedings—18th IEEE/ACIS International Conference on Computer and Information Science, ICIS 2019*. IEEE. doi:10.1109/ICIS46139.2019.8940162

Yalagi, P. S., and P. V. Mane. 2021. "Smart Library Automation Using Face Recognition." *Journal of Physics: Conference Series* 1854. doi:10.1088/1742-6596/1854/1/012041/meta

Yao, Fei, Chengyu Zhang, and Wu Chen. 2015. "Smart Talking Robot Xiaotu: Participatory Library Service Based on Artificial Intelligence." *Library Hi Tech* 33(2):245–260. doi:10.1108/LHT-02-2015-0010

Younis, M. I. 2012. "SLMS: A Smart Library Management System Based on an RFID Technology." *International Journal of Reasoning-based Intelligent Systems* 4(4):186–191. doi:10.1504/IJRIS.2012.051717

Yu, K., and G. Huang. 2020. "Exploring Consumers' Intent to Use Smart Libraries with Technology Acceptance Model." *Electronic Library* 38(3):447–461. doi:10.1108/EL-08-2019-0188

Yuke, Li. 2017. "Library New Directions in Knowledge Management: Recycling of Technical Design Research of Virtual and Reality Knowledge Space." *Ifla Wlic*:1–12.

Zheng, Wang. 2019. "How Do Library Staff View Librarian Robotics? Librarian Staff's Ignored Humanistic Views on the Impact and Threat of Robotics Adoption." *Ifla Wlic*: 1–17.

Zhao, J. et al. 2019. "Research on Smart Library Big Service Application in Big Data Environment." Pp. 242–243, in *Advances in Computational Science and Computing* (877). doi:10.1007/978-3-030-02116-0_28.

Zimmerman, T., and H.-C. Chang. 2018. "Getting Smarter: Definition, Scope, and Implications of Smart Libraries." Pp. 403–404, in *Proceedings of the ACM/IEEE Joint Conference on Digital Libraries*. IEEE. doi:10.1145/3197026.3203906

14 AI-Based Smart Education System for a Smart City Using an Improved Self-Adaptive Leap-Frogging Algorithm

Ahmed Muayad Younus, Mohanad S.S. Abumandil, Veer P. Gangwar, Shashi Kant Gupta

CONTENTS

14.1 INTRODUCTION

Smart cities' main goal is to give inhabitants comfort, security, and a healthy life-style through intelligent collaboration. People gain from smart cities because they make use of the opportunities presented by widespread and cooperative computer technology (Alhathli et al., 2020).

Digital environments are expected to play a crucial role in addressing issues in such a situation. They are transportation, care services, consumables, learning and education, sensing metropolitan movements, calculating with several databases, handling massive amounts of municipal data, and environmental challenges such as pollution reduction (Chauhan and Rani, 2021).

Due to growing urbanization and population increase, present city services and administration would be unable to provide enough added value to inhabitants. Improving the methods and processes that improve the community's quality of life is critical, and this necessitates the provision of these services powerfully and effectively (Rani, Chauhan, Kataria, and Khang, 2021; Rani, Kataria et al., 2021).

With the tremendous advancements in computer and wired/wireless communication technology in so-called smart cities, the idea of incorporating various tiers of smartness and intelligence has emerged.

Education has traditionally been offered in classrooms across the world, using blackboards, chalkboards, textbooks, exams, and assignments. Students enrolled in these physical schools and were required to attend sessions five or six days a week. Professors led these classes, standing at the front of the room and giving the students orders to follow.

In a smart education system, the instructor serves as a guide who assists students in navigating their learning process using self-contained learning systems that are personalized and adjusted to the student's learning style and pace. The smart classroom makes learning pleasant and exciting for kids while also allowing teachers to successfully educate, as shown in Figure 14.1.

As a result, the smart class is a technological benefit to education. It gives pupils the chance to assist in better comprehending the idea and achieving academic achievement (Rani, Mishra et al., 2021).

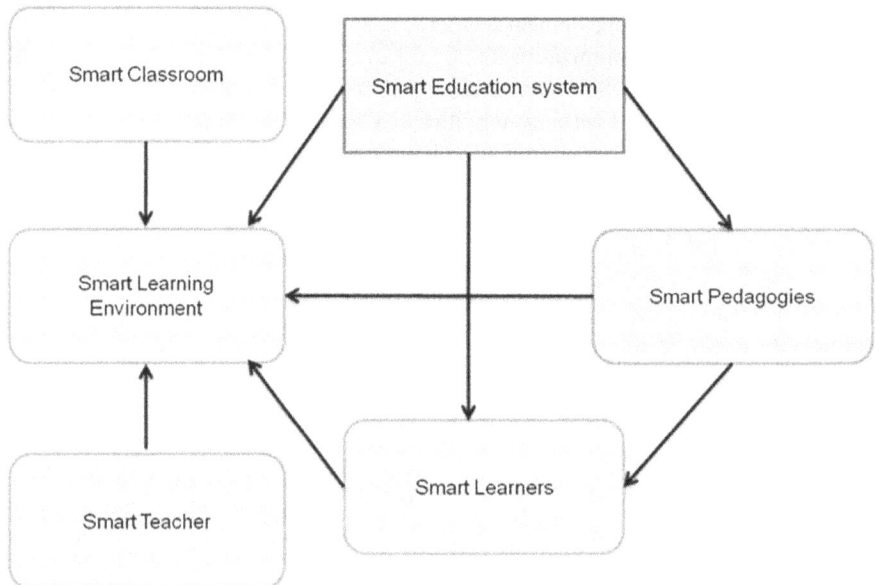

FIGURE 14.1   General framework of the smart education system.

We may use several digital instruments to access the digital/smart education system, such as mobiles, tablets, computers, projectors, whiteboards, digital textbooks, and so on.

The main components involved in the smart education framework are smart classrooms, smart learning environments, smart pedagogies, and smart learners.

Education is more important than ever, especially public education assisted by AI (artificial intelligence). This is because we must not only reposition and enhance our core competencies, but we must also use this great technology to gather knowledge, learn from everyone else, and collaborate with all abilities around the globe in order to get new knowledge. The advantages of collaboration are beneficial not only to individuals but also to mankind as a whole.

This chapter starts with the advancements in smart education during the last few decades. After that, the smart education model is investigated, which is a four-tier structure of smart approaches and key components of a smart learning environment.

Then we look at two of the most current R&D AI-based smart education applications, AI language learning robots and VR (virtual reality) and AR (augmented reality) instructors, to illustrate how AI technologies, such as ML (machine learning), NLP (natural language processing), ontological knowledge bases, VR, and AR, may be merged to create a new era of smart education.

In the field of developing smart digital education, AI has increasingly received a lot of attention. Researchers used computational intelligence (CI) and machine learning technology to create a smart tutoring system (STS).

The convergence of AI, information science, and the Internet of Things (IoT), on the other hand, is allowing the design and creation of web-based smart solutions for all educational and learning tasks (Khang et al., IoT & Healthcare, 2021).

This chapter discusses the CI and knowledge engineering approaches for creating smart educational and learning systems.

In the 21st century, AI has become a prominent study issue in a variety of fields, encompassing technology, research, educational, healthcare, industry, auditing, economics, advertising, economics, stock market, and law, to mention a few.

The scope of AI has evolved dramatically since the intelligence of computers with machine learning skills has always had enormous repercussions for industry, governments, and society. They have an impact on global environmental trends as well.

AI has the potential to assist in the resolution of critical challenges in the realm of environmental protection (e.g., optimization of energy resources, logistics, supply chain management, waste management, etc.).

In this regard, there is a trend in smart production to incorporate AI into environmental manufacturing procedures in order to comply with increasingly stringent environmental standards (Rana et al., AI & Blockchain, 2021). AI involved in the smart education system is depicted in Figure 14.2.

In the education sector, AI is used to give smart content for everyone, better assessment of skills and weaknesses, immediate attention, feedback to teachers, customized learning, automating admin tasks, tutoring, and guidance outside the classroom.

Furthermore, as Hendrik Fink, head of PricewaterhouseCoopers' Sustainability Services, stated in March 2019 that if we appropriately incorporate AI, then we may

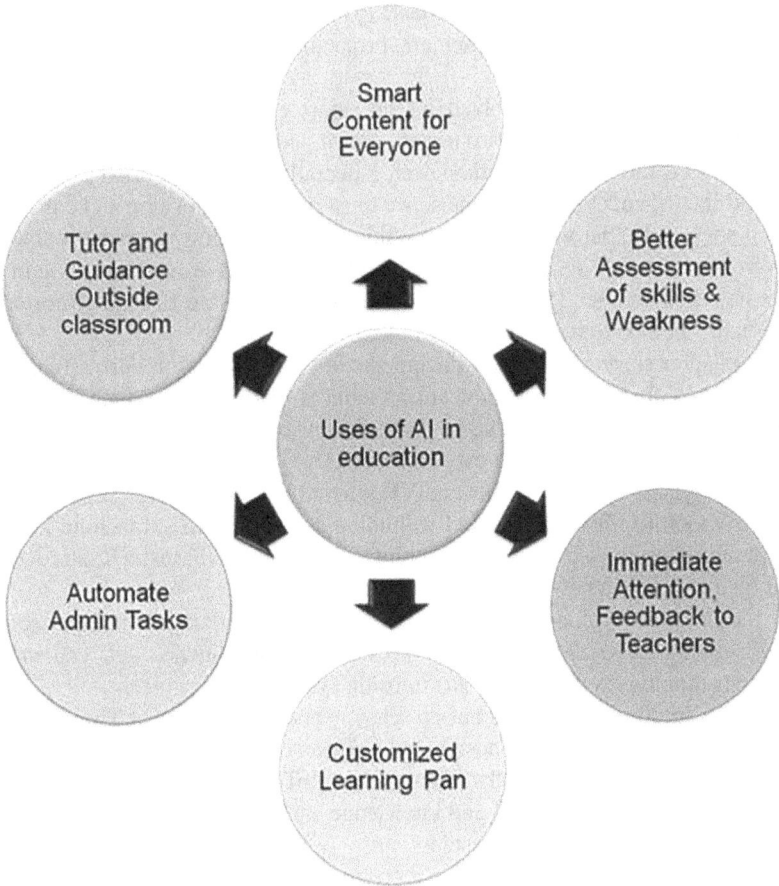

FIGURE 14.2   Artificial intelligence in the smart education system.

be able to produce a revolutionary in terms of sustainable development. "AI will drive the fourth industrial revolution."

As a result, AI subfields, such as computer vision, natural language processing, image recognition, and data mining, have risen to prominence among today's computer companies. Because of the rapid advancement of technology available today, the issue of AI has sparked a lot of interest in the scientific community.

In this chapter, we will discuss an AI-based smart education system for the smart city using the improved self-adaptive leap-frogging algorithm (ISALFA). Students can understand and retain material more easily using smart education technologies.

Suppose you feel that we remember many things from our lives that we have experienced or seen rather than merely reading. As a result, these sessions are a good study innovation. Furthermore, students do not need to physically visit digital learning classrooms because most learning takes place online.

14.2   PROBLEM STATEMENT

The main drawback of a smart education system is the usage of electricity is extremely important. It is tough to create presentations, visualizations, and programs for every class.

High devices are needed, such as computers, digital boards, tablets, and other electronic devices. So the maintenance cost is high, and it requires proper network connection such as LAN, WAN, and the Internet, among other things.

In the existing method, smart education systems using blockchain have some issues related to security, scalability, and efficiency. The main challenge in the blockchain is how to integrate blockchain with its legacy system(s).

14.3   PROPOSED WORK

This chapter gives a new method to overcome the drawbacks of the smart education system by the existing method. This chapter discusses the AI-based smart education system for a smart city using the ISALFA.

AI provides its knowledge to achieve scalability, and AI technology is a way to solve problems with scheduling and routing. The schematic representation of the suggested optimization is depicted in Figure 14.3.

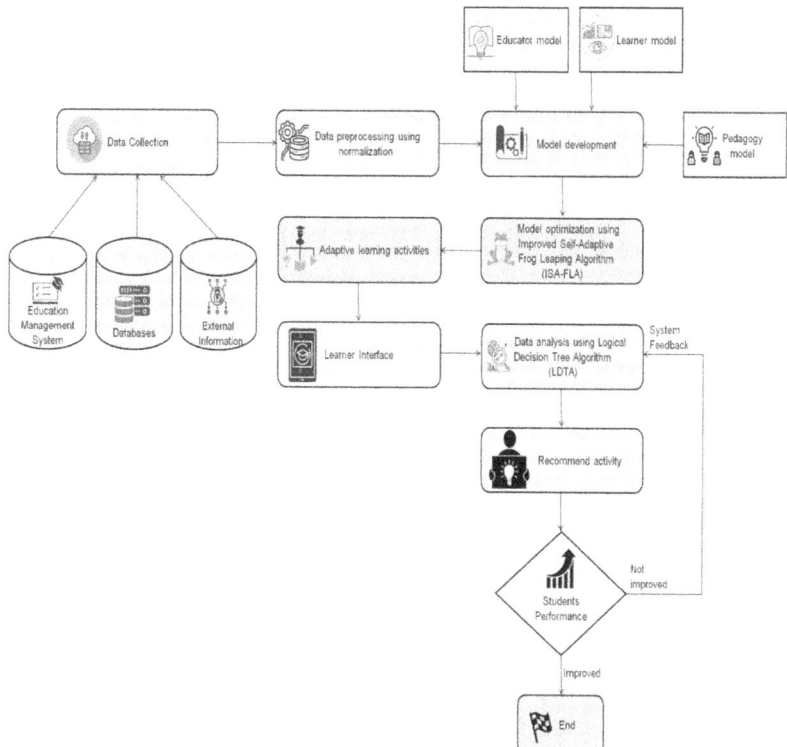

FIGURE 14.3   Schematic representation of the suggested optimization.

14.3.1 DATA COLLECTION

In the collection of data, there are three important types of data collection presented.

- Education management system
- Databases
- External information

Education management system collects and stores a large quantity of information, such as lectures, material, assignments, and student submissions.

Schools employ a database to keep track of information on their students, such as how many days they've missed due to illness and their performance records. External information contains other personal information about students.

14.3.2 DATA PREPROCESSING

Data preprocessing is a job that entails the preparation and transformation of data into an appropriate format. Preprocessing data seeks to minimize data size, establish data relationships, normalize data, eliminate outliers, and extract data characteristics (Bhambri et al., Cloud & IoT, 2022).

Data cleansing, integration, transformation, and minimization are some of the strategies used.

- The primary goal of data normalization is to reduce or eliminate redundant data. Min-max normalization is a method for performing linear transformations on a set of data.
- This is a technique for keeping the original data connected. Min-max normalization is a simple method for fitting data into a predefined boundary.

$$S' = \left(\frac{S - minvalue\,of\,S}{maxvalue\,of\,S - min\,value\,of\,S} \right) * (Z - W) + W \qquad (14.1)$$

Here,

- S,—has min-max normalized data one
- $[Z, W]$ is a predefined boundary
- S—original data range

K-score normalization is a method that uses concepts like quantitative variables to obtain normalized values or sets of information from unstructured data.

As a result, the K-score parameter may be used to normalize unstructured data, as presented in the following equation:

$$qa' = std^{-qa^-}(B^{\bar{B}}) \qquad (14.2)$$

- q_d-K score normalized one's value
- q_a_the value of row $\bar{B}$

$$std^{(B)} = \sqrt{\frac{1}{(r-1)}\sum_{j=1}^{r}\left(q_a - \bar{B}\right)^2} \tag{14.3}$$

$$\bar{B} = \frac{1}{r}\sum_{j=1}^{r} q_a \tag{14.4}$$

The technique that gives a scale from −1 to 1 is known as decimal scaling. The decimal scaling procedure, as a consequence,

$$q^a = {}^{-q}x \tag{14.5}$$

10

- q^a—scaled value
- q—range of values x- smallest integer Max($|q^a|$) <1

Using min-max normalization, the original data is converted linearly. Assume that min_x and max_x are the lowest and maximum values for variable E.

By computing [new-min_x, new-max_x], A value e of E is transferred to e' in the range through min-max normalization. [new-min_x, newmax$_x$].

$$E, = ((e - min_x)/(max_x - min_x)) * (new - max_x - new - min_x) + new - min_x \tag{14.6}$$

The values for variable E are normalized using the mean and standard deviation of E in w-score normalization. E value e of E is normalized to e' using the following formula:

$$e, = ((e - \bar{X})/\sigma X) \tag{14.7}$$

where e and σX are the mean and standard deviation of variable e, respectively. When the real minimum and maximum of variable e are unknown, this approach of normalizing is beneficial.

The decimal point of values of variable e is moved during normalization using decimal scaling. The number of decimal points shifted is determined by e' absolute maximum value. A value e of E is normalized to e' using the following formula:

$$e, = (e/10^i) \tag{14.8}$$

where i is the smallest integer such that Max (|e'|) < 1.

14.3.3 MODEL DEVELOPMENT

A computer system that tries to assist students in understanding the topic and offer teaching, an exercise, or feedback to pupils (students) without human interaction is known as a model development for a smart education system (Rana et al., 2021).

It was created by combining three scientific fields: computer science (AI), psychology, and studies. If this strategy is followed, the effectiveness of each instructional classroom session will be led and systematic.

14.3.3.1 Educator Model

The educator model represents the instructor of a smart class. They conduct lectures, tests, or quizzes for the learners.

14.3.3.2 Leaner Model

The learner model contains information about the learner (student), such as the student's data, degree of skill, and learning style.

14.3.3.3 Pedagogical Model

The information regarding both the instructor's and student's strategies are contained in the pedagogical model. As a result, the instructor's teaching strategy and the student's learning strategy are inextricably linked.

14.3.4 MODEL OPTIMIZATION

The smart education system for a smart city is a popular method that faces issues in terms of heterogeneity, scalability, security, dependability, and some legacy systems. AI has the benefits of problem-solving skills, a quick convergence time, a strong optimization ability, and a high level of resilience.

Therefore, to optimize some issues in the existing smart education method with the AI-based smart education system for the smart city, developers should use the ISALFA.

14.3.5 IMPROVED SELF-ADAPTIVE LEAP-FROGGING ALGORITHM

The ISALFA is an AI-based periodic basic search approach that was developed by simulating the information sharing and communication mechanisms in frog searching.

The technique, which combines the benefits of the memetic algorithm (MA) with the particle swarm optimization (PSO) methods, was initially proposed by Eusuff and Lansey. It offers fewer parameter settings, a quick solution speed, and a great global optimization capability.

There are mainly three types of operations involved in ISALFA:

- Initialization operation
- Grouping operation
- Local update operation

14.3.5.1   Initialization Operation

Initially, we can take the population of frogs as F= K1, K2, . . . KSK.

There are SK people in all, and each one is a randomly generated potential solution.

$$K_a{}^b, = K_{min}{}^b + (0, 1) (K_{max}b - K_{min}{}^b) b$$
$$= 1, 2, \ldots D \; a = 1, 2, \ldots, NP \tag{14.9}$$

The jth individual is $K_{i,t} = (k^1{}_{i,t}, k^2{}_{i,t}, \ldots, k^j{}_{i,t}, \ldots, k^D{}_{i,t}) \; 1 \le j \le SK$. The evolutionary number of each meme group is denoted by t. The solution's dimension is denoted by D.

14.3.5.2   Grouping Operation

The population may be evenly divided into M subpopulations using the grouping procedure. The population's SK candidate solutions are first sorted in decreasing order by Eq. 14.10's fitness value, and then the first candidate solution is classified into the first subgroup, and the second set of solutions is classified into the second subgroup.

By similarity, the rest is calculated using Eq. 14.11, and the SK design parameters are subdivided into M subgroups, with s being the subgroup number to which Hsu et al. (2020) allocate the jth candidate solution.

$$(K)K \subseteq [K_{min}],= (K_1, K_2, \ldots, K_n) \tag{14.10}$$
$$0 = a\%M, a = 1, 2, \ldots, NP \tag{14.11}$$

14.3.5.3   Local Update Operation

$Fw_s{}^o$ is the candidate solution with the least populated attribute in the tth evolutionary generation as the worst solution, $Fb_s{}^o$ is the candidate solution with the largest value in the sth subgroup in the tth developmental source as an optimal solution, and Fg is the candidate solution with the highest density attribute in all previous generations as an optimal solution.

$$F\dot{w} = Fw_s o + (0, 1). (Fb_s{}^o - Fw_s{}^o) \tag{14.12}$$

Algorithm 14.1. represents the ISALFA.

Algorithm 14.1—ISALFA

Create frog populations, with each frog's location information represented as a linear combination (h, σ). U frogs create a matrix T of $U \times 2$.

for i = 1: G do for u = 1: F do

Applying AI to identify initial centroids;

$$\{g_1, g_2, \ldots, g_h\} = Artificalintellgince\ (S, T(u), H = h(s_j, s_p));$$

To every data point, figure out (j, p), (j, p) using equations
   For every data points Compute *ISAFLA* and *AVG_ISAFLA*;

   End for

Arrange the fit and save the global ideal fitness as S_m, along with the location data.

   for $j = 1: n$ for $p = 1: k$

Identify the group's best and worst fitness, as well as the position information S_v and S^a.
   Upgrade the worst fitness as well as the Sa that goes with it;
   Upgrade the global optimal fitness as well as the Sv that goes with it;

   End for
   End for
   End for

14.3.6   ADAPTIVE LEARNING ACTIVITIES AND LEARNER INTERFACE

Making adaptations in an educational environment to suit individual variances is what adaptation is smart education systems are all about. It is divided into two components such that adaptively and adaptability.

   An e-learning system can be adaptable depending on who has control over or initiates the adaptation, the student or the system.

   Adaptively refers to the procedures through which the system adjusts to the learner's facts or knowledge in a system-controlled manner. Adaptability is enabled, on the other hand, when the technology allows for end-user modification and student control.

   The user of an adaptable system can change the system's settings to meet their own requirements.

   As a result, flexibility gives learners and educators the capacity to personalize their learning experiences. In an adaptive system, the system assumes the learner's needs and adapts its behavior appropriately.

14.3.7   DATA ANALYSIS USING A DECISION TREE ALGORITHM

Ross Quinlan created the ID3 and C4.5 algorithms, which are based on information gain. The ID3 approach uses information from the training data to create decision trees, but the C4.5 algorithm uses a variety of information called a gain ratio. The test data is subsequently classified using the decision tree that has been generated.

The dataset used to train the decision tree is comprised of objects and other properties, and it is used as input to the algorithms.

These methods use the same decision tree construction procedure. The information gain is computed for all the attributes after the class entropy and the entropy of each attribute are determined, as shown in Equations 14.13, 14.14, and 14.15.

The attribute with the maximum information gain is deemed the most informative attribute in the ID3 algorithm and is chosen as the root node. The technique is continued until the tree contains all of the properties.

$$(O) = - \sum NJ = 1 \; tj * log2 \; (tj) \tag{14.13}$$
$$(O) = \sum ep = 1 \; |-|OOp|| * inf \; (Op) \tag{14.14}$$
$$(L) = (O) - (O) \tag{14.15}$$

The C4.5 algorithm was created to address some of ID3's shortcomings. The IGR, which is used in the C4.5 technique, is a less biased selection criterion that normalizes information gain using split entropy.

The split entropy and IGR, as indicated in Equations 14.16 and 14.17, are two additional parameters derived in C4.5 in addition to those of Equations 14.13, 1.14, and 14.15.

$$SplitInfo_L O = - \sum_{p=1}^{e} |{-}|^0{}_0{}^p| * log_2 |{-}|^0 O^p| \tag{14.16}$$

$$IGA(L) = \frac{Gain(L)}{SplitInf_{OL}O} \tag{14.17}$$

Algorithm 14.2 represents the data analysis using a logical decision tree algorithm.

Algorithm C4.5

Input: Example, Destination Attribute, Attribute
Output: Instances that have been classified

- Look for the most basic case.
- Create a TD based on the training data.
- Determine the attribute that has the most information gain.
- (A_Best) Entropy minimization is defined as A Best.
- Divide X into X1, X2, and X3.
- Based on the value of A_Best.
- Rep procedure 1–3 for X1, X2, and X3.
- Apply the TD to each ti D.

The following are examples of base cases:

- The examples in the training set are all from the same class (a tree leaf identified with that class is returned).
- There is no data in the training set (it gives failed tree leaf).
- There are no attributes in the attribute list (produces a leaf with the most common class named or the destruction of all classes).

Output: a decision tree that appropriately classifies the data

The IGR is used to determine the best divide attribute and to assess the information worth of various features, after which the decision tree is built using the depth-first method. Every node in the tree does the same thing, initializing its data and producing child nodes.

The IGR is a kind of information gain that is used to eliminate feature bias in characteristics with multiple branches. If the data is uniformly distributed, the gain ratio is high; however, if all data enters one branch, the gain ratio is low (Khang, Bhambri et al., Big Data, Cloud Computing & IoT, 2022).

14.4 PERFORMANCE ANALSYSIS

This chapter gives an idea about how to improve the issues involved in the existing smart education system. This chapter discusses the AI-based smart education system for the smart city using an ISALFA. We examined the performance of this system like student satisfaction, efficiency, and average score.

14.4.1 STUDENT SATISFACTION

In this chapter, we have analyzed students' feedback or satisfaction with the smart education system. The satisfaction level is observed by teaching level, communication response, level of interest, and overall education, as shown in Figure 14.4.

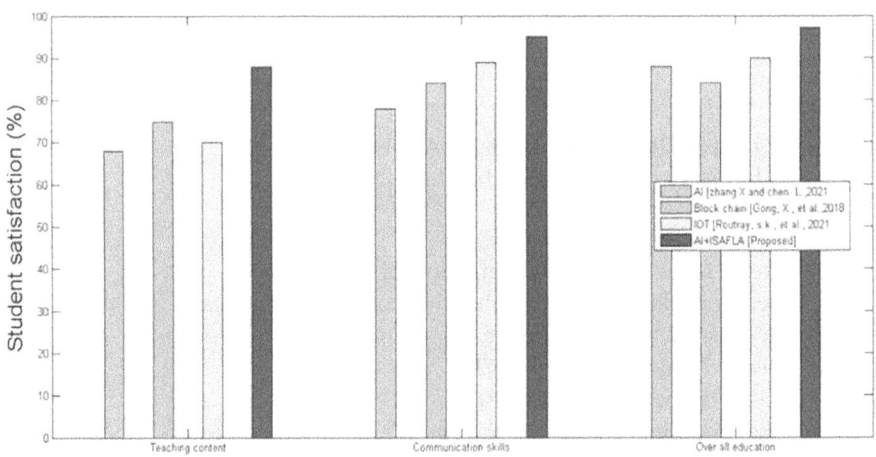

FIGURE 14.4 Comparison of student satisfaction.

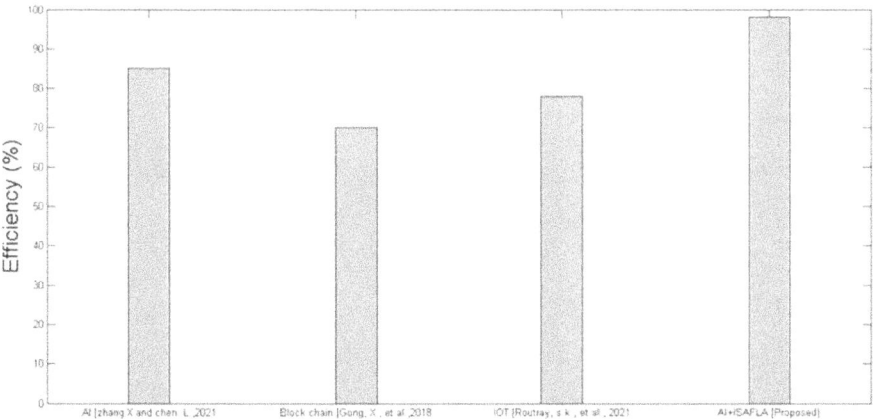

FIGURE 14.5   Comparison of efficiency.

In Figure 14.4, it is clearly displayed in the graph that the proposed AI+ISALFA method has a higher level of student satisfaction when compared to the existing three smart education systems: AI, blockchain, and IoT (Hasan et al., 2022).

14.4.2 EFFICIENCY

We have compared some other existing smart education systems with the proposed method. The existing methods have a few drawbacks, such as inappropriate teaching methods, misleading information, additional student distractions, students learning inefficiently, technology that occasionally fails, and some difficult legacy systems, as shown in Figure 14.5.

The comparative analysis of the efficiency of the suggested optimization (AI+ISALFA) is displayed in Figure 14.5 when compared to the existing three smart education systems: AI, blockchain, and IoT (Tailor et al., 2022). In a comparative analysis, the proposed method gives more efficiency to the described issues.

14.4.3 AVERAGE SCORE

A major number of students believe that the learning style and substance of learning resources provided by instructors on the suggested system are valuable and efficient for learning because the proposed method gives more uninterrupted communication between learners and educators. Finally, we have analyzed the average score by the experiment, as shown in Figure 14.6.

The comparative analysis of the average score for the suggested optimization (AI+ISALFA) is displayed in Figure 14.6 when compared to the existing three smart education systems: AI, blockchain, and IoT. It clearly shows the great score after the implementation of the proposed method.

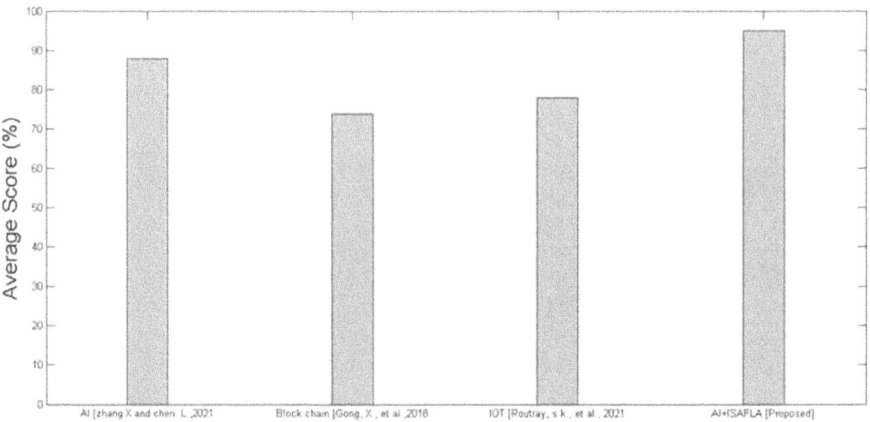

FIGURE 14.6 Comparison of average score.

14.5 CONCLUSION

Technology is being more widely used in today's educational environment. Using smart gadgets to better teaching and learning, modern technologies are strengthening today's education.

The new optimization method mentioned in this chapter is ideal for creating a smart education ecosystem. Smart education makes teaching, learning, communication, and cooperation easier with the use of current technologies. To optimize the efficiency and advantages, this optimization is an effective use of technology that creates a scalable and cost-efficient smart education system.

The proposed optimization will be most useful for educational administrators, e-learners, and other end users for emerging technologies in entrepreneurial solutions, as well as the existing employment of this mode of emerging technologies in the educational environment.

REFERENCES

Alhathli, M., Masthoff, J. and Beacham, N., 2020, Adapting learning activity selection to emotional stability and competence. *Frontiers in Artificial Intelligence*, 3, p. 11. doi:10.3389/frai.2020.00011.

Bhambri, P., Rani, S., Gupta, G. and Khang, A., 2022, *Cloud and fog computing platforms for internet of things*. CRC Press, ISBN: 978-1-032-101507, doi:10.1201/9781032101507.

Chauhan, M. and Rani, S., 2021, COVID-19: A revolution in the field of education in India. In *Learning How to Learn Using Multimedia* (pp. 23–42). Springer, Singapore.

Hsu Tsu-An, Yang Jinn-Moon, Nikhil Pathak, Yi-Ping Kuo, Teng-Yuan Chang, Chin-Ting Huang, Hui-Chen Hung, John Guann-Yi Yu, 2020. *Zika Virus NS3 Protease Pharmacophore Anchor Model and Drug Discovery*, Article number: 8929, https://www.nature.com/articles/s41598-020-65489-w.

Khang, A., Bhambri, P., Rani, S. and Kataria, A., 2022. *Big data, cloud computing and internet of things*. CRC Press, ISBN: 978-1-032-284200, doi:10.1201/978103228420.

Khang, A., Chowdhury, S. and Sharma, S., 2022. *The data-driven blockchain ecosystem: Fundamentals, applications and emerging technologies.* CRC Press, ISBN: 978-1-032-21624, doi: 10.1201/9781003269281.

Khang, A., Rani, S., Chauhan, M. and Kataria, A., 2021. IoT equipped intelligent distributed framework for smart healthcare systems. *Networking and Internet Architecture*, 2021, https://arxiv.org/abs/2110.04997v2, doi:10.48550/arXiv.2110.04997.

Hasan, S., Khang, A. and Sivakumar, T.B., 2022. Cryptocurrency Methodologies and Techniques. In *The Data-Driven Blockchain Ecosystem: Fundamentals, Applications, and Emerging Technologies* (1st ed., pp. 27–37). CRC Press, https://doi.org/10.1201/9781003269281

Rana, G., Khang, A. and Khanh, H.H., 2021. The role of artificial intelligence in blockchain applications. In *Reinventing Manufacturing and Business Processes Through Artificial Intelligence* (p. 20). CRC Press, doi:10.1201/9781003145011.

Rani, S., Kataria, A., Sharma, V., Ghosh, S., Karar, V., Lee, K. and Choi, C., 2021, Threats and corrective measures for IoT security with observance of cybercrime: A survey. *Wireless Communications and Mobile Computing*, 2021, pp. 1–30.

Rani, S., Mishra, R.K., Usman, M., Kataria, A., Kumar, P., Bhambri, P. and Mishra, A.K., 2021, Amalgamation of advanced technologies for sustainable development of smart city environment: A review. *IEEE Access*, 9, pp. 150060–150087.

Tailor, R.K, Khang, A. and Pareek, R. (Eds.)., 2022. RPA in blockchain. In *The Data-Driven Blockchain Ecosystem: Fundamentals, Applications, and Emerging Technologies* (1st ed., pp. 149–164). CRC Press, https://doi.org/10.1201/9781003269281.

15 Smart Healthcare Solutions for Smart Cities

Meetali Chauhan

CONTENTS

15.1 INTRODUCTION

In the traditional healthcare system, it was mandatory for patients to visit hospitals for regular checkups and medical treatments. Doctors and nurses had to meet patients face to face. But such physical interactions of doctors with patients were not sufficient to correctly diagnose the parameters related to hypertension, blood pressure, sugar levels, and cardiovascular disorders.

DOI: 10.1201/9781003252542-15

A smart healthcare system provides an ease to people by introducing health-care apps via mobile phones and laptops. This system has reduced the need for patients to visit hospitals and clinics to consult doctors. Various services are provided by these healthcare apps, such as booking an ambulance in emer-gencies, consulting doctors online via healthcare apps, and gathering data of patients.

Various important components can be tracked online, such as ambulance and blood supply from the blood bank. In addition to this, the biggest advantage is that senior citizens, for whom it is not feasible to visit doctors on a regular basis, can consult doctors via the e-healthcare system.

Personal data or patient-related information can be shared to doctors, medical specialists, insurance companies, and pharmacies over the network. This informa-tion helps to facilitate patients with a systematic workflow pattern and on-time response. Smart healthcare is a kind of movement that focuses on affordable tech-nologies like mobile phones, computers, and laptops which connect citizens to e-healthcare services.

With the usage of the traditional healthcare system, it was difficult for doctors to provide treatment to long-distance patients in case of emergency (Islam et al., 2015; Rani, Mishra et al., 2021). But healthcare apps and services have resolved this major issue by providing medical facilities to the people in their homes (Bhatt & Chakraborty, 2021).

Smart healthcare system uses technologies, such as artificial intelligence, 5G, big data, microelectronics, cloud computing, and IoT-based wearable and non-wear-able devices. Healthcare enhances communication between people such as doctors, patients, and nurses (Tian et al., 2019).

15.1.1 SMART HEALTHCARE FRAMEWORK

A smart healthcare system is a collaboration of healthcare and medical devices. It consists of medical staff, patients, medical equipment, medical records, hospitals, and clinics. A smart healthcare system is a framework that involves various compo-nents, such as Io—based medical devices (IOMT), cloud computing, edge comput-ing, and 4G/5G/6G wireless communication, as shown in Figure 15.2.

15.1.2 SMART HEALTHCARE DOMAINS

A smart healthcare system is divided into three basic healthcare units: personal care, home care, and hospital care (Algarni, 2019). Each healthcare unit can use IoT devices, which consist of sensors and vibrators (Bedón-Molina et al., 2020).

These devices can be wearable, such as smart watches and smart bands, or non-wearable, such as blood pressure machines and weight-checking machines. The following table is the detailed description of various smart healthcare domains (Algarni, 2019).

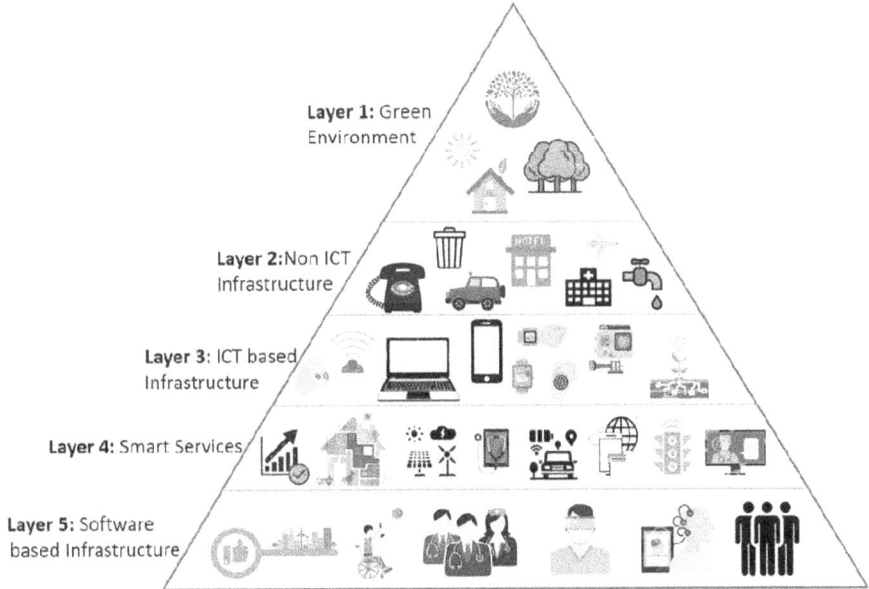

FIGURE 15.1   Architecture of a smart city.

Smart Healthcare Domains	Features
Self-care	• Preventing disease with necessary precautions • Checking health parameters using smart devices on a regular basis • Creating a personal healthy diet • Undergoing medical checkups on a regular basis
Home care	• Monitoring a patient remotely at home • Attending to patients in their homes • Using alarms by patients at home for medical requirements • Online appointment of patients with the doctors • Medical services provided to patients at home • E-medical prescriptions suggested to patients
Hospital care	• Managing chronic diseases of patients • Presence of the patient mandatory at hospital/clinic for treatment • Presence of medical staff is mandatory for attending to patients • Urgent requirement of ambulance, medical equipment, medications, medical staff, and specialists for critical patients • Healthcare providers required to care for the patient 24/7 • Hospital care required in case of surgery or operation of patients

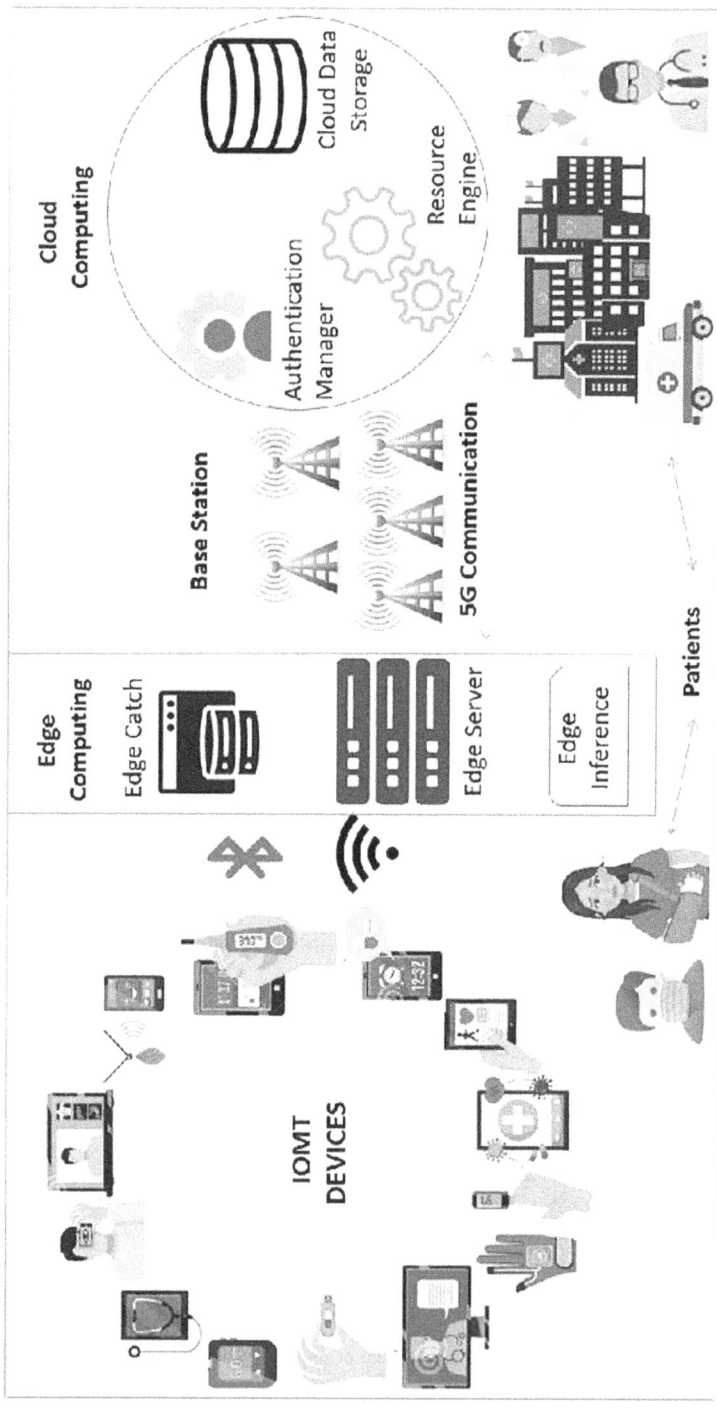

FIGURE 15.2 Smart healthcare framework.

15.2 ADVANTAGES OF SMART HEALTHCARE

15.2.1 REMOTE MONITORING

Remote monitoring via IoT devices, such as sensors, actuators, and smart alarms, has saved the lives of patients who are critically ill. Patients suffering from chronic diseases for whom it is not feasible to move out of their homes to visit doctors due to certain constraints can easily rely on e-medical checkups and e-medical prescriptions.

15.2.2 REDUCTION OF COST

Machines measuring various parameters, such as blood pressure, heart rate, and sugar level, have helped patients to early diagnose their parameters. These machines consist of sensors and actuators, which help measure these health parameters. This has reduced the cost of visiting doctors for minor checkups and made medical testing more affordable.

15.2.3 ACCESS TO MEDICAL INFORMATION

Patients have access to medical records, such as blood tests and medical reports, comprising various body parameters specifying the range for each of those parameters. Such information mentioned in reports helps the patients to consult specialists according to their requirements. In addition to this, these records help doctors to analyze patient body and provide them best remedial measures with minimal complications.

15.2.4 TRACK OF HEALTH STATUS

Smart devices, such as mobile phones and laptops, comprise e-medical healthcare apps, which are used for providing medical facilities. These devices help the patients to track the record of ongoing treatment and administer medication suggested to them for improving health conditions. Various other devices, such as smart watches and smart bands, track various parameters, such as the number of steps taken, blood pressure, and heart rate.

15.2.5 EFFICIENT DATA MANAGEMENT

With the usage of smart devices, the healthcare authorities keep records of all types of data, such as staff members, patients, equipment used, and medical records. Earlier, healthcare centers used to store all records manually, but now smart devices have made it feasible for healthcare centers to store medical data online.

15.3 MAJOR CHALLENGES

15.3.1 FINANCIAL ASPECTS (INTERNET)

Cost is one of the major concerns when it comes to planning healthcare apps in the smart healthcare sector. There is a lot of expenditure on resources to develop

IoT-based apps, but once developed, they are useful for citizens in the smart health-care sector.

15.3.2 PRIVACY AND SECURITY

The advancements in computing, artificial intelligence, and machine learning tech-nologies have made it possible for the rise of the smart healthcare sector. But at the same time, citizens face privacy and security issues in terms of data collection and leakage (Hasan et al., 2022; Tailor et al., 2022).

For example, citizens use smart healthcare monitoring apps for patients residing at home. But the hackers might hack the system to keep an eye on someone's home, which reveals privacy in front of the hacker (Vladimir Hahanov et al., Cybersecurity, 2022). Another problem arises with monitoring devices that lack prescribed stan-dards to reveal information. This might risk patient health in critical situations.

The use of mobile healthcare apps is also not safe as a hacker can track the loca-tion of the user with the help of information provided by the citizens on the app and stalk that citizen (Cook et al., 2018).

The exchange of patients' data takes place over a network via Bluetooth and Wi-Fi connection. This communication over a network can be easily interrupted by a man-in-middle attack (Muhammad et al., 2021).

15.3.3 UNIFORMITY

There is no standardization in rules and regulations for healthcare products and devices. This causes interoperability issues. Mismanagement of electronic health-care records is another major issue. The records are collected from various resources. The standardization needs to be dependent on a variety of factors, such as protocol stacks, communication layer, physical media access control layer (MAC), gateway interface, and data aggregation interface.

15.3.4 HARDWARE PLATFORM

Healthcare-based IoT hardware is difficult to use more than IoT devices. There is a requirement for libraries, frameworks, and specialized platforms that can be used by designers and software developers to access important codes, documents, message templates, classes, and documents.

15.3.5 TECHNOLOGY TRANSITION

The healthcare sector can be updated by modernizing devices and sensors using the latest IoT-based approaches. But there occur compatibility and flexibility issues with existing devices to cope with advanced technologies (Hasan et al., 2022).

15.3.6 LOW POWER PROTOCOL

An IoT device faces issues regarding power consumption requirements, as it is not feasible to provide such services every time. For example, finding a device with such a protocol that requires low power consumption with the availability of services at the MAC layer is a difficult process.

15.3.7 CONTINUOUS MONITORING

In the healthcare sector, there are critical situations where patients who are suffering from chronic diseases require continuous monitoring. In such a situation, there is a requirement for constant monitoring and logging in to the screens.

15.3.8 DATA PROTECTION

There is a risk to data shared with authorized users, organizations, and applications. Shared patient data might be accessed by hackers in between communication for retrieving personal information (Khang, Rana et al., AI & Blockchain, 2021). So data exposed to hackers for attacks, threats, and vulnerabilities is quite a big challenge (Khang, Chowdhury et al., Blockchain, 2022).

15.4   TECHNOLOGIES USED IN SMART HEALTHCARE

In healthcare-based applications, the transmission of data generated to the Internet by sources or devices is mandatory. Data collection done by a user is analyzed for further processing (Arunachalam et al., 2021; Gupta & Rani, 2010; Rani & Kaur, 2012). This processing provides an enhancement to their devices.

With the advancement in technologies, better communication and processing of data are required. Thus, instead of preferring other networks, the best alternative is to prefer a privileged network to fulfill the requirements of its own applications.

In recent times, most industries have been actively involved in the development of wired or wireless communication channels or protocols. But cost and infrastructure are two components of the development of technology (Balaji et al., 2019).

Though multiple connectivity options are available for communication technologies. It depends on products and the associated systems (Khanna & Kaur, 2020). The smart healthcare domain consists of IoT-based standard communication protocols, which are based on the broadband network.

IoT consists of smart equipment, such as GPS, mobile phones, foodstuff, healthcare, appliances, and sensors, which try to achieve a common objective of satisfying consumers in all domains, including the smart healthcare sector (Rani, Bhambri et al., 2021).

The cabling cost is expensive, so wireless communication using sensors is preferred. Low power type of standard communication is preferred for connectivity of devices. Given here are some of the networks comprising location and area of distance:

A. HAN (HOME AREA NETWORK)

It consists of short-range standards, such as Wi-Fi, Dash 7, and Zigbee. In homes, the components that need to be monitored are controlled are connected using HAN.

B. WAN (WIDE AREA NETWORK)

It provides a broader area of coverage than HAN in terms of communication between customers and their distributors. For this setup, a fiber cable or wireless broadband is required, such as 3G and LTE (Rani, Chauhan et al. 2021).

C. FIELD AREA NETWORKS

These are mainly used for the interconnection of customers with substations (Talari et al., 2017). Given her are different types of *technologies* used in smart healthcare (Dachyar et al., 2019):

15.4.1 Software Technologies

Technology	Description
1. IoT platform	It is software that allows the connection of IoT devices. It combines components of the healthcare system with a single product. Services provided are device monitoring, security management, data acquisition, application development, programming, visualization, and analytics.
2. Real-time database	A real-time database does processing for handling workloads.
3. IoT edge analytics	Data is analyzed at the sensor and sent to the remote cloud (Bhambri et al., Cloud & IoT, 2022).
4. IoT-based streaming analytics	It involves creating stream-processing real-time logical data from devices.
5. Unsupervised machine learning	This comprises discovery patterns and information from data that was previously undetected. In this type of learning, the training data of the algorithm does not include the desired type of output.
6. Deep learning	It brings a generation of capable applications that perform complex sensing and recognition tasks for supporting interaction between humans and surroundings.
7. Supervised learning	This type of model uses a training set model, and it yields the desired output.
8. Container security	It comprises software that holds dependencies of applications to run quickly and makes them reliable to run on different computing environments.
9. Security platforms	Security platforms protect electronic devices from physical tampering and attacks on information, using encrypted texts and communication.
10. Cloud computing	It uses networks or servers to manage, process, and store data.

15.4.2 Hardware Technologies

Technology	Description
1. Cloud-connected sensors	These are the sensors that send the data directly to the cloud.
2. Edge gateways	These serve as a point of connection for sensors, devices, and controllers.
3. Quantum computing	This type of computation consists of quantum mechanics phenomena.

15.4.3 CONNECTIVITY TECHNOLOGIES

Technology	Description
1. Cellular IoT 2G/3G/4G	All these provide connectivity to smart healthcare apps. They provide reliable communication and support voice and video calls.
2. Sub	It is a kind of asynchronous service communication used in messaging protocols.
3. WLAN	This includes a Wi-FI network.
4. WPAN	It includes short-term connectivity up to 100 m.
5. Bluetooth	Bluetooth is the key solution for the connectivity of wireless headphones and sensors with smartphones. It is designed using Bluetooth Low-Energy (BLE) protocol, which requires less power consumption from the device.
6. RFID	It is used in logistics, where the main requirement is to determine the position of the object inside buildings. One of the applications of RFID is its ability to track patients in hospitals for improvement in healthcare centers and fulfilling its requirements.
7. Wi-Fi	It is a wireless protocol used for communication. Mostly used in the world of IoT with higher power consumption. It has high signal strength and is reliable for fast transfer of data comprising good connectivity. Its development is based on IEEE 802.11.
8. Zigbee	Zigbee is a wireless standard of network. Mostly preferred in traffic management systems, machines, industrial areas, and electronics of the household. It offers a low exchange rate of data and power consumption. In addition to this, it is less reliable and secure.
9. Thread	Threads have been designed mainly for smart homes. It has IPV6 connectivity, which provides a facility of communication between two connected devices, cloud access services, and user interaction using thread mobile applications.
10. NB-IoT	Narrow-band IoT is based on the latest radio technology. It assures low power consumption and grants the provision of quality signal.
11. LTE-CAT M1	LTE-CAT M1 is a low-power wide area (LPWA) connection standard. It is used for connecting IoT and M2M devices, which have a requirement for a medium rate of data. It supports a long range of battery life and is used in cellular technologies, such as 3G and 2G.
12. LoRaWAN	LoRaWAN stands for low-power long-range wide area networking protocol. It is used for less consumption of power along with multiple devices. Focusing on WAN applications, LoRaWAN was initially designed for furnishing low-power WAN. This provides features, such as low-cost secure communication with IoT, M2M, smart city, and industrial applications.
13. Sigfox	The motive of Sigfox is to provide an alternative for connectivity of low-power M2M applications. This requires reduced levels of data transfer, but for this, Wi-Fi range is short and the cellular range is expensive. So Sigfox provides UNB as a technology that enables handling low data transfer rates. Sigfox has proved to be a suitable option for various M2M apps, such as security devices, intelligent meters, patient monitors, and environmental sensors.
14. e-SIM	It is used in mobile devices that provide SIM provision
15. WNAN	The wireless neighborhood area network is a medium range of technology and is based on IEEE 802.15.4 standard.

15.4.4   OTHER TECHNOLOGIES

15.4.4.1   Big Data

Big data is one of the best methods used for capturing and analyzing the data of patients. Earlier, medical data was stored physically by maintaining manual records with additional costs. But big data technology has feasible conditions to maintain data in digital form with great ease.

This technology has helped healthcare management to maintain all types of records for patients, such as billing system, clinical records, and medical and personal information records of patients (Khang, Bhambri et al., Big Data, Cloud Computing & IoT, 2022).

The systematic storage of data helps to provide quick access to the patient's record when required on an urgent basis.

15.4.4.2   Cloud Computing

Cloud computing technology helps to store on-demand data. The computing resources help to maintain and describe data with the Internet. It provides required information by searching the bulk amount of data with great ease whenever required on an urgent basis (Bhambri et al., Cloud & IoT, 2022).

In addition to this, it also shares proprietary resources of data which helps medical team to perform their job efficiently (Kaur et al., 2015; Rani, Kataria, Chauhan et al., 2022).

15.4.4.3   Smart Sensors

Smart sensors are used to monitor and control patients' health parameters. These sensors help to communicate over a digital network by producing accurate results (Singh et al., 2017; Rani, Kataria et al., 2021).

Smart sensors can easily monitor various parameters of patients such as blood pressure, oxygen level, temperature, sugar level, and fluid management system. It is very effective to collect necessary information on defective bones, the health status of patients, and information related to biological tissues (Rani & Kumar, 2022).

15.4.4.4   Software

Software is used to store patient data, which helps to provide on-time treatment, improves patient care, and helps in easy diagnosis. Medical software provides great ease with effective performance and improves communication between doctors and patients.

Software helps to store confidential details of the patient, disease-related reports of the patients, medical history records of patients, and other necessary information. The best part about using software for accessing medical records is that it is safe and secure to use, as only the authenticated user can access the required details of the patients (Tailor et al., 2022).

15.4.4.5   Artificial Intelligence

Artificial intelligence (AI) is the most renowned technology in the healthcare sector. It helps to perform, evaluate, predict, validate, and analyze the data in a predefined manner.

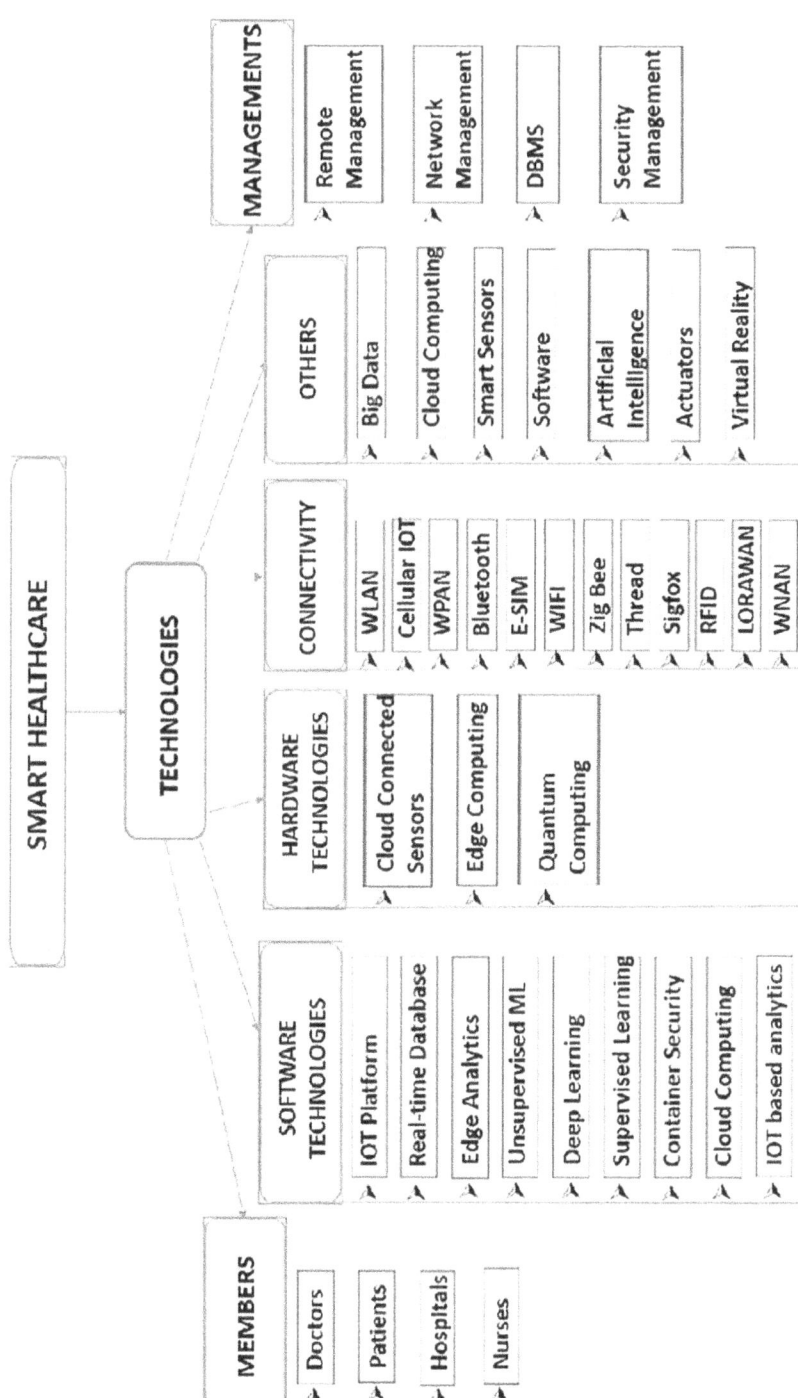

FIGURE 15.3   Smart healthcare technologies.

AI provides the efficient capability to control and predict viral or bacterial infections in advance. Doctors and surgeons achieve accurate and effective results using this technology. It helps to measure the pain of the patient and provide effective medication to the patient accordingly.

15.4.4.6 Actuators

Actuators are mechanisms used for controlling the system, and they provide motion to the system. The purpose of the actuator is to control certain parameters and provides accuracy. Actuators help to design hospital beds & adjust them according to the patient's requirements.

15.4.4.7 Virtual or Augmented Reality

The best way to collaborate with human beings with an electronic system is a virtual or augmented reality system that provides real-time information. Virtual reality helps to improve the safety of patients, quality of planning, and effective treatment.

It provides information to the doctors for improving the quality of surgery undertaken by the patients, as shown in Figure 15.3. It also provides digital information in image or sound form, as shown in Figure 15.2. (Rani, Kataria, Chauhan et al., 2022).

15.5 CONCLUSION

With the evolution of technology, people are switching toward better education, employment, transportation, and healthcare services in a smart city ecosystem. The healthcare domain is evolving rapidly in a smart city environment.

A variety of technologies are used in the area of smart healthcare. With the amalgamation of these technologies, healthcare services have become more convenient, accurate, and easily accessible.

REFERENCES

Algarni, A., "A Survey and Classification of Security and Privacy Research in Smart Healthcare Systems," *IEEE Access*, vol. 7, pp. 101879–101894, 2019. doi:10.1109/ACCESS.2019.2930962.

Arunachalam, P., N. Janakiraman, Arun Kumar Sivaraman, A. Balasundaram, Rajiv Vincent, Sita Rani, Barnali Dey, A. Muralidhar, and M. Rajesh. "Synovial Sarcoma Classification Technique Using Support Vector Machine and Structure Features," *Intelligent Automation & Soft Computing*, vol. 32, no. 2, pp. 1241–1259, 2021. doi:10.32604/iasc.2022.022573.

Balaji, S., K. Nathani, and R. Santhakumar, "IoT Technology, Applications and Challenges," *Wireless Personal Communications*, no. 0123456789, 2019. doi:10.1007/s11277-019-06407-w.

Banerjee, K., V. Bali, N. Nawaz, S. Bali, S. Mathur, R. K. Mishra, and S. Rani, @A Machine-Learning Approach for Prediction of Water Contamination Using Latitude, Longitude, and Elevation," *Water*, vol. 14, no. 5, p. 728, 2022. doi:10.3390/w14050728.

Bedón-Molina, J., M. J. Lopez, and I. S. Derpich, "A Home-based Smart Health Model," *Advances in Mechanical Engineering*, vol. 12, no. 6, pp. 1–16, 2020. doi:10.1177/16878140 20935282.

Bhambri, P., Sita Rani, Gaurav Gupta, and Khang, A., *Cloud and Fog Computing Platforms for Internet of Things*. CRC Press, 2022, doi:10.1201/9781032101507, ISBN: 978-1-032-101507.

Bhatt, V., and S. Chakraborty, "Real-time Healthcare Monitoring Using Smart Systems: A Step Towards Healthcare Service Orchestration," *MDPI*, pp. 772–777, 2021. doi:10.1109/ICAIS50930.2021.9396029.

Cook, D. J., G. Duncan, G. Sprint, and R. L. Fritz, "Using Smart City Technology to Make Healthcare Smarter," *Proc. IEEE*, vol. 106, no. 4, pp. 708–722, 2018. doi:10.1109/JPROC.2017.2787688.

Dachyar, M., T. Y. M. Zagloel, and L. R. Saragih, "Knowledge Growth and Development : Internet of Things (IoT) Research, 2006–2018," *Heliyon*, vol. 5, no. June, p. e02264, 2019. doi:10.1016/j.heliyon.2019.e02264.

Gupta, O. P., and S. Rani, "Bioinformatics Applications and Tools: An Overview," *CiiT-International Journal of Biometrics and Bioinformatics*, vol. 3, no. 3, pp. 107–110, 2010.

Hasan, S., Khang, A., and T. B. Sivakumar, "Cryptocurrency Methodologies and Techniques," in *The Data-Driven Blockchain Ecosystem: Fundamentals, Applications, and Emerging Technologies* (1st ed., pp. 27–37). CRC Press, 2022, https://doi.org/10.1201/9781003269281.

Islam, S. M. R., D. Kwak, M. H. Kabir, M. Hossain, and K. S. Kwak, "The Internet of Things for Health Care: A Comprehensive Survey," *IEEE Access*, vol. 3, pp. 678–708, 2015. doi:10.1109/ACCESS.2015.2437951

Kaur, G., R. Kaur, and S. Rani, "Cloud Computing-A new Trend in IT Era," *International Journal of Science Technology & Management*, Vol. 1, pp. 1–6, 2015.

Khang, A., Pankaj Bhambri, Sita Rani, Aman Kataria, *Big Data, Cloud Computing and Internet of Things*. CRC Press, 2022, doi:10.1201/9781032284200, ISBN: 978-1-032-284200.

Khang, A., Subrata Chowdhury, and Seema Sharma, *The Data-Driven Blockchain Ecosystem: Fundamentals, Applications and Emerging Technologies*. CRC Press, 2022, doi:10.1201/9781003269281, ISBN: 978-1-032-21624.

Khang, A., Geeta Rana, Ravindra Sharma, Alok Kumar Goel, and Ashok Kumar Dubey, "The Role of Artificial Intelligence in Blockchain Applications," in *Reinventing Manufacturing and Business Processes Through Artificial Intelligence* (p. 20). CRC Press, 2021, doi:10.1201/9781003145011.

Khanna, A. and S. Kaur, *Internet of Things (IoT), Applications and Challenges*, No. 0123456789. Springer, 2020, doi:10.1007/s11277-020-07446-4.

Muhammad, G., F. Alshehri, F. Karray, A. El Saddik, M. Alsulaiman, and T. H. Falk, "A Comprehensive Survey on Multimodal Medical Signals Fusion for Smart Healthcare Systems," *Information Fusion*, vol. 76, no. July, pp. 355–375, 2021. doi:10.1016/j.inffus.2021.06.007.

Rani, S., P. Bhambri, and M. Chauhan, "A Machine Learning Model for Kids' Behavior Analysis from Facial Emotions using Principal Component Analysis," In *2021 5th Asian Conference on Artificial Intelligence Technology (ACAIT)*. IEEE, pp. 522–525, 2021 October. doi:10.1109/ACAIT53529.2021.9731203.

Rani, S., M. Chauhan, A. Kataria, and Khang, A., "IoT Equipped Intelligent Distributed Framework for Smart Healthcare Systems," *arXiv preprint arXiv:2110.04997*, 2021.

Rani, S., A. Kataria, and M. Chauhan, "Fog Computing in Industry 4.0: Applications and Challenges—A Research Roadmap," in *Energy Conservation Solutions for Fog-Edge Computing Paradigms* (pp. 173–190). Springer, Singapore, 2022.

Rani, S., A. Kataria, M. Chauhan, P. Rattan, R. Kumar, and A. K. Sivaraman, "Security and Privacy Challenges in the Deployment of Cyber-Physical Systems in Smart City Applications: State-of-Art Work," in *Materials Today: Proceedings*. Elsevier, Amsterdam, 2022.

Rani, S., A. Kataria, V. Sharma, S. Ghosh, V. Karar, K. Lee, and C. Choi, "Threats and Corrective Measures for IoT Security with Observance of Cybercrime: A Survey," *Wireless Communications and Mobile Computing*, vol. 2021, 2021. doi:10.48550/arXiv.2010.08793.

Rani, S., and S. Kaur, "Cluster Analysis Method for Multiple Sequence Alignment," *International Journal of Computer Applications*, vol. 43, no. 14, pp. 19–25, 2012. doi:10.5120/6171-8595.

Rani, S., and R. Kumar, "Bibliometric Review of Actuators: Key Automation Technology in a Smart City Framework," *Materials Today: Proceedings*. Elsevier, Amsterdam, 2022.

Rani, S., R. K. Mishra, M. Usman, A. Kataria, P. Kumar, P. Bhambri, and A. K. Mishra, "Amalgamation of Advanced Technologies for Sustainable Development of Smart City Environment: A Review," *IEEE Access*, vol. 9, pp. 150060–150087, 2021.

Shrestha, N. M., A. Alsadoon, P. W. C. Prasad, L. Hourany, and A. Elchouemi, "Enhanced e-Health Framework for Security and Privacy in Healthcare System," *2016 6th International Conference on Digital Information Processing Communications. ICDIPC 2016* (pp. 75–79). IEEE, 2016. doi:10.1109/ICDIPC.2016.7470795.

Singh, P., O. P. Gupta, and S. Saini, "A Brief Research Study of Wireless Sensor Network," *Advances in Computational Sciences and Technology*, vol. 10, no. 5, pp. 733–739, 2017. https://www.ripublication.com/acst17/acstv10n5_07.pdf.

Talari, S., M. Shafie Khah, P. Siano, V. Loia, A. Tommasetti, and J. Catalao, "A Review of Smart Cities Based on the Internet of Things Concept," *Innovative Methods for Smart Grids Planning and Management*, pp. 1–23, 2017. doi:10.3390/en10040421.

Tian, S., W. Yang, J. M. Le Grange, P. Wang, W. Huang, and Z. Ye, "Smart Healthcare: Making Medical Care More Intelligent," *Journal of Global Health*, vol. 3, no. 3, pp. 62–65, 2019. doi:10.1016/j.glohj.2019.07.001.

Tailor, R. K., Khang, A., and Ranu Pareek (Eds.), "RPA in Blockchain," in *The Data-Driven Blockchain Ecosystem: Fundamentals, Applications, and Emerging Technologies* (1st ed., pp. 149–164). CRC Press, 2022. https://doi.org/10.1201/9781003269281.

Vladimir Hahanov, Khang, A., Gardashova Latafat Abbas, and Vugar Abdullayev Hajimahmud, "Cyber-Physical-Social System and İncident Management," in *AI-Centric Smart City Ecosystems: Technologies, Design and Implementation* (1st ed.). CRC Press, 2022. https://doi.org/10.1201/9781003252542.

16 Advanced Technologies and Data Management in the Smart Healthcare System

Alex Khang, Nazila Ali Ragimova,
Vugar Abdullayev Hajimahmud,
Abuzarova Vusala Alyar

CONTENTS

16.1 INTRODUCTION

Many areas must constantly be evolving for world prosperity—for example, technology, industry, science, education, and health sectors. In fact, the list goes on and on. However, the recent era—the digital age—requires further development of technology. However, it is known that with the digital age, many areas have begun to digitalize and enter the digital world.

Given the situation of the last two years, we can see that the health sector is now keeping pace with the technological era in a more sustainable way. Although the pandemic is perceived by people as a catastrophe—an economic or population catastrophe—it has shown that this world, this humanity, must embrace and adapt to a new era (Arunachalam et al., 2021).

DOI: 10.1201/9781003252542-16

For example, a few years ago, distance learning, distance education, and even distance treatment were relevant in very few countries, but now very few countries do not practice it. That is, the vast majority of countries are already moving to remote living (Banerjee et al., 2022).

However, it should be noted that compared to other areas, countries have paid more attention to improving their health systems. In this sense, today's health has become a new cornerstone for the future healthcare system (Chauhan and Rani, 2021).

16.2 TODAY'S HEALTHCARE AS THE FOUNDATION OF THE FUTURE

"A healthy body has a healthy spirit." Although this expression comes from the past, it proves the importance of human health today.

While it is important that many areas are constantly evolving, perhaps the most important sector is the health sector, which must constantly evolve (Gupta and Rani, 2013).

Certain diseases have occurred in each period. The most dangerous of these is the so-called Black Death or Black Plague. The disease, which dates back to the 1300s, is a population disaster that has killed millions of people (Rani, Kataria et al., 2022).

The Spanish flu or the flu epidemic caused by the H1N1 virus between 1918 and 1920 is considered another disaster. And finally, the coronavirus (COVID-19 pandemic) has engulfed our world in the new era (Rani, Kataria, Sharma et al., 2021).

The spread of such diseases has led to the discovery of health innovations for each period. Today's health innovations are at a different level than in the past.

We can look at this in two ways:

1. Structure (construction of new hospitals, provision of technological equipment, design, etc.)
2. Treatment/medicine (impact of technology on drugs for new diseases, advances in the treatment of other ongoing diseases, technological innovations affecting them, the impact of biotechnology, etc.)

Recently, one of the most focused projects is the smart hospital project.

Smart hospitals are part of the e-health phenomenon. The smart hospital is improving patient care using optimized and automated processes, especially the Internet of Things (IoT), to connect medical devices to artificial intelligence and data analysis (Rani, Khang et al., 2021).

Many technologies are used in the smart hospital, such as the IoT, big data, cloud computing, artificial intelligence (AI) (Rana et al., 2021), robotics, 3D printing, mobile health (MHealth), RFID, biosensors, integration platforms, and wearable devices (Khang, Bhambri et al., Big Data, Cloud Computing & IoT, 2022).

And it should be noted that no matter how strong the hospital staff, at the same time, patients now prefer hospitals with a high level of comfort. In this regard, new types of hospitals should pay special attention to the proper use of technology.

If the world's oldest doctor is not equipped with such necessary equipment, he may have difficulty treating the patient in front of him. Therefore, hospital managers should not be afraid to introduce the right technology to their staff. The smart hospital project is an ideal model for innovators, medical professionals, patients, and the development of technology (Rani, Mishra et al., 2021).

Due to the current pandemic, some technologies in healthcare have become more advanced than others.

Examples of this are these technologies:

1. Telemedicine (telehealth)
2. New methods of drug development
3. Information-based health
4. Nanomedicine (based on nanotechnology)
5. Devices that support 5G
6. Tricorders
7. Digital health assistants
8. More intelligent pacemakers
9. Laboratory on chip
10. Wearable devices

In addition, 3D printing, virtual reality, augmented reality, artificial intelligence, big data, remote control, blockchain, and mobile-based technologies can also be noted (Sudevan et al., 2021).

In particular, telemedicine has been further improved. In January 2020, about 24% of healthcare organizations had an existing telehealth program. Many regulatory barriers to forced healthcare have been removed, and healthcare organizations now have nearly a year of information on how to evaluate and improve telehealth services.

These are the main components of healthcare for the new era. For future health, specialists continue to work.

16.3 TECHNOLOGICAL PROGRESS TOWARD THE FUTURE WORLD OF HEALTHCARE

The world of future healthcare will be better than it is today, just as today's healthcare has improved in the last 30 years ago.

In general, according to experts, there are three factors that can change health:

1. **Demographic structure:** The number of elderly people around the world continues to grow as the demand for health services accelerates.
2. **Benefit:** The financial burden of the national budget and employer-funded health insurance will push countries to the necessary reforms.
3. **Technology:** Technological changes in medicine and technology/information continue to bring new ideas, bringing new approaches to disease management. These technologies improve the results of the treatment of patients in a more cost-effective way.

As mentioned, the smart hospital project is also an area that needs to be further developed in the future.

Given the current situation, as well as new experiences in healthcare, we can say that the demand for smart hospitals will increase in the near future.

Undoubtedly, patients will prefer hospitals that will serve them safer and older. Eventually, this will push other hospitals to become smart hospitals.

Hospitals have long been reluctant to adopt evolving technologies and practices, primarily due to high investments.

However, as the cost of technology declines sharply and hospitals use new technologies, they will be forced to adopt new technologies.

In addition to helping patients, these changes will benefit hospitals themselves by digitizing asset tracking, employee management, and planning for better operational efficiency.

In addition, the technologies mentioned earlier will affect the current state and the future of healthcare.

16.3.1 TELEMEDICINE

Telemedicine—with patients from a distance, from home or even from work—allows patients to communicate with their doctor. The best part of this is that it improves access to healthcare for those who live in remote areas or have difficulty accessing physical services.

16.3.2 INFORMATION-BASED HEALTH

With information-based health, we can link this to big data and electronic medical records. It is known that big data is the most popular technology in healthcare. Through electronic medical records, patient information is now more accessible to physicians and patients themselves. Big data technology is also responsible for the proper management of this data.

16.3.3 AUGMENTED REALITY

Augmented reality is an ideal technology, especially for medical students, allowing them to engage in operations that are closer to reality. This helps not only students but also medical staff and doctors to develop their skills without getting involved in real surgery.

16.3.4 3D PRINTING

Researchers are exploring the potential of 3D printing in medicine. For example, the use of 3D printers to replicate multidimensional models of problem areas is being improved. Surgeons can manage the models and simulate various possible surgical replicas before performing the actual operation.

Alternatively, 3D printing can be used to create models of bones or other organs in the human body.

16.3.5 Nanomedicine

Nanomedicine is an application of nanotechnology in health. Nanomedicine is used for the early detection and prevention of diseases and the diagnosis of many diseases, and it has the potential to significantly improve. This technology is still being tested.

In general, nanomedicine is understood as a key tool for personalized, targeted, and rehabilitative medicine by delivering patients to the next level of new drugs, treatments, and implantable devices for real progress in health (Rani, Khang et al., IoT & Healthcare, 2021).

In addition, one of the most widely used technologies in recent times is e-health applications.

Nowadays, with almost everyone having a smartphone, e-health apps make it more accessible to focus on your health and well-being. You will have a personal doctor based on artificial intelligence who is closer to you. However, it should be noted that these applications cannot replace a real doctor. It may happen in the future, but it is not for the present.

16.4 THE CONCEPT OF DATA MANAGEMENT

It is not meaningless to call this century the information age. The number of data in the common database of the world's internet continues to grow. For the past 20 years, the number of current data has increased dramatically. Unfortunately, the abundance of information has both positive and negative effects. The most important of these is the proper management of that data.

Data management, in a narrow sense, refers to the process of receiving, storing, processing, organizing, and transmitting data in the warehouse of any enterprise.

The data management process involves a combination of different functions in corporate systems to ensure that information is accurate, relevant, and accessible.

Proper data management helps to make the right decisions within the enterprise. In addition, data management is important for businesses to secure the data of both their employees and customers.

There are a number of concepts related to data management:

1. Clearing the data
2. Data transmission
3. Data storage
4. Data directories
5. Data architecture
6. Data security
7. Data modeling

Clearing raw data and converting it into the correct form and format for analysis, as well as making adjustments and merging datasets, happens with data cleaning.

Data transmission channels allow the automatic transfer of data from one system to another.

Databases are used to connect different data sources, manage many types of data stored by enterprises, and provide a clear route for data analysis. Here, there is the concept of ETL (extract, transform, load), which is built to take data from a system, transform it and upload it to the organization's database.

Data directories help manage metadata to create a complete description of the data and provide a summary of changes, locations, and quality, while also making it easier to find the data (Bhambri et al., Cloud & IoT, 2022).

The data architecture provides a formal approach to creating and managing data flows.

Data security is the process of preventing unauthorized third-party access to data in an enterprise's warehouse.

Data modeling is the process of documenting data flows.

It is also possible to face a number of problems in data management. The most important of these is that the amount of data is too large. It is in this respect that big data management is then implemented.

A possible risk in an environment where data cannot be managed centrally is data leakage. As a countermeasure, it is necessary to manage the data in a centralized manner and to understand which data belongs to whom and when it is extracted.

The concept of data management can be summarized in this form.

16.5   THE IMPORTANCE OF DATA MANAGEMENT

Data management is a very important process at the enterprise level. In general, we can show it as in the following graph.

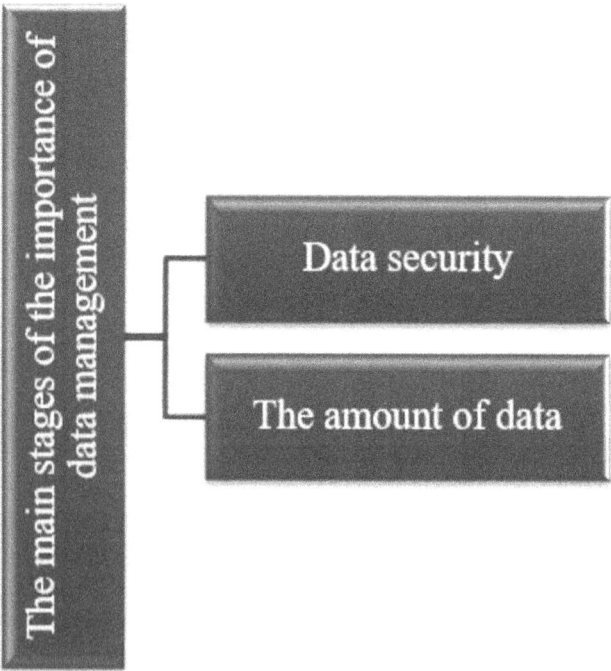

FIGURE 16.1   Importance of data management.

1. Enterprise data is a very important resource. In this regard, their safety is an important factor.
2. Excessive amount of data complicates management and decision-making processes.

Loss of data can lead to disasters within the enterprise. It also damages the company's image.

16.6   TECHNOLOGICAL PROGRESS IN DATA MANAGEMENT

The concept of data management is not so old. This concept was put forward in the 1980s. Therefore, the term data management can be taken as a young enough concept.

Along with the constant growth of different types of data, their management is also evolving in different ways.

Today, there are a number of trends in data management. These trends will also affect the future development of data management.

1. Artificial intelligence and machine learning
2. Augmented data analytics
3. Blockchain technology
4. DataOps
5. Data factory

Artificial intelligence (AI) and machine learning (ML) provide high-value automation for manual processes that are prone to human error. Basic data management tasks, such as data identification and classification, can be managed more efficiently and accurately by advanced technologies in the AI/ML space (Rana et al., AI & Blockchain, 2021).

ML will enhance the ability to manage a variety of data, including data cataloging, metadata management, data mapping, anomaly detection, and other key processes. At the same time, AI will contribute to the proposed actions, the automatic detection of metadata, and the automatic monitoring of controls (Bartley, 2021).

Increased data analytics is aimed at reducing the number of manual tasks. Thus, in the near future, the percentage of manual work will decrease. Instead, the percentage of automation of data management will increase. Here, too, AI and ML technologies will be used: for data management—collection, storage, processing, and so on.

In blockchain technology, distributed stationery systems allow businesses to keep transaction logs, asset tracking, and audit trails more securely. Together with blockchain technology, this technology keeps data in a decentralized form that cannot be changed, financial transaction data, sensitive data retrieval activities, and so on. It also enhances the accuracy and precision of data processing records (Tabatabaei, 2019).

DataOps applies DevOps principles to data management. DataOps provides data and analytics within the company that integrates practices such as flexible development, technology, processes, and statistical process control. Instead of creating barriers to data management between different commands, tools, and processes, DataOps aims to break down those barriers and build a company-wide data operation whose components are integrated and optimized. It plays a crucial role in the democratization of information within an organization (Bartley et al., 2016).

Data factory manages the collection, management, integration, and distribution of data according to a single architecture. Data structures offer a number of benefits, including the elimination of data silos, simplification of data management, the launch of hybrid cloud and local infrastructures, and scalability. In addition to solutions such as ETL, ELT, and enhanced data management, companies also use capabilities such as graphics technology and semantic standards to build data structures (Tabatabaei, 2019). Data management is also important for the health sector, as mentioned earlier. In particular, the proper management of patient records (i.e., patient information) is critical to both the patient and the hospital (medical staff).

Properly managed data cannot be viewed by a third party. This plays an important role in protecting the patient's privacy. On the other hand, the advantage of this is that patient information is easily accessible to both the doctor and the patient. This is especially true with the ability to easily find any information we have mentioned in the data catalogs section (Tabatabaei, 2019).

The problem of data management in smart hospitals can also be solved in a more convenient way than in other hospitals (Mann et al., 2020).

16.7 CONCLUSION

With the current pandemic situation in the world, the demand for technology has increased. Education, industry, production, health, and so on have become key drivers in meeting the demand for technology.

Remote work has become a way of life due to pandemic conditions. It is true that normalization has already taken place in many areas, but it has become a kind of practice for mankind. People will be more experienced in similar events that may happen in the future.

In turn, one of the most affected areas was health. Over the past ten years, the health sector has changed its development to some extent. Smart hospital projects have become more relevant.

At the same time, as a result of increasing people's use of Internet resources in the current situation, there has been an increase in the amount of data. This, in turn, has had an impact on the data management process.

In general, we can note the following results from the theoretical knowledge mentioned in the article:

1. The current pandemic has led to some radical changes in health.
2. Many technologies in the field of healthcare, including artificial intelligence, machine learning, big data, blockchain technology, cloud computing, and so on (Rana et al., AI & Blockchain, 2021), play a big role.

3. Smart hospitals will be implemented in many countries in the near future and will provide better services to patients.
4. A number of the mentioned technologies (artificial intelligence, machine learning, big data, etc.) play an important role in the implementation of the smart hospital project.
5. Data management remains relevant in many areas, including health (Rani, Khang et al., IoT & Healthcare, 2021).
6. In data management, many technologies help also. The most important of these are artificial intelligence and machine learning. Because with their help, a large percentage of manual work will be automated.

In healthcare, data management is important. It is known that the health sector (hospitals, the Ministry of Health, etc.) contains more information than many small organizations. And it is important to manage this information properly.

REFERENCES

Arunachalam, P., N. Janakiraman, A.K. Sivaraman, A. Balasundaram, R. Vincent, S. Rani, B. Dey, A. Muralidhar, and M. Rajesh, "Synovial Sarcoma Classification Technique Using Support Vector Machine and Structure Features," *Intelligent Automation & Soft Computing*, 32(2), 1241–1259, 2021. doi:10.32604/iasc.2022.022573.

Banerjee, K., V. Bali, N. Nawaz, S. Bali, S. Mathur, R.K. Mishra, and S. Rani, "A Machine-Learning Approach for Prediction of Water Contamination Using Latitude, Longitude, and Elevation," *Water*, 14(5), 728, 2022. doi:10.3390/w14050728.

Bartley, Kevin, "Data Management 2021: Trends Technology That Will Define the Year," *e-Book, Rivery*, 2021. https://rivery.io/blog/data-management-2021-trends-technology-that-will-define-the-year.

Bhambri, Pankaj, Sita Rani, Gaurav Gupta, and Khang, A., *Cloud and Fog Computing Platforms for Internet of Things*. CRC Press, 2022. ISBN: 978-1-032-101507, doi:10.1201/9781032101507.

Briney, Kristin, *Data Management for Researchers: Organize, Maintain and Share Your Data for Research Success (Research Skills)* (p. 250), Pelagic Publishing, September 1, 2015. https://www.amazon.com/Data-Management-Researchers-Organize-maintain/dp/1784270113. Pelagic Publishing.

Chauhan, M., and S. Rani, "Covid-19: A Revolution in the Field of Education in India," in *Learning How to Learn Using Multimedia* (pp. 23–42). Springer, Singapore, 2021.

Gupta, O.P., and S. Rani. "Accelerating Molecular Sequence Analysis Using Distributed Computing Environment," *International Journal of Scientific & Engineering Research— IJSER*, 4(10), 262–266, 2013.

Khang, A., Pankaj Bhambri, Sita Rani, and Aman Kataria, *Big Data, Cloud Computing and Internet of Things*. CRC Press, 2022. ISBN: 978-1-032-284200, doi:10.1201/9781032284200.

Khang, A., Subrata Chowdhury, and Seema Sharma, *The Data-Driven Blockchain Ecosystem: Fundamentals, Applications and Emerging Technologies*. CRC Press, 2022. ISBN: 978-1-032-21624, doi:10.1201/9781003269281.

Mann, S., Y. Arora, and S. Anand, "Smart Hospitals with the Use of' Internet of Things' and Artificial Intelligence," Available at SSRN 3569591, 2020. doi:10.2139/ssrn.3569591.

Meskó, Bertalan, *The Guide to the Future of Medicine: Technology and the Human Touch* (p. 228). Webicina Kft, August 25, 2014. https://www.amazon.com/Guide-Future-Medicine-Technology-Human/dp/9630898020.

Rana, G., Khang, A., Ravindra Sharma, Alok Kumar Goel, and Ashok Kumar Dubey, "The Role of Artificial Intelligence in Blockchain Applications," in *Reinventing Manufacturing and Business Processes Through Artificial Intelligence*. CRC Press, 2021, doi:10.1201/9781003145011.

Rani, S., A. Kataria, and M. Chauhan, "Fog Computing in Industry 4.0: Applications and Challenges—A Research Roadmap," in *Energy Conservation Solutions for Fog-Edge Computing Paradigms* (pp. 173–190). Springer, Singapore, 2022.

Rani, S., A. Kataria, M. Chauhan, P. Rattan, R. Kumar, and A.K. Sivaraman, "Security and Privacy Challenges in the Deployment of Cyber-Physical Systems in Smart City Applications: State-of-Art Work," in *Materials Today: Proceedings*. Elsevier, Amsterdam, 2022.

Rani, S., A. Kataria, V. Sharma, S. Ghosh, V. Karar, K. Lee, and C. Choi, "Threats and Corrective Measures for IoT Security with Observance of Cybercrime: A Survey," *Wireless Communications and Mobile Computing*, 2021, 1–30, 2021.

Rani, Sita, Khang, A., Meetali Chauhan, and Aman Kataria, "IoT Equipped Intelligent Distributed Framework for Smart Healthcare Systems," *Networking and Internet Architecture*, 2021, https://arxiv.org/abs/2110.04997v2, doi:10.48550/arXiv.2110.04997.

Rani, S., and R. Kumar, "Bibliometric Review of Actuators: Key Automation Technology in a Smart City Framework," in *Materials Today: Proceedings*. Elsevier, Amsterdam, 2022.

Rani, S., R.K. Mishra, M. Usman, A. Kataria, P. Kumar, P. Bhambri, and A.K. Mishra, "Amalgamation of Advanced Technologies for Sustainable Development of Smart City Environment: A Review," *IEEE Access*, 9, 150060–150087, 2021.

Sudevan, S., B. Barwani, E. Al Maani, S. Rani, and A.K. Sivaraman, "Impact of Blended Learning during Covid-19 in Sultanate of Oman," *Annals of the Romanian Society for Cell Biology*, 14978–14987, 2021.

Tabatabaei, Seyyed Mohammad, "Medical Big Data Analytics: An Interesting but Challenging Interdisciplinary Field of Study," *EC Clinical and Experimental Anatomy ECO*, 1, 22–23, 2019.

17 The Key Assistant of Smart City
Sensors and Tools

*Alex Khang, Vladimir Hahanov, Eugenia Litvinova,
Svetlana Chumachenko, Vugar Abdullayev
Hajimahmud, Abuzarova Vusala Alyar*

CONTENTS

17.1 INTRODUCTION

It is possible to encounter dubious situations at any time, whether it is in a "smart" city or a "normal" city. Since a smart city is a "people + technology + knowledge" system, in addition to the risks of an ordinary city, there will be cyber risks. In general, a city (assuming it is smart) may have the following (suspicious) situations: crime (theft, murder, armed conflict, physical conflict), bags left by suspects within the city that could be a bomb threat, traffic accidents, and physical risks. Also, more intelligent cities, given their technology, have become an integral part of human life but also increase the risk of theft of personal information, including bank accounts. In addition, it is possible to note the moral risks associated with the theft of data, such as identification and biometric data, with advanced cyberattacks (Arunachalam et al., 2021).

If we accept smart cities as a "human + technology + knowledge" system, the transmission of information (in the correct form) during the risks and concerns that occur here can lead to physical or spiritual salvation (Bhambri et al., 2022).

In general, proper data management (collection, storage, processing, and transmission, which are very important events for timely transmission of information) is

DOI: 10.1201/9781003252542-17

very important both in the smart city life system and in other areas (military, industry, corporate relations, etc.).

Sensors are the biggest helper in the implementation of processes, such as data collection and transmission (Rani, Chauhan, Kataria, and Khang, 2021).

In addition, these processes can be considered as integrated processes under the term cyber-physical system. In this regard, according to the topic, we will look at the follows:

- Smart cities
- Cyber-physical systems
- Smart processes in smart cities
- Sensors
- Principle of operation of sensors (detection systems)
- Sensors in solving smart city problems (security)

17.2 SMART CITIES

It is relatively difficult to describe a smart city in the current situation. In general, a smart city is a cyber-physical system that makes extensive use of IT infrastructure in the implementation of intra-city processes, solves the problems of people related to the city, ensures that humans and technology coexist constantly, allows the expression to become a reality, and is equipped with the technological capabilities designed for the survival of such a society (Rani, Arya et al., 2022).

Smart (human + technology) = Smart society

The idea of a smart city is relatively old, dating back to the last century (1974). Formally, its application in reality is relatively new.

In reality, however, the first smart city to be created was Songdo, South Korea. After that, different countries joined the project and began to implement this project in their territories, in different cities (Rani, Kataria et al., 2022).

Although the concept of a smart city is limited to the word *city*, it is a relatively broad concept. It is enough to briefly mention the mentioned definition and look at its components in accordance with our topic (Rani, Kataria, Chauhan et al., 2022; Rani, Kataria et al., 2021).

In addition to the "human + technology + knowledge" components of a smart city, its location is also one of its key components (Rani and Kumar, 2022). In short, in addition to the main factors of the smart city (people, technology, and knowledge), we can add the geographical-position factor, as shown in Figure 17.1.

In this regard, Mr. Kasarda's opinion is very interesting (Vallianatos, 2015). Thus, he noted in his book (*Aerotropolis: The Way We'll Live Next*), "The 18th century was really the century of water, the 19th century the century of railways, the 20th century the century of highways, cars, trucks, and the 21st century will be an increasingly century of aviation."

He claims that the next station of smart cities is the airport.

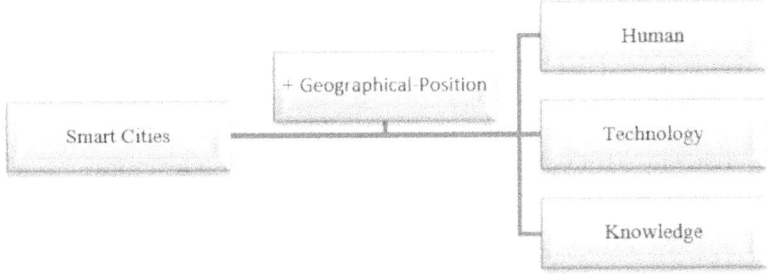

FIGURE 17.1 Smart city = (H + T + K) + geographical position.

According to him, in the 21st century, it is important to build cities in areas with ideal airports. When we talk about smart cities, many technologies come to mind, such as big data, blockchain, Internet of things (Khang et al., IoT & Healthcare, 2021), and so on.

One of them is the cyber-physical systems. We will look at this in the next section. Finally, in the simplest case, a smart city can be thought of as a model above the concept of a city that comes to life in the human brain (Khang, Bhambri et al., Big Data, Cloud Computing & IoT, 2022).

A very top model of this, as in science fiction movies, could be places where flying robots act as police, ambulances, restaurant owners, CEOs, and so on. Maybe in the future, these films may become a reality, like many films (Rani, Mishra et al., 2021).

17.3 CYBER-PHYSICAL SYSTEMS

Cypber-physical systems (CPSs) are systems that connect the physical world with the virtual processing world. CPSs have different components. These components include modern preventive information technologies

In other words, CPS collects information in the real (physical) world, analyzes this information using digital technology in the virtual (cyber) world, facilitates the use of information and knowledge, and returns that information to the physical side to create added value.

In a CPS, various information in the real world (physical space) is collected by IoT devices and processed and analyzed in the virtual world (cyberspace).

As mentioned, the processes within the smart city are generally equivalent to the purpose of the existence of CPSs.

In terms of the function and structure of the ICT system that builds a smart city, it can be said that it is the CPS itself. CPS is a system that optimizes activities in real space by collecting and analyzing information in virtual space that reflects the state and movement of real space, such as residential areas and streets.

In addition to various data terminals, IoT devices must be used to collect data. Big data analysis using cloud or artificial intelligence (AI) is required for analysis and optimization in virtual space (Rana et al., AI & Blockchain, 2021).

Moreover, a data terminal is required to receive the analysis results and provide some information to the residents.

In the future, control data will be transferred to infrastructure equipment, self-propelled (driverless) cars, robots, and so on will be returned and will become an autonomous system that operates optimally according to the conditions and movements of the city (Ochoa et al., 2017).

17.4   SMART PROCESSES IN SMART CITIES

When we talk about what smart processes can be in smart cities, we can basically mean the digitization of processes (connections) in normal cities.

The components that exist in a city, and the relationships between them can be shown as follows:

1. Power supply (control of light energy transmitted to homes, businesses in the city, streetlights, etc.)
2. Transport (control of parking systems, traffic lights, etc.)
3. Gas supply (gas supply to both enterprises and homes, control of meters)
4. Internet services
5. Water supply systems (control of used water, wastewater, etc.)
6. Security systems (surveillance of homes, city, traffic cameras, etc.)
7. Filling stations
8. Accident warning systems

All processes of these systems within the smart city will be digitized.

We take a closer look at these processes, such as intelligent transport, intelligent parking, and security detection systems.

The most common technology used by these systems is sensors.

After first looking at the sensors and how they work, we will look at the detection distances and operation of the sensors in these systems.

17.5   SENSORS AND TOOLS

Sensors are devices that measure an event or change in the environment and convert it into an electronic signal that can be read and calculated. Sensors can detect any aspect of the physical environment and turn it into useful information. Thus, sensors convert stimuli such as heat, light, sound, and movement into electrical signals. These signals pass through an interface that converts them into binary code and transmits them to a computer for processing.

A sensor network is a group of sensors where each sensor monitors information in a different location and sends that information to a central location for storage, viewing, and analysis.

There are two types of sensor networks, wired and wireless. Sensor network components include sensor nodes, sensors, gateways, and control nodes. The four topologies of sensor networks are point-to-point, star, tree, and mesh. The architecture of the wireless sensors is shown in Figure 17.2:

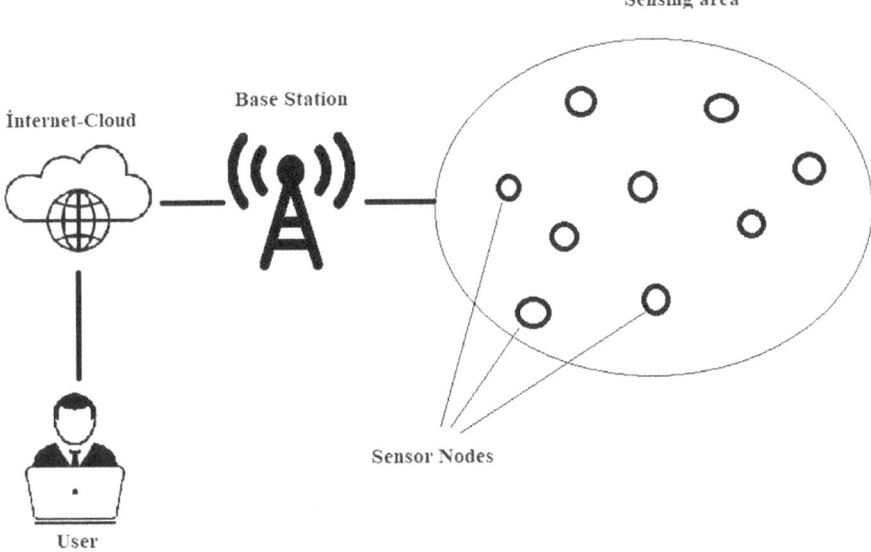

FIGURE 17.2 Wireless sensor architecture.

Inside the wireless sensors, there are original sensory nodes (nodes) distributed in space, which can work in different conditions: noise, temperature, or pressure. As can be seen from the previous image, the sensor nodes must be directly connected to the base (i.e., the destination) in order to transmit information. Among them, the router acts as a router interface. In general, such networks—which are router nodes—are known as multi-hop networks.

In the previous picture, instead of a "user," there can be a "control center." The main sensors used in smart cities include biosensors, electronic sensors, chemical sensors, and smart grid (network) sensors (Alotaibi et al., 2019):

- **Biosensors:** The current function of biosensors is to detect anolyte in biomedicine. Existing biosensors include sensors used for ionization and atomic sensors, such as neutron and MEMS sensors, among many other biosensor devices.
- **Electronic sensors:** The main function of electronic sensors is to detect various forms of energy using electroscopes, magnetic anomalies, and voltage detectors, among others. Electronic sensors include, for example, environmental monitoring sensors, parking sensors, and speedometer sensors, among others. The wireless nature of the sensors results in limitations such as high power consumption, low agility, and high complexity.
- **Chemical sensors:** The chemical sensor currently includes sensors such as carbon dioxide sensors, oxygen sensors, electronic nozzles, and catalytic bead sensors. When working with chemicals, they are sufficient to identify any problems that may be related to the chemicals.

- **Smart grid sensors:** Smart network technologies are classified into five important segments according to their roles, sensors, advanced components, communication, decision support systems, and other highly advanced components.

In general, the most helpful sensors in detecting suspicious situations are electronic sensors. In this regard, in the next section, we look at the principle of operation of these sensors.

17.6   WORKING PRINCIPLE OF SENSORS

The working principle of sensors (detection systems) electronic sensors, as noted, include sensors such as parking sensors and monitoring control sensors. There are two main types of parking sensors: ultrasonic and electromagnetic.

The ultrasonic wave generated by the ultrasound sensor must be evenly distributed in any direction, and the ultrasonic wave returning from any direction must be detected (Alharbi and Soh, 2019).

Like ultrasonic parking sensors, electromagnetic parking sensors work by emitting a wave and then setting its return time.

Electromagnetic sensors emit radio waves, not sound waves. The radio wave is reflected from the object and then returns to the sensor again—a change in frequency is used to calculate the distance from the sensor to the object (Iwaya et al., 2015).

Electromagnetic sensors offer a larger area of detection than ultrasonic sensors, but they also bring with them a number of related problems.

They only detect moving objects, which pose certain danger. The detection distance of such sensors varies from three to five meters. In simple terms, it is as follows:

FIGURE 17.3   Electromagnetic sensors in cars.

Sensors are also used in many areas to detect suspicious situations. In the military, in order to maintain public order, within public transport, in intercity areas (specially installed cells), as well as in civilian areas such as houses.

For example, in the military, it is very useful to identify the enemy, study the physical condition of the soldier, and so on. Such sensors are a key part of health monitoring systems.

In addition, sensors increase their capabilities in conjunction with technologies such as radars and motion and sound detectors. With the help of sensors, a lot of information can be obtained in air, sea, and land without losing coverage and accuracy.

However, given the general urban life, the number of control sensors should be increased, especially in terms of security.

In general, we must pay attention to the detection distance of the most important sensors to be installed in the city.

17.7   SENSORS FOR SOLVING SMART CITY PROBLEMS (SAFETY)

Basically, sensors for solving smart city problems promote safety. There must be some nuances that distinguish smart cities from ordinary cities. As noted, the smart city project focuses on improving people's quality of life, as well as solving urban problems.

Smart cities can consist of hundreds of smart homes. While smart homes are owned by one person, smart cities are owned by everyone. In this regard, many problems in the smart home and their solutions should be applied to smart cities.

A (smart) house needs light, energy, gas, water, waste systems, and a safe environment.

The biggest challenges facing large cities in the future are environmental threats, governance, resources, security, technology, and inequality.

This is also true for smart cities. However, one of the biggest problems in the city is the lack of security.

Within the city, car accidents (which are most frequent accidents in the world), theft, murder, and so on occur frequently. Timely prevention of such events will affect the course of subsequent events.

In general, biometric data and surveillance cameras are widely used around the world to detect violence and illegal activities. Intelligent lighting systems are a useful and cost-effective tool that uses common sensors such as light and motion detectors and can improve safety tasks. Surveillance cameras, face recognition systems, and global positioning systems (GPS), along with data management systems, are increasingly common tools in the hands of law enforcement.

Sensors should be used more for such activities. Thus, there is a great need for such systems, especially in high-risk areas. On the other hand, it is known that as society develops, unfortunately, the number of violence increases instead of decreasing. There are cases of theft and murder.

On the other hand, there are some accidents that occur, such as traffic accidents, which can be caused by human negligence or simple carelessness. In such cases, not only the surveillance cameras but also the sensors placed inside cars can play an important role. There will be a slight difference of one second between a person calling an ambulance as a result of an accident and the sensor sending an ambulance alert at that moment, which is very important during life-threatening situations.

Surveillance camera alerts can be more useful at other times.

For example, the average time for an ambulance to arrive at the scene of an accident (although it varies from country to country and from area to area) is assumed to be between six and nine minutes.

Considering the area in which an injured person is injured and other factors, a person may not be able to call an ambulance after an injury, or it may take some time for someone else to find him or her and call an ambulance. This can increase the risk of life-threatening and even death with excessive blood loss.

However, with the help of sensors installed in the smart city, it is possible for the medical team to quickly find the injured person by collecting and sending information to the center in accordance with the changes in the environment (a wounded person collapsing, conflict between two people, etc.). Also, if anyone has injured that person, it is possible to alert the police center with the information collected by the sensors (the person's face, general appearance, etc.).

In terms of maintaining public order, sensors are an indispensable tool for smart cities.

As we have mentioned, in the military field, sensors are also used to recognize the enemy in wartime. However, since these are technologies based on artificial intelligence, it can be difficult to recognize a person (i.e., an enemy). AI can determine who is an enemy and who is not and send a warning to the command center. In this regard, the lack of sensors can lead to some dangerous consequences.

In short, the city should give priority to such activities and create a large connected network system. A (simple) description is as follows:

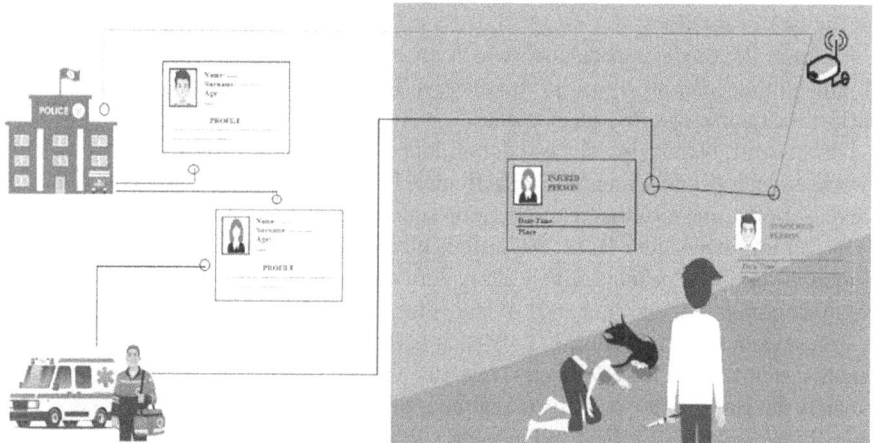

FIGURE 17.4 Connected network.

There are more cost-effective sensors that have a range of 300 to 3,000 meters, but sensors for normal value observation have a maximum range of ≈25 m. It is better to use short-sighted sensors instead of long-sighted sensors within the city because there are many buildings in cities that will prevent the sensor from seeing. In this regard, the use of short-range sensors will give better results.

In addition, sensors can be used to improve the quality of life in smart cities:

1. Automatic lighting of traffic lights as a result of lowering the temperature (or dimming the light)
2. Informing the driver without entering the parking area through parking sensors to indicate which space is empty or full (e.g., marking on a sign in front of the parking area)
3. Quickly detecting suspicious situations by increasing the detection distance on sensors installed for security purposes (which will definitely lead to an increase in power)

17.8 CONCLUSION

The number of smart cities is growing every year. A smart society needs smart cities, homes, and networks.

There are technologies that are indispensable for any smart city infrastructure, one of which is the sensors and their data we mentioned in our article (Moreno et al., 2021).

Today, there are very few areas where sensors are not used. There are even fewer devices.

1. Sensors are tools that help and will solve many problems. Control, monitoring, data collection—in short, proper management is possible through sensors.
2. Sensors are also increasingly being used to solve security problems, especially in smart cities.

Most importantly, it is important to increase the detection distances of the sensors to a certain extent and, of course, to put them into operation at a reasonable price. Thus, it is possible to increase the detection distance, but it will require additional energy. In this chapter, we mainly looked at the use of sensors in terms of public order.

REFERENCES

Alharbi, N., B. Soh, "Roles and Challenges of Network Sensors in Smart Cities," in *International Conference on Smart Power & Internet Energy Systems.* IOP Publishing Ltd, 2019, doi:10.1088/1755-1315/322/1/012002.

Alotaibi, Majid, "Security to Wireless Sensor Networks Against Malicious Attacks Using Hamming Residue Method," *EURASIP Journal on Wireless Communications and Networking*, Volume 2019, 8 January 2019, doi:10.1186/s13638-018-1337-5.

Arunachalam, P., N. Janakiraman, A.K. Sivaraman, A. Balasundaram, R. Vincent, S. Rani, B. Dey, A. Muralidhar, M. Rajesh, "Synovial Sarcoma Classification Technique Using Support Vector Machine and Structure Features," *Intelligent Automation & Soft Computing*, Volume 32(2), 2021, pp. 1241–1259, doi:10.32604/iasc.2022.022573.

Bhambri, Pankaj, Sita Rani, Gaurav Gupta, Alex Khang, *Cloud and Fog Computing Platforms for Internet of Things.* CRC Press, 2022. ISBN: 978-1-032-101507, doi:10.1201/9781032101507.

Iwaya, Kazuki, Riichi Murayama, Takahiro Hirayama, "Study of Ultrasonic Sensor That Is Effective for All Direction Using an Electromagnetic Force," *Proceedings of the SPIE*, Volume 9302, March 2015, id. 93022U 7 p, doi:10.1117/12.2077337.

Khang, A., Pankaj Bhambri, Sita Rani, Aman Kataria, *Big Data, Cloud Computing and Internet of Things*. CRC Press, 2022. ISBN: 978-1-032-284200, doi:10.1201/9781032284200.

Khang, A., Subrata Chowdhury, Seema Sharma, *The Data-Driven Blockchain Ecosystem: Fundamentals, Applications and Emerging Technologies*. CRC Press, 2022. ISBN: 978-1-032-21624, doi:10.1201/9781003269281.

Khang, A., Sita Rani, Meetali Chauhan, Aman Kataria, "IoT Equipped Intelligent Distributed Framework for Smart Healthcare Systems," *Networking and Internet Architecture*, 2021, https://arxiv.org/abs/2110.04997v2, doi:10.48550/arXiv.2110.04997.

Moreno, Ramírez, Mauricio Adolfo, Sajjad Keshtkar, Diego A. Padilla Reyes, Edrick Ramos-López, "Sensors for Sustainable Smart Cities: A Review," *ResearchGate*, September 2021, doi:10.3390/app11178198.

Ochoa, Sergio F., Giancar lo Fortino, Giuseppe Di Fatta, "Cyber-physical Systems, Internet of Things and Big Data," *Future Generation Computer Systems*, Volume 75, October 2017, pp. 82–84, doi:10.1016/j.future.2017.05.040.

Rana, G., Khang, A., Khanh, H.H., "The Role of Artificial Intelligence in Blockchain Applications," in *Reinventing Manufacturing and Business Processes Through Artificial Intelligence* (p. 20). CRC Press, 2021, doi:10.1201/9781003145011.

Rani, S., V. Arya, A. Kataria. "Dynamic Pricing-Based E-commerce Model for the Produce of Organic Farming in India: A Research Roadmap with Main Advertence to Vegetables," in *Proceedings of Data Analytics and Management* (pp. 327–336). Springer, Singapore, 2022.

Rani, S., A. Kataria, M. Chauhan, "Fog Computing in Industry 4.0: Applications and Challenges—A Research Roadmap," in *Energy Conservation Solutions for Fog-Edge Computing Paradigms* (pp. 173–190). Springer, Singapore, 2022.

Rani, S., A. Kataria, M. Chauhan, P. Rattan, R. Kumar, A.K. Sivaraman, "Security and Privacy Challenges in the Deployment of Cyber-Physical Systems in Smart City Applications: State-of-Art Work," in *Materials Today: Proceedings*. Elsevier, Amsterdam, 2022.

Rani, S., A. Kataria, V. Sharma, S. Ghosh, V. Karar, K. Lee, C. Choi, "Threats and Corrective Measures for IoT Security with Observance of Cybercrime: A Survey," *Wireless Communications and Mobile Computing*, Volume 2021, 2021, pp. 1–30.

Rani, S., R. Kumar, "Bibliometric Review of Actuators: Key Automation Technology in a Smart City Framework," in *Materials Today: Proceedings*. Elsevier, 2022, doi:10.1016/j.matpr.2021.12.469.

Rani, S., R.K. Mishra, M. Usman, A. Kataria, P. Kumar, P. Bhambri, A.K. Mishra, "Amalgamation of Advanced Technologies for Sustainable Development of Smart City Environment: A Review," *IEEE Access*, Volume 9, 2021, pp. 150060–150087, doi:10.1109/ACCESS.2021.3125527.

Vallianatos, Mark, "Uncovering the Early History of 'Big Data' and the 'Smart City' in Los Angeles," *Boom California*, June 16, 2015. https://boomcalifornia.org/2015/06/16/uncovering-the-early-history-of-big-data-and-the-smart-city-in-la/.

Index

Note: Page numbers in *italics* indicate a figure on the corresponding page.